The Handbook of Environment

Editor-in-Chief: O. Hutzinger

Volume 5 Water Pollution
Part O

The Handbook of Environmental Chemistry

Recently Published and Forthcoming Volumes

Environmental Specimen Banking
Volume Editors: S. A. Wise and P. P. R. Becker
Vol. 3/S, 2006

Polymers: Chances and Risks
Volume Editors: P. Eyerer, M. Weller and C. Hübner
Vol. 3/V, 2006

The Rhine
Volume Editor: T. P. Knepper
Vol. 5/L, 03.2006

Persistent Organic Pollutants in the Great Lakes
Volume Editor: R. A. Hites
Vol. 5/N, 2006

Antifouling Paint Biocides
Volume Editor: I. Konstantinou
Vol. 5/O, 2006

Estuaries
Volume Editor: P. J. Wangersky
Vol. 5/H, 2006

The Caspian Sea Environment
Volume Editors: A. Kostianoy and A. Kosarev
Vol. 5/P, 2005

Marine Organic Matter: Biomarkers, Isotopes and DNA
Volume Editor: J. K. Volkman
Vol. 2/N, 2005

Environmental Photochemistry Part II
Volume Editors: P. Boule, D. Bahnemann and P. Robertson
Vol. 2/M, 2005

Air Quality in Airplane Cabins and Similar Enclosed Spaces
Volume Editor: M. B. Hocking
Vol. 4/H, 2005

Environmental Effects of Marine Finfish Aquaculture
Volume Editor: B. T. Hargrave
Vol. 5/M, 2005

The Mediterranean Sea
Volume Editor: A. Saliot
Vol. 5/K, 2005

Environmental Impact Assessment of Recycled Wastes on Surface and Ground Waters
Engineering Modeling and Sustainability
Volume Editor: T. A. Kassim
Vol. 5/F (3 Vols.), 2005

Oxidants and Antioxidant Defense Systems
Volume Editor: T. Grune
Vol. 2/O, 2005

Antifouling Paint Biocides

Volume Editor:
Ioannis K. Konstantinou

With contributions by
T. A. Albanis · A. Baart · D. Barceló · J. Boon · C. Brunori · K. Fent
A. R. Fernández-Alba · B. van Hattum · M. D. Hernando · I. Ipolyi
I. K. Konstantinou · P. Massanisso · M. Mezcua · H. Mieno
R. Morabito · H. Okamura · I. Omae · J. W. Readman
V. A. Sakkas · N. Voulvoulis · H. Yamada

Springer

Environmental chemistry is a rather young and interdisciplinary field of science. Its aim is a complete description of the environment and of transformations occurring on a local or global scale. Environmental chemistry also gives an account of the impact of man's activities on the natural environment by describing observed changes.
The Handbook of Environmental Chemistry provides the compilation of today's knowledge. Contributions are written by leading experts with practical experience in their fields. The Handbook will grow with the increase in our scientific understanding and should provide a valuable source not only for scientists, but also for environmental managers and decision-makers.
The Handbook of Environmental Chemistry is published in a series of five volumes:

Volume 1: The Natural Environment and the Biogeochemical Cycles
Volume 2: Reactions and Processes
Volume 3: Anthropogenic Compounds
Volume 4: Air Pollution
Volume 5: Water Pollution

The series Volume 1 The Natural Environment and the Biogeochemical Cycles describes the natural environment and gives an account of the global cycles for elements and classes of natural compounds.
The series Volume 2 Reactions and Processes is an account of physical transport, and chemical and biological transformations of chemicals in the environment.
The series Volume 3 Anthropogenic Compounds describes synthetic compounds, and compound classes as well as elements and naturally occurring chemical entities which are mobilized by man's activities.
The series Volume 4 Air Pollution and Volume 5 Water Pollution deal with the description of civilization's effects on the atmosphere and hydrosphere.
Within the individual series articles do not appear in a predetermined sequence. Instead, we invite contributors as our knowledge matures enough to warrant a handbook article.
Suggestions for new topics from the scientific community to members of the Advisory Board or to the Publisher are very welcome.

ISSN 1433-6863

ISBN 978-3-642-06846-1 e-ISBN 978-3-540-32844-5

DOI 10.1007/11555148

Springer is a part of Springer Science+Business Media

springer.com

Softcover reprint of the hardcover 1st edition 2006

Cover design: E. Kirchner, Springer-Verlag

Printed on acid-free paper 02/3141 YL – 5 4 3 2 1 0

Editor-in-Chief

Volume Editor

Dr. Ioannis K. Konstantinou

Department of Environmental
and Natural Resources Management
University of Ioannina
Seferi 2
30100, Agrinio, Greece
iokonst@cc.uoi.gr

Advisory Board

Prof. Dr. D. Mackay
Department of Chemical Engineering
and Applied Chemistry
University of Toronto
Toronto, ON, M5S 1A4, Canada

Prof. Dr. A. H. Neilson
Swedish Environmental Research Institute
P.O. Box 21060
10031 Stockholm, Sweden
ahsdair@ivl.se

Prof. Dr. J. Paasivirta
Department of Chemistry
University of Jyväskylä
Survontie 9
P.O. Box 35
40351 Jyväskylä, Finland

Prof. Dr. Dr. H. Parlar
Institut für Lebensmitteltechnologie
und Analytische Chemie
Technische Universität München
85350 Freising-Weihenstephan, Germany

Prof. Dr. S. H. Safe
Department of Veterinary
Physiology and Pharmacology
College of Veterinary Medicine
Texas A & M University
College Station, TX 77843-4466, USA
ssafe@cvm.tamu.edu

Prof. P. J. Wangersky
University of Victoria
Centre for Earth and Ocean Research
P.O. Box 1700
Victoria, BC, V8W 3P6, Canada
wangers@telus.net

The Handbook of Environmental Chemistry Also Available Electronically

For all customers who have a standing order to The Handbook of Environmental Chemistry, we offer the electronic version via SpringerLink free of charge. Please contact your librarian who can receive a password or free access to the full articles by registering at:

springerlink.com

If you do not have a subscription, you can still view the tables of contents of the volumes and the abstract of each article by going to the SpringerLink Homepage, clicking on "Browse by Online Libraries", then "Chemical Sciences", and finally choose The Handbook of Environmental Chemistry.

You will find information about the

- Editorial Board
- Aims and Scope
- Instructions for Authors
- Sample Contribution

at springer.com using the search function.

Preface

Environmental Chemistry is a relatively young science. Interest in this subject, however, is growing very rapidly and, although no agreement has been reached as yet about the exact content and limits of this interdisciplinary discipline, there appears to be increasing interest in seeing environmental topics which are based on chemistry embodied in this subject. One of the first objectives of Environmental Chemistry must be the study of the environment and of natural chemical processes which occur in the environment. A major purpose of this series on Environmental Chemistry, therefore, is to present a reasonably uniform view of various aspects of the chemistry of the environment and chemical reactions occurring in the environment.

The industrial activities of man have given a new dimension to Environmental Chemistry. We have now synthesized and described over five million chemical compounds and chemical industry produces about hundred and fifty million tons of synthetic chemicals annually. We ship billions of tons of oil per year and through mining operations and other geophysical modifications, large quantities of inorganic and organic materials are released from their natural deposits. Cities and metropolitan areas of up to 15 million inhabitants produce large quantities of waste in relatively small and confined areas. Much of the chemical products and waste products of modern society are released into the environment either during production, storage, transport, use or ultimate disposal. These released materials participate in natural cycles and reactions and frequently lead to interference and disturbance of natural systems.

Environmental Chemistry is concerned with reactions in the environment. It is about distribution and equilibria between environmental compartments. It is about reactions, pathways, thermodynamics and kinetics. An important purpose of this Handbook, is to aid understanding of the basic distribution and chemical reaction processes which occur in the environment.

Laws regulating toxic substances in various countries are designed to assess and control risk of chemicals to man and his environment. Science can contribute in two areas to this assessment; firstly in the area of toxicology and secondly in the area of chemical exposure. The available concentration ("environmental exposure concentration") depends on the fate of chemical compounds in the environment and thus their distribution and reaction behaviour in the environment. One very important contribution of Environmental Chemistry to the above mentioned toxic substances laws is to develop laboratory test methods, or mathematical correlations and models that predict the environmental fate of new chemical compounds. The third purpose of this Handbook is to help in the basic understanding and development of such test methods and models.

The last explicit purpose of the Handbook is to present, in concise form, the most important properties relating to environmental chemistry and hazard assessment for the most important series of chemical compounds.

At the moment three volumes of the Handbook are planned. Volume 1 deals with the natural environment and the biogeochemical cycles therein, including some background information such as energetics and ecology. Volume 2 is concerned with reactions and processes in the environment and deals with physical factors such as transport and adsorption, and chemical, photochemical and biochemical reactions in the environment, as well as some aspects of pharmacokinetics and metabolism within organisms. Volume 3 deals with anthropogenic compounds, their chemical backgrounds, production methods and information about their use, their environmental behaviour, analytical methodology and some important aspects of their toxic effects. The material for volume 1, 2 and 3 was each more than could easily be fitted into a single volume, and for this reason, as well as for the purpose of rapid publication of available manuscripts, all three volumes were divided in the parts A and B. Part A of all three volumes is now being published and the second part of each of these volumes should appear about six months thereafter. Publisher and editor hope to keep materials of the volumes one to three up to date and to extend coverage in the subject areas by publishing further parts in the future. Plans also exist for volumes dealing with different subject matter such as analysis, chemical technology and toxicology, and readers are encouraged to offer suggestions and advice as to future editions of "The Handbook of Environmental Chemistry".

Most chapters in the Handbook are written to a fairly advanced level and should be of interest to the graduate student and practising scientist. I also hope that the subject matter treated will be of interest to people outside chemistry and to scientists in industry as well as government and regulatory bodies. It would be very satisfying for me to see the books used as a basis for developing graduate courses in Environmental Chemistry.

Due to the breadth of the subject matter, it was not easy to edit this Handbook. Specialists had to be found in quite different areas of science who were willing to contribute a chapter within the prescribed schedule. It is with great satisfaction that I thank all 52 authors from 8 countries for their understanding and for devoting their time to this effort. Special thanks are due to Dr. F. Boschke of Springer for his advice and discussions throughout all stages of preparation of the Handbook. Mrs. A. Heinrich of Springer has significantly contributed to the technical development of the book through her conscientious and efficient work. Finally I like to thank my family, students and colleagues for being so patient with me during several critical phases of preparation for the Handbook, and to some colleagues and the secretaries for technical help.

I consider it a privilege to see my chosen subject grow. My interest in Environmental Chemistry dates back to my early college days in Vienna. I received significant impulses during my postdoctoral period at the University of California and my interest slowly developed during my time with the National Research Council of Canada, before I could devote my full time of Environmental Chemistry, here in Amsterdam. I hope this Handbook may help deepen the interest of other scientists in this subject.

Amsterdam, May 1980 *O. Hutzinger*

Twenty-one years have now passed since the appearance of the first volumes of the Handbook. Although the basic concept has remained the same changes and adjustments were necessary.

Some years ago publishers and editors agreed to expand the Handbook by two new open-end volume series: Air Pollution and Water Pollution. These broad topics could not be fitted easily into the headings of the first three volumes. All five volume series are integrated through the choice of topics and by a system of cross referencing.

The outline of the Handbook is thus as follows:

1. The Natural Environment and the Biochemical Cycles,
2. Reaction and Processes,
3. Anthropogenic Compounds,
4. Air Pollution,
5. Water Pollution.

Rapid developments in Environmental Chemistry and the increasing breadth of the subject matter covered made it necessary to establish volume-editors. Each subject is now supervised by specialists in their respective fields.

A recent development is the accessibility of all new volumes of the Handbook from 1990 onwards, available via the Springer Homepage springeronline.com or springerlink.com.

During the last 5 to 10 years there was a growing tendency to include subject matters of societal relevance into a broad view of Environmental Chemistry. Topics include LCA (Life Cycle Analysis), Environmental Management, Sustainable Development and others. Whilst these topics are of great importance for the development and acceptance of Environmental Chemistry Publishers and Editors have decided to keep the Handbook essentially a source of information on "hard sciences".

With books in press and in preparation we have now well over 40 volumes available. Authors, volume-editors and editor-in-chief are rewarded by the broad acceptance of the "Handbook" in the scientific community.

Bayreuth, July 2001 *Otto Hutzinger*

Contents

Foreword

The need for effective antifoulants, which prevent the settlement and growth of marine organisms on submerged structures, such as ship's hulls, oil rig supports, buoys and fish cages is recognised worldwide as being of significant economic importance. Tributyltin (TBT)-based antifouling paints have been widely acclaimed as the most effective antifoulants ever devised and consequently were the most widely used active ingredients in paint formulations for many years. However, since 1990 they have been internationally regulated due to their severe impact on the aquatic ecosystem. Although much research has been performed in the field of TBT monitoring, environmental fate and effects, a large number of environmental scientists around the world continue to engage in this issue since ongoing sources of TBT remain. The ecotoxicological problems associated with the use of TBT have led to policy actions. Legislation in many countries banned the application of TBT-based paints to small vessels (< 25 m) and the International Maritime Organization (IMO), Marine Environment Protection Committee (MEPC) endorsed the ban on the use of TBT as an antifouling agent on ships by January 2003 and the presence of such paints on vessel hulls by January 2008.

This imminent ban of the TBT-based paints has been the cause of a major change in the antifouling paint industry and led to an increase in vessels using alternative TBT-free coatings containing copper combined with organic booster biocides, the majority of which are already used in agriculture. Biocide-containing coatings are already used and applied to the hulls of ships and boats in order to prevent the growth of marine species. Worldwide around 18 compounds like Irgarol 1051, sea-nine 211, dichlofluanid, chlorothalonil, zinc pyrithione, diuron, TCMS pyridine, TCMTB, zineb, etc., are currently used as antifouling biocides. These biocides are also the most frequently used in many countries. As a result, important levels of contamination have been observed in the aquatic environment worldwide, especially in coastal areas with high yachting activity, particularly in marinas and sportive harbours. Since these alternatives to TBT are also toxic and their putative impact on non-target organisms is poorly known in some cases, their contamination in the aquatic environment has been a topic of increasing importance over the last few years. Already, many countries have reached on agreement on the restriction of biocides such as Irgarol 1051 and diuron.

Therefore, environmentally safe biocidal additives that will perform equally as well or even better than the currently used substances are sought. This search led investigators to study natural products. The development of antifoulants containing environmentally safe natural products has anticipated the conservation of the marine environment.

This volume represents a comprehensive coverage of the antifouling biocides field and addresses a broad spectrum of the environmental issues. It reviews systematically the currently available data, results and discussion on topics such as the occurrence of TBT-based and alternative antifouling biocides in the aquatic environment, trace analytical techniques for the determination of biocide residues in various matrices, the environmental fate and behaviour, inputs estimation, the toxic effects and the risk assessment, with an emphasis on the last 10-year period. It also highlights the gaps in scientific knowledge where more research and monitoring efforts are needed, especially in the fields of ecotoxicology and long-term risk assessment.

I would like to thank all the contributors to this volume and all experts in the field, who have shared their expertise and experience with the reader. I greatly appreciate their efforts and believe that they will be rewarded by the production of such an interesting volume. In particular, I would like to extend my thanks to Prof. O. Hutzinger for inviting me to coordinate the preparation of this book and to Springer for their advice and assistance. Finally, it is my hope that readers will enjoy the reading of this book and the content will constitute an important resource for researchers, students, environmental managers and professionals interested in this interdisciplinary field.

Ioannina, July 2005 Ioannis K. Konstantinou

Hdb Env Chem Vol. 5, Part O (2006): 1–15
DOI 10.1007/698_5_047

Published online: 9 December 2005

Development, Occurrence and Regulation of Antifouling Paint Biocides: Historical Review and Future Trends

James W. Readman

Plymouth Marine Laboratory, Prospect Place, The Hoe, Plymouth PL1 3DH, UK
jwre@pml.ac.uk

Abstract Antifouling agents have been used on ships since the ancient Roman and Greek civilizations. A brief history is provided through to the demise of tributyltin (TBT) and the introduction of organic "booster" biocides. It is upon these latter compounds that the chapter is focused. A broad description of published data and of work undertaken through the Assessment of Antifouling Agents in Coastal Environments (ACE) project is provided to afford an overview of levels, behaviours and potential threats posed by the compounds. Legislative measures that influence and control usage are described. Finally, options for antifouling measures projected into the future are described and discussed.

Keywords Antifouling · Booster biocides · Environmental behaviour · History · Regulations

1 Background and Historical Perspectives

Antifouling of boats and ships is not a new concept. The history of antifouling has recently been reviewed [1]. The ancient civilizations of the Romans and the Greeks coated their vessels with lead sheathing secured by cooper nails. Columbus' ships are thought to have been coated with pitch and tallow. In the UK, lead sheathing was abandoned by the Navy in the late 1600s and antifouling paints containing tar, grease, sulphur pitch and brimstone were developed. One hundred years later, copper sheathing was used, which prevented fouling through dissolution of the toxic metal ions. It was in the mid-1800s that antifouling paints really began to develop. This was attributed to the introduction of iron ships on which copper sheathing caused corrosion of the iron. Paints were prepared by adding toxicants such as copper oxide, arsenic, and mercury oxide to resin binders. These proved to be effective. Following the Second World War, the introduction of petroleum-based resins and health and safety concerns relating to organo-arsenicals and mercurials meant that synthetic copper based paints became most popular. In the late 1950s and early 1960s, a new formulation using tributyltin (TBT) proved to be excellent in the prevention of fouling. This is where our story begins.

The efficiency of TBT, especially in "self-polishing" formulations, was remarkable, and the application of TBT-based paints rapidly expanded. Added bonuses also included the fact that it did not cause galvanic corrosion on aluminium hulls, it was colourless, and periods between dry-docking were extended. Whilst this appears ideal, unfortunately, the use of the compound had environmental consequences. As the popularity of TBT grew, oyster producers in France were reporting shell malformations, which rendered their produce worthless. This effect was traced to TBT in the water. In Arcachon Bay (France) alone, it has been estimated that TBT provoked a loss in revenue of 147 million US dollars through reduced oyster production [2]. Wild populations of other mollusc species were also found to be affected at very low concentrations ($< 10\,\mathrm{ng\,L^{-1}}$) [3]. Female dog whelks (*Nucella* sp.) were shown to develop male characteristics (termed imposex) at these levels [4]. Imposex was also reported in the open North Sea [5]. Although dealt with in more detail in a later section, national and international legislation was introduced to restrict the use of TBT. In 1989, the European Community introduced a directive to prevent the use of TBT on boats under 25 m [6]. This provoked paint manufacturers and chemical companies to develop and sell a range of agents for new antifouling paints for the "small boat" market. Although usually added to copper-based formulations, they were also added to TBT-based paints to enhance efficacy for larger vessels. These compounds have since been termed "booster biocides". Examples of the types of compounds that were used or promoted for use included:

- 2-methylthio-4-tertiary-butylamino-6-cyclopropylamino-s-triazine (Irgarol 1051);
- 1-(3,4-dichlorophenyl)-3,3-dimethylurea (diuron);
- 4,5-dichloro-2-n-octyl-4-isothiazolin-3-one (SeaNine 211);
- *N*-dichlorofluoromethylthio-*N*′, *N*′-dimethyl-*N*-phenylsulphamide (dichlofluanid);

- 2,4,5,6-tetrachloro iso phthalo nitrile (chlorothalonil);
- bis(1hydroxy-2(1H)-pyridethionato-O,S)-T-4zinc (zinc pyrithione);
- 2-(thiocyanomethyl thio)benzthiazole (TCMBT);
- 2,3,5,6-tetrachloro-4-(methyl sulphonyl) pyridine (TCMS pyridine);
- cuprous thiocyanate;
- 4-chloro-meta-cresol;
- arsenic trioxide;
- cis1-(3-chloroallyl)-3,5,7-triaza-1-azonia adamantane chloride;
- zineb;
- folpet;
- thiram;
- oxy tetracycline hydrochloride;
- ziram;
- maneb.

Many of these compounds were known to be highly toxic. Negligible data, at that time, was, however, available concerning contamination, and (potential) effects/risks of these compounds in coastal and marine environments. Towards the end of the 1990s, this lack of information was addressed through the "Assessment of Antifouling Agents in Coastal Environments (ACE)" project of the European Commission (MAS3-CT98-0178) (1999–2002).

This project was designed to provide:

- information on usage and geographical differences in usage of antifouling agents and products in Europe;
- suitably sensitive analytical (chemical) techniques for quantifying antifouling "booster" biocides;
- an assessment of the extent of contamination of European harbours and marinas and coastal waters through chemical surveys;
- information concerning the dissipation, transport and ecotoxicological effects of "booster" biocides (through experimentation under laboratory and field conditions);
- models that predict environmental concentration and impact;
- a critical comparison of products regarding environmental impact.

Partners in the project were: J.W. Readman (Plymouth Marine Laboratory, UK); B. van Hattum, and M. L'Amoree (Institute for Environmental Studies, Vrije Universiteit, Amsterdam, Netherlands); D. Barcelo (CID-CSIC, Barcelona, Spain); T.A. Albanis (University of Ioannina, Greece); B. Riemann (National Environmental Research Institute, Denmark); H. Blanck and F. Gronvall (Botanical Institute, Göteborg University, Sweden); K. Gustavson (DHI Water & Environment, Denmark); J. Tronczynski and C. Munschy (IFREMER, Centre de Nantes, France); A. Jacobson (Rohm & Haas).

Full details of the ACE project (including reports) are available at www.pml.ac.uk/ace. Much of the data briefly summarised in the present chapter relate to the substantial efforts of all partners within the ACE project.

2 Usage of Antifouling Agents

Usage of antifouling paints differs regionally according to legislation, location of the manufacturer, marketing and consumer preferences. Whilst the list of potential booster biocides provided above is substantial, not all compounds are marketed. For example in the UK, although recent legislative changes have occurred (as discussed in Sect. 5), during the last decade usage of antifouling agents was massively dominated by copper(1)oxide followed by (in order of usage) diuron, Irgarol 1051, zinc pyrithione and dichlofluanid [7]. This will change with the newly introduced legislation (see Sect. 5). Within ACE, similar investigations on usage were undertaken in all the partner's member states. Table 1 summarises the booster biocides that are permitted on yachts less than 25 m in length. Investigations revealed that of these products,

Table 1 Usage of booster biocides: ingredients permitted for use on yachts < 25 m

	UK[a]	France[b]	Greece[b]	Spain[b]	Sweden	Denmark[c]	Neth[a,c]
Copper(1) oxide	+	+	+	+	+[d]	+	+
Copper thiocyanate	+	+			+[d]	+	+
Cu powder					+[d]	+	
Chromium trioxide							+
Diuron	–	+	+	+		–	+
Irgarol 1051	–	+	+	+	+	–	+
Zinc pyrithione	+	+	+	+	–	+	
Dichlofluanid	+	+	+	+			+
TCMTB	–						
Chlorothalonil	–	+	+				
TCMS pyridine	–						
Sea-Nine 211	–			+		+[e]	
Ziram			+				+
Zineb	+						+
Folpet			+				
Total (booster biocides)	3	5[b]	7[b]	5[b]	1	2	5

[a] UK = United Kingdom and Neth = The Netherlands
[b] very limited/no approval scheme (in principle, all can be used)
[c] regulations currently under debate
[d] leach rate regulated on west coast; banned on east coast
[e] although approved, product not used on pleasure craft

booster biocides that were the most used were diuron, Irgarol 1051, dichlofluanid, chlorothalonil and SeaNine 211. For this reason, research within ACE was focused on these compounds. Whilst zinc pyrithione was also considered to be of emerging importance, usage, then, was comparatively low. In addition the compound is difficult to analyse. For these reasons, very little data is available for this biocide.

3 Extent of Contamination

Concentrations of booster biocides in coastal environments are a function of the inputs from vessels, dilution/flushing of the systems, and degradation of the compounds.

The first reported contamination of coastal waters by booster biocides was for Irgarol 1051 on the Cote d'Azur [8]. Substantial concentrations (up to approximately 1700 ng L^{-1}) were recorded in marinas of the region. Subsequent papers confirmed broad contamination in other areas of high boating activity in Europe. More recently, booster biocide contamination has been reported in waters from Japan, the United States, Singapore, Australia and Bermuda. Several reviews have described and compared the extent of contamination (e.g. [9–12]) (see also other chapters within this book).

Critical to monitoring of the extent of contamination is the development of suitably sensitive analytical techniques. This topic is dealt with in detail in the chapter by Barcelo and Fernandez-Alba. Within ACE, several highly sensitive chromatographic methods for the analysis of the selected booster biocides and their metabolites in environmental waters and sediments were developed. Methods were directed towards: Irgarol 1051, its metabolite 2-methylthio-4-tert-butylamino-s-triazine; diuron and its byproducts dimethyl diuron and 1-(3,4-dichlorophenyl)urea; chlorothalonil; vinclozolin; dichlofluanid; and SeaNine 211. Extractions employed on-line and off-line solid-phase extraction (SPE) cartridges and disks, solid-phase micro-extraction (SPME), headspace-SPME, XAD-2 resin and liquid-liquid techniques. Sediment analyses used an ultrasonication extraction protocol. A comparative ELISA method was also developed for trace level determinations. Quantification was carried out by gas chromatography (GC) with electron capture (ECD), nitrogen phosphorus (NPD), flame photometric (FPD) and mass spectrometric (MS) (including ion-trap tandem MS) detection. High-performance liquid chromatography was also used in quantification with detection using electrospray MS/MS and atmospheric chemical ionization mass spectrometry (HPLC-ACPI-MS).

Approximately 800 water samples (and sediments from some areas) were collected within ACE from the areas shown in Fig. 1. These included marinas, harbours, estuaries and coastal waters and covered diverse European

Fig. 1 Location of sampling areas (indicated by *squares*) investigated during the ACE Project

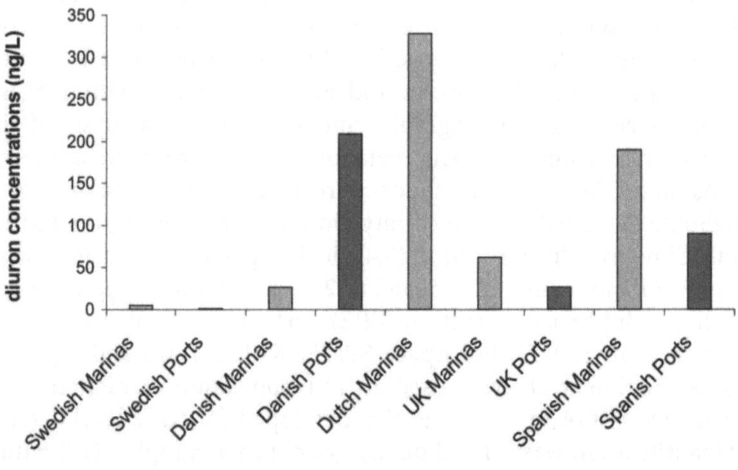

Fig. 2 Mean concentrations ($ng\,L^{-1}$) of diuron in samples taken from marinas and ports

coastal systems. Results from analyses are summarised in Table 2. They indicate that of the major booster biocides, highest mean concentrations of diuron were encountered. The distribution of this compound is shown in Fig. 2 and indicates highest levels in North Western Europe. Irgarol 1051 tended to be present at lower mean concentrations than diuron, although for

Table 2 Concentrations ($ng\,L^{-1}$) of antifouling booster biocides measured in European coastal waters

Country	Site Description	No. of samples analysed		Irgarol 1051	Diuron	Dichlo-fluanid	Chloro-thalonil	Seanine
Sweden	Marinas	10	range	2–364	< 1–35	< 1	< 1	< 1–3
			mean	61	5	< 1	< 1	< 1
			median	16	3	< 1	< 1	0
	Ports	8	range	< 1–6	< 1–3	< 1	< 1	< 1–1
			mean	2	1	< 1	< 1	< 1
			median	1	0	< 1	< 1	< 1
	Coastal	19	range	< 1–36	< 1–7	< 1	< 1	< 1
			mean		2	< 1	< 1	< 1
			median	0	2	< 1	< 1	< 1
Denmark	Marinas	21	range	4–9	37–174	n/a	n/a	n/a
			mean	2	27			
			median	0	0			
	Ports	3	range	< 1–68	< 1–628	n/a	n/a	n/a
			mean	23	209			
			median	0	0			
Nether-lands	Marinas	26	range	< 1–87	< 1–1129	n/a	n/a	n/a
			mean	20	328			
			median	17	233			
	Coastal	12	range	< 1–39	< 1–282	n/a	n/a	n/a
			mean	4	51			
			median	0	19			

Table 2 (continued)

Country	Site Description	No. of samples analysed		Irgarol 1051	Diuron	Dichlo-fluanid	Chloro-thalonil	Seanine
UK	Marinas	168	range	< 1–621	< 1–685	< 1–390	< 1–30	< 1
			mean	52	62	8	1	< 1
			median	19	< 1	< 1	< 1	< 1
	Ports	47	range	< 1–208	< 1–110	< 1–26	< 1–20	< 1
			mean	10	27	1	1	< 1
			median	4	20	< 1	< 1	< 1
	Estuaries	64	range	< 1–47	< 1–438	< 1–40	< 1	< 1
			mean	9	43	1	< 1	< 1
			median	7	20	< 1	< 1	< 1
	Coastal	49	range	< 1–92	< 1–465	< 1–7	< 1–26	< 1
			mean	6	23	1	1	< 1
			median	2	7	< 1	< 1	< 1
France	English channel Marinas	3	range	6–23	n/a	< 1	8–11	n/a
			mean	15		< 1	9	
			median	17		< 1	9	
	Atlantic coast Marinas	14	range	3–491	n/a	< 1	< 1	n/a
			mean	55		< 1	< 1	
			median	18		< 1	< 1	
	Atlantic Coastal	19	range	< 1–21	n/a	< 1	< 1	n/a
			mean	5		< 1	< 1	
			median	2		< 1	< 1	

Table 2 (continued)

Country	Site Description	No. of samples analysed		Irgarol 1051	Diuron	Dichlo-fluanid	Chloro-thalonil	Seanine
	Mediterranean Marinas	18	range	11–244	n/a	< 1	< 1–27	n/a
			mean	67		< 1	9	
			median	33		< 1	6	
	Mediterranean Coastal	32	range	< 1–11	n/a	< 1	< 1–2	n/a
			mean	1		< 1	1	
			median	1		< 1	< 1	
Spain	Marinas	112	range	< 1–670	< 1–2190	< 1–760	< 1	< 1–3700
			mean	80	190	30	< 1	110
			median	40	80	< 1	< 1	< 1
	Ports	11	range	30–323	< 1–240	< 1	< 1	< 1
			mean	100	90	< 1	< 1	< 1
			median	80	60	< 1	< 1	< 1
Greece	Marinas	58	range	< 1–90	n/a	< 1–284	< 1–63	< 1
			mean	18		61	16	< 1
			median	15		38	16	< 1
	Ports	27	range	< 1–24	n/a	< 1–88	< 1–35	Detected
			mean	6		25	10	< 1
			median	< 1		< 1	11	< 1

"n/a" indicates "not analysed"

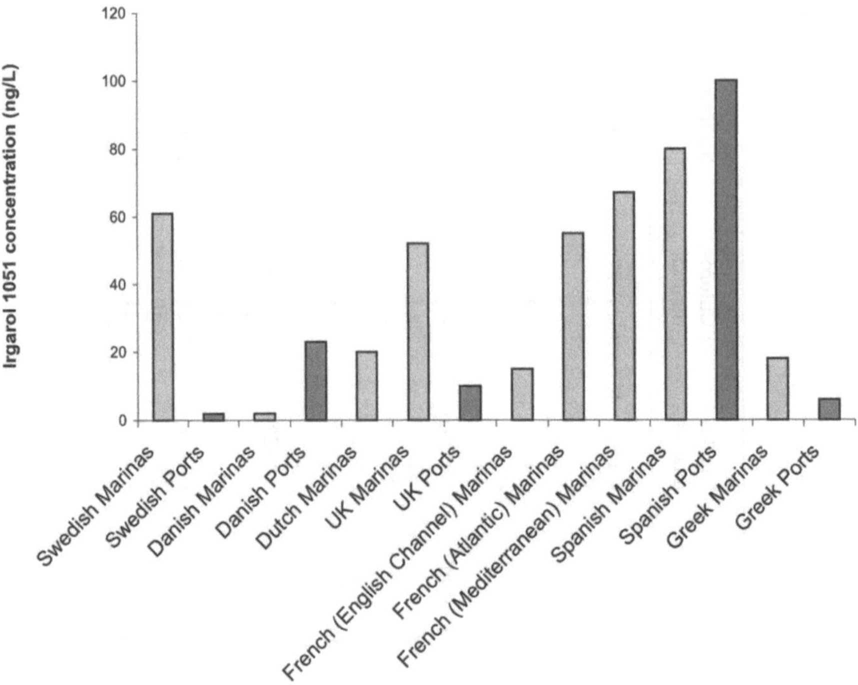

Fig. 3 Mean concentrations ($ng\,L^{-1}$) of Irgarol 1051 in samples taken from marinas and ports

this compound, Mediterranean coastal environments were the most contaminated (Fig. 3). Chlorothalonil, dichlofluanid and SeaNine 211 were sporadically encountered, primarily in the Mediterranean. In isolated cases, however, high concentrations of these were recorded. Measurable concentrations of the degradation products of Irgarol 1051 and diuron were also recorded, albeit at lower levels than those of the parent compounds.

In summary, on a global scale, data are available for the biocides most commonly used in Europe, North America and Japan (i.e. diuron, Irgarol 1051 and SeaNine 211), whilst negligible data are available for the others.

4 Fates, Effects and Environmental Risks

The behaviour and fates of booster biocides in coastal waters have been reviewed [9, 10] and further details are provided in chapters within this volume.

Removal of the compounds from the water column can occur through biotic degradation, photo-degradation, chemical hydrolysis, sorption to particulates followed by sedimentation, volatilization, or bioaccumulation. A comparative study of the disappearance of booster biocides from natural sea water containing the diatom *Amphora coeffeaeformis* [13] concluded that Irgarol 1051 and diuron were not easily degraded even after 8 weeks, whereas chlorothalonil was unlikely to persist and SeaNine 211 was easily degraded. In another study [14], the authors demonstrated that the toxicity of SeaNine 211 and zinc pyrithione to *Acartia tonsa* declined rapidly through either rapid degradation or partitioning to sediments. Half-lives of Irgarol 1051 (100 d), dichlofluanid (18 h), chlorothalonil (1.8 d), SeaNine 211 (< 24 h), zinc pyrithione (< 24 h), TCMTB (740 h) and zineb (96 h) have been reported [10]. Further details on the mechanisms of degradation are provided in the reviews [9, 10] and the other chapters of this book.

Zinc pyrithione (the zinc chelate of 2-pyridinethiol-1-oxide) is unusual as it is reported to transchelate with copper to form copper pyrithione (which also acts as a booster biocide) [15]. Both zinc and copper pyrithiones are considered to degrade rapidly.

Toxicological properties of the compounds have been reviewed [9, 11] and are detailed in the chapter by Yamada within this volume. Marine plants appear particularly vulnerable to many of these biocides. The first published study on the herbicidal properties of the booster biocides was by Dahl and Blanck [16] on the toxicity of Irgarol 1051 to periphyton communities. Long-term effects were detected at 0.25 to 1 nM (63 to 250 ng L^{-1}), which is within the range of concentrations reported for coastal waters. Later studies [17–19] have confirmed the vulnerability of algae/phytoplankton to booster biocides, and especially corals through damage to their endosymbiotic microalgae (zooxanthellae) [20]. Subsequent to the review by Konstantinou and Albanis [9], some other papers address algal toxicity [21–23]. Using natural populations of phytoplankton, Readman et al. [21] report toxic effects of Irgarol at low concentrations (Fig. 4) with an EC50 (72 h) of 70 ng L^{-1}. Again, this concentration is well within the range of concentrations reported in coastal waters.

Within the ACE Project, endocrine disruption was also assessed. Based on an evaluation of potentially suitable test systems, it was decided to apply the ER-CALUX (estrogen responsive-chemically activated luciferase expression) assay to determine (anti) estrogenicity. None of the antifoulants selected (Irgarol 1051, SeaNine, chlorothalonil, diuron, dichlofluanid, maneb and ziram) showed a strong estrogenic response.

The critical feature in risk evaluation of the booster biocides relates to persistence and toxicity. Although substantial information has been accrued, some authors [24] consider that additional data is still required to properly evaluate the risks associated with the widespread use of Irgarol 1051, diuron, SeaNine 211 and chlorothalonil. These authors caution against the use of TCMS pyridine, TCMTB and dichlofluanid and, again, identify a lack of appropriate data. In their initial risk evaluation, zinc pyrithione and zineb appear the least hazardous options for the aquatic environment.

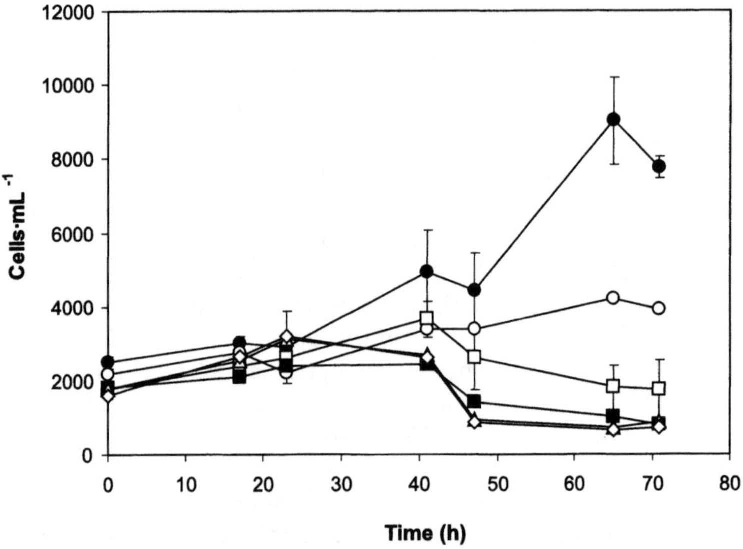

Fig. 4 Cell numbers (measured by flow cytometry) of marine eukaryotic phytoplankton (natural population) exposed to different concentrations of Irgarol 1051 over a 72-h period. Symbols indicate: • 0 ng L^{-1} (control); ○ 112 ng L^{-1}; □ 331 ng L^{-1}; ■ 465 ng L^{-1}; △ 1082 ng L^{-1}; ◇ 2020 ng L^{-1} (adapted from [21])

5 Recent Legislation

The permitted use of booster biocides in selected European Countries (those of the ACE partners) has already been provided in Table 1. The legislative position from a global perspective (a key points summary) is available *via* the Internet [25].

In the UK, diuron, Irgarol 1051, TCMTB, chlorothalonil, TCMS pyridine, and Sea-Nine registrations have recently been removed for boats less than 25 m (however, Sea-Nine is still registered on ships > 25 m) [26]. The currently registered biocides for boats less than 25 m are zinc pyrithione, dichlofluanid, and zineb [26]. In the Netherlands, files for antifouling agents are being reviewed. Current information can be obtained from the "College vor Toelating van Bestrijdingsmiddelen" (CTB), the regulatory authority. In Spain, Greece, and France, there are very limited registration schemes and, in principle, all can be used. In Denmark, diuron and Irgarol 1051 were banned for use on pleasure craft in 2000. Results from an environmental risk analysis of Sea-Nine and zinc pyrithione in Danish waters demonstrated that, in most cases examined, the PEC/PNEC (predicted no effect level/predicted no effect level) ratio was less than 1, indicating an acceptable risk. The European Union has instituted the Biocidal Products Directive (BPD) (98/8/EC) for authorisation of biocidal products within the European Union. The BPD harmonises the data requirements for existing and new biocides within the EU. Antifouling agents are included in this directive (Product Type 21). Any antifouling agent seeking registration was required to notify in 2002 and provide a base set of data. Time-scales for submission of additional necessary data have not been established; however they will be part of the 10-year plan to fully implement the directive. It has been suggested that antifouling agents are likely to be reviewed in 2006 [25].

6
Future Developments

Clearly there is a need to protect non-target organisms in the environment from antifouling agents. However, antifoulants are critical for shipping, and with an imminent ban on TBT, alternative strategies must be found. Yebra et al. [1] have reviewed this topic.

In the short term, the most feasible option is to utilize copper-based formulations with the most environmentally friendly booster biocides. Whilst substantial data on the biocides is presented within this volume, there remain deficits in our knowledge concerning the environmental behaviour and effects of these compounds. Persistence is an important feature of the agents, but needs to be tempered with longevity of antifouling performance. In the slightly longer term, some companies are addressing this issue through polymer research, which controls the release of active compounds and affords self-polishing capabilities to maintain the active ingredient in the surface layer. This challenge is substantial, as it needs to maintain a constant, and ideally a small release, that is adequate for antifouling the vessel (even during stationary periods), whilst maximizing the lifetime of the coating.

An alternative strategy to the polymer option relates to the use of natural products (or synthesized analogues), which prevent fouling of organisms in the wild. Usually, these compounds act enzymatically by interfering with the metabolism, or dissolving adhesive materials of the fouling organisms. Structures include terpenoids, steroids, heterocyclics, alkaloids and polyphenolics. This approach to the prevention of fouling is, however, far from commercial exploitation.

Another important area of technological development of antifouling relates to non-stick fouling-residue coatings that prevent adhesion by the fouling organisms. The demands of this type of material, presently, restricts choice to two families of compounds: the fluoro-polymers and silicones. This option, from an environmental viewpoint, is appealing. Results from performance tests for currently available coatings indicate modest performance, requiring comparatively high speeds to remove fouling organisms. Coupled with expense, poor adhesion qualities to the hull and susceptibility to damage, further advances will be necessary [1].

Finally, it would appear logical that production of engineered highly reactive nano-particles would afford an alternative, high-tech option. To the knowledge of the author, however, this has not, as yet, been addressed.

7
Conclusions

The use of antifouling materials has been pertinent since ancient times. The removal of TBT from control options has provoked change, primarily through accelerated introduction of organic booster biocides. Through the ACE project, the most popular biocides were evaluated and, together with information in the literature, combined usage, persistence and toxicity of some products (e.g. Irgarol 1051 and diuron) were shown to be likely to cause damage to non-target organisms, especially phytoplankton, periphyton and corals. Data necessary for the risk assessments of other booster biocides is lacking. Options for the future will, in the short term, probably be directed towards copper-based paints combined with the most environmentally friendly booster biocides. In the medium term, improvements in polymer synthesis to control release is likely. Research into natural toxin products and their synthetic analogues, together with investigations into non-stick coatings will continue, aiming to afford a longer-term solution to fouling. Although not presently addressed, nano-particle technology might also afford an alternative option to antifouling in the future.

Acknowledgements Much of the reported research was supported by the European Commission under the ACE (Assessment of Antifouling Agents in Coastal Environments) contract (MAS 3-CT98-0178) (a component of the EC IMPACTS cluster). The author is most grateful to all the scientists who participated in the ACE project and who contributed to the data described herein.

References

1. Yebra DM, Kiil S, Dam-Johansen K (2004) Prog Org Coat 50:75
2. Alzieu C (1991) Mar Environ Res 32:7
3. Evans SM, Leksono T, McKinnell PD (1995) Mar Pollut Bull 30:14
4. Bryan GW, Gibbs PE, Burt GR, Hummerstone LG (1986) J Mar Biol Assoc UK 66:611
5. Ten Hallers-Tjabbes CC, Kemp JF, Boon JP (1994) Mar Pollut Bull 28:311
6. Communauté Européenne (1989) J Off Communautes Eur L398/19
7. Environment Agency (1998) Environmental Problems from Antifouling Agents. R & D Technical Report P215 EA. Bristol, UK
8. Readman JW, Long Wee Kwong L, Grondin D, Bartocci J, Villeneuve J-P, Mee LD (1993) Environ Sci Technol 27:1940
9. Konstantinou IK, Albanis TA (2004) Environ Intern 30:235
10. Thomas KV (2001) Biofouling 17:73

11. Hall LW, Giddings JM, Solomon KR, Balcomb R (1999) Crit Rev Toxicol 29:367
12. Vouvoulis N, Scrimshaw MD, Lester JN (1999) Appl Organomet Chem 13:135
13. Callow ME, Finlay JA (1996) Biofouling 10:239
14. Madsen T, Samsoe-Petersen L, Gustavson K, Rasmussen D (2000) Danish EPA Report No. 531 2000
15. Turley PA, Fenn RJ, Figura PM, Ritter JC (2000) Biofouling 15:175
16. Dahl B, Blanck H (1996) Mar Pollut Bull 32:342
17. Okamura H, Aoyama I, Takami T, Maruyama T, Suzuki Y, Matsumoto M, Katsuyama I, Hamada J, Beppu T, Tanaka O, Maguire RJ, Liu D, Lau YL, Pacepavicius GJ (2000) Mar Pollut Bull 40:754
18. Fernandez-Alba AR, Hernado MD, Piedra L, Chisti Y (2002) Anal Chim Acta 456:303
19. Jacobson AH, Willingham GL (2000) Sci Total Environ 258:103
20. Owen R, Knap A, Toaspern M, Carbery K (2002) Mar Pollut Bull 44:623
21. Readman JW, Devilla RA, Tarran G, Llewellyn CA, Fileman TW, Easton A, Burkhill PH, Mantoura RFC (2004) Mar Environ Res 58:353
22. Devilla RA, Brown MT, Donkin M, Tarran GA, Aiken J, Readman JW (2005) Mar Ecol Prog Ser 286:1
23. Devilla RA, Brown MT, Donkin M, Readman JW (2005) Aquatic Tox 71:25
24. Voulvoulis N, Scrimshaw MD, Lester JN (2002) Chemosphere 47:789
25. International Coatings Ltd (2004) http://www.yachtpaint.com/superyacht/sy/pdf/antifouling_legislation.pdf
26. Health and Safety Executive (2002) Information Document HSE 730/15:13

Hdb Env Chem Vol. 5, Part O (2006): 17–50
DOI 10.1007/698_5_048

Published online: 22 December 2005

Chemistry and Fate of Organotin Antifouling Biocides in the Environment

Iwao Omae

Omae Research Laboratories, 335-23, Mizuno, Sayama, 350-1317 Saitama, Japan
um5i-oome@asahi-net.or.jp

Abstract Organotin antifouling paints are degraded relatively rapidly in seawater mainly by microorganisms, but they have high affinities to sediment and organisms in the sea. The concentrations of tin in seawater have been observed to decrease in many harbors in the world since the ban to use the paints for small vessels came into force, however, in a few areas frequented by large ships, organotin compounds continued to be input onto the surface of the sediment. Therefore, recently, the number of articles on the investigation of depuration of organotin compounds by organisms have increased. For example, pyroverdins, azotochelin and ornibactin obtained from microorganisms such as *Pseudomonas chlororaphis* and *Burkholderia cepacia*, and organisms such as dinoflagellates, diatoms, mussel, green algae and chrysophytes are able to depurate organotin compounds of TBT and TPT in seawater. Furthermore, with the help of these depuration technologies, we hope that the marine environment will improve in the world, at least up to the point in time where environmentally friendly best alternative antifouling paints are used routinely for all vessels.

Keywords Bioaccumulation · Biomethylation · Degradation · Fate · Organotin antifouling biocides

Abbreviations

DBT	dibutyltin compounds
DPT	diphenyltin compounds
IMO	International Maritime Organization
MBT	monobutyltin compounds
MEPC	Maritime Environmental Protection Committee
MPT	monophenyltin compounds
PVC	polyvinyl chloride
TBT	tributyltin compounds
TBTCl	tributyltin chloride
TBTF	tributyltin fluoride
TBTO	tributyltin oxide
TPT	triphenyltin compounds

1 Introduction

The historical development of organotin antifouling paints has two big steps. The first step was the discovery of tributyltin (TBT) compounds in the 1950s by the research group of van der Kerk in the Netherlands. The second step was the discovery of self-polishing antifouling paints in the 1970s. The first step involved the discovery of the most effective antifouling compounds, although they exhibit very low toxicity towards mammals, among the many kinds of antifoulants. The second step involved the discovery of polymers with very high performance as the paints containing the antifoulants and they were used for a long time. These polymers easily allowed a constant release rate of the antifoulants into seawater and maintained a low frictional resistance to seawater by keeping a polished smooth surface. This meets the need to reduce the ecological impact of components of antifouling paints that are being released into the aqueous environment. Leaching occurs by hydrolysis of the painted polymer at the surface of the ships hull by seawater. The self-polishing antifouling paints have very long lives and are the most cost effective. Almost all vessels in the world used these antifouling paints. The leaching rates of organotin compounds from the antifouling paint film into seawater from one vessel are very low. However, as too many vessels release organotin compounds into seawater, the amount of organotin compounds in seawater in closed sea areas such as harbors and marinas, gradually increases year by year.

The first use of the organotin-based antifouling boat-bottom paints began in the early 1960s. In 1974, oyster growers first reported the occurrence of abnormal shell growth in *Crassostrea gigas* (the Pacific oyster) along the east coast of England. However, it was not until the mid 1980s that researchers in France and the UK began to suggest that the use of TBT in the antifouling paints was adversely impacting a number of marine species other than the fouling organisms [1–3].

Legislation to ban the use of the organotins for small boats was first introduced in France in 1982 and followed by the UK in 1987. Similar legislations have been introduced worldwide since then, and led to a significant fall in the concentrations of organotins in seawater.

The "International Convention on the Control of Harmful Antifouling Systems on Ships" adopted on 5 October 2001 by the International Maritime Organization (IMO) stated that: (1) by 1 January 2003, a ban on the application of TBT-based antifouling paints should be introduced; and (2) 1 January 2008 is the last date for having a TBT-based antifouling paint on a vessel. In addition, the MEPC also proposed that the IMO promotes the use of environmentally safe antifouling technologies to replace TBT [1].

However, by November 2005, two years later, there were only 12 countries including Antigua & Barbuda, Bulgaria, Denmark, Japan, Latvia, Nigeria, Norway, Poland, Romania, Saint Kitts and Nevis, Spain and Sweden, who had ratified this control. Since at least 25 countries are required to ratify the control it has not become effective yet.

There are three issues regarding the control: The first is that the organotin antifouling paints have too high a performance, as some researchers suggest that organotin antifouling paints have ten years of life and they are too cost effective. The second is that alternative antifouling paints do not inspire confidence in providing high antifouling activity and environmental problems yet. The third is that many articles have reported that concentrations of the organotins in marinas decreased after regulation of small vessels.

Although the control was approved in 1999—by the IMO conference—by 2005 it had still not come into effect. It has been stated that world opinion is gradually changing from supporting organotin antifouling paints to supporting alternative antifouling paints. Hence, if good alternative antifouling paints were to be found, organotin antifouling paints would be easily and completely replaced by the alternative antifouling paints. However, at this time, the best alternative antifouling paints have not been determined. Hence, there is a need for reliable data regarding both the fate of organotin compounds and the performance of the most cost-effective and environmentally friendly alternative antifouling paints.

2 Organotin Compounds and Their Properties

2.1 Organotin Compounds

There are two kinds of organotin compounds, i.e., bivalent compounds and tetravalent compounds. The bivalent compounds, which have two organic groups and one lone pair of electrons, are shown in structural formula 1.

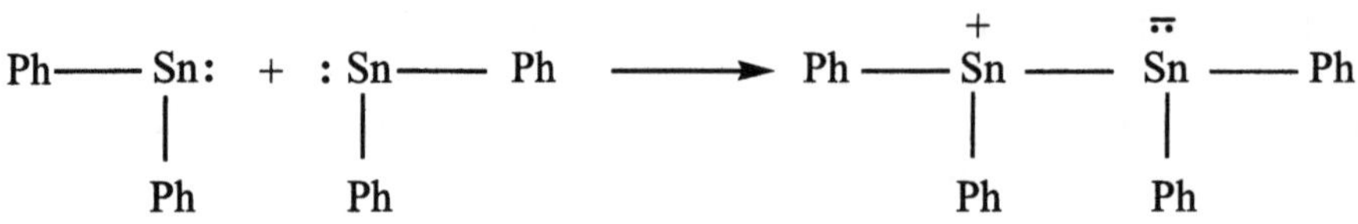

$\longrightarrow$ higher polymers (1)

These compounds are usually polymeric substances; and only in a few cases, monomeric compounds. They slowly polymerize on standing [4]. The bond angle is approximately 90° (as p^2 bonding), for example, 94(5)° (SnF_2) [5, 6]. In structural formula 2, a molecule with sp^2 hybridization, a vacant *p* orbital and bulky groups [4, 7, 8] is shown. In this case, the bond angle is for example 117.4° in $(C_5H_5)SnCl$ [9].

$$R_3SnX \rightarrow R_2SnX_2 \rightarrow RSnX_3 \rightarrow SnX_4 \qquad (2)$$

It is particularly to be noted that these molecules exist in a singlet state with their nonbonding electron paired. Since the central tin atom in either of the structural formulaes, 1 and 2, has only six electrons in the valence shell, polymerization forming a stable octet is expected to occur [4]. When freshly prepared, diphenyltin is monomeric, but it readily polymerizes to reach a pentamer or greater based on its molecular weight as shown in Eq. 1 [10].

These bivalent tin compounds, wherever possible, adopt structures in which the metal achieves a coordination number higher than two either by complexation, chelation or by bridging. The bivalent tin compounds having active lone pair electrons are utilized as catalysts [11]. The potential for bivalent organometallic tin species in organic syntheses has also been studied. However, as the organotin compound, tetravalent compounds (R_nSnX_{4-n}, R is alkyl, aryl, etc.; X is halogen, hydroxyl, etc.) are used, since they are very stable and easy to handle.

The organotin compounds were first used as stabilizers for polyvinyl chloride (PVC) in 1940 [12]. Although the organotin compounds have a variety of uses, their use as stabilizers for PVC has been the most important application of the organotin compounds.

In 1950, the research team under the leadership of Prof. G.J.M. van der Kerk made important contributions to the study of organotin chemistry as described above. In particular, the powerful biocidal properties of trialkyltin and triaryltin derivatives were established [13, 14]. The introduction of biocidal uses of the triorganotin compounds began in the late 1950s, when they were used as agricultural fungicides; their uses in wood-preserving compounds began in 1960 and antifouling paints in the early 1960s [15]. Other biocidal uses include use as antifeedants, acaricides, stone preservants, disinfectants and molluscicides.

The organotin compounds are used mainly by utilizing the following two characteristics. The first is the strong affinity of the tin atom with a donor ligand atom such as sulfur, oxygen or nitrogen. The second comprises physiological activities such as biocidal activities (e.g., bactericidal, fungicidal, acaricidal,

Table 1 Industrial applications of organotin compounds [16–22]

Application	Compound
	R_3SnX
Agriculture	
fungicides	Ph_3SnX (X = OH, OAc), $(Cy)_3SnOH$
antifeedants	Ph_3SnX (X = OH, OAc),
acaricides	$(Cy)_3SnOH$, $((Ph-C(CH_3)_2CH_2)_3Sn)_2O$
Antifouling paint biocides	Ph_3SnX (X = OH, OAc, F, Cl, SCS · NMe_2), $(Bu_3Sn)_2O$, Bu_3SnX (X = F, Cl, OAc) $(-CH_2C(CH_3)(COOSnBu_3)-)_n$
Wood preservative fungicides	$(Bu_3Sn)_2O$, $(Bu_3Sn)_3PO_4$, Bu_3Sn(naphthalenate)
Stone preservation	$(Bu_3Sn)_2O$
Disinfectants	$Bu_3SnOCOPh$, $(Bu_3Sn)_2O$
Molluscicides (field trials)	Bu_3SnF, $(Bu_3Sn)_2O$
	R_2SnX_2
Heat and light stabilizers for rigid PVC	$R_2Sn(SCH_2COO-1-Oct)_2$ (R = Me, Bu, $BuOCOCH_2CH_2$), $(R_2)Sn(OCOCH=CHCOO)_n$ (R = Bu, Oct), $(R_2)Sn(OCOCH=CHCOOOct)_2$, $(R_2)Sn(OCOC_{11}H_{23})_2$, $(R_2)Sn(SC_{12}H_{25})_2$
Homogeneous catalysts for RTV silicon, polyurethane foam and transestserification reactions	$R_2Sn(OCOCH_3)_2$, $R_2Sn(OCO-i-Oct)_2$, $(R_2)Sn(OCOC_{11}H_{23})_2$, $(R_2)Sn(OCOC_{12}H_{25})_2$, $(Bu_3Sn)_2O$
Precursor for forming SnO_2 films on glass	Me_2SnCl_2
Anthelmintics for poultry	$Bu_2Sn(OCOC_{11}H_{23})_2$
	$RSnX_3$
Heat stabilizers for rigid PVC	$RSn(SCH_2COO-1-Oct)_3$ (R = Me, Bu, Oct, $BuOCOCH_2CH_2$), $(BuSnS_{1.5})_4$ $R_2Sn(SCH_2COO-i-Oct)_2$ (R = Me, Bu, Oct)
Homogenous catalysts for transesterification reactions	$[BuSn(O)(OH)]_n$, $BuSn(OH)_2Cl$
Precursor for SnO_2 films on glass	$BuSnCl_3$, $MeSnCl_3$ [a], Me_2SnCl_2

[a] These compounds are used in combination with the corresponding R_2SnX_2 derivatives

insecticidal and molluscicidal activities) and repellent activities against wood-destroying organisms, marine animals, marine plants, rodents, etc.

The industrial uses of the organotin compounds, utilizing the first characteristic, are as stabilizers for PVC and the other vinyl polymers, and as catalysts. The industrial uses utilizing the second characteristic are as wood preservatives, antifouling agents, agrochemicals, pharmaceuticals, disinfectants, rodent repellents, masonry and stonework-protecting agents, and slime-preventing agents, etc. Other minor uses of organotin compounds are in glass applications (such as easy handling of raw materials by using inorganic tin compounds (SnO_2), flame retardants, etc.) [16].

The industrial applications of the organotin compounds are shown in Table 1 [16–22].

In 1950, the annual production of organotin compounds was only around 50 tons. This rose to 2000 tons in 1960 and 16 000 tons in 1970, and production in the mid-1980s was estimated at over 40 000 tons [15].

As discussed by Blunden et al. [18] in 1985, the most important use of the organotins was as PVC stabilizers (ca. 70%); biocidal uses such as antifoulants, agrochemicals and wood preservatives, are ca. 20%. Around 3500 tons per year were used as organotin antifoulants.

2.2 Properties of Organotin Compounds

Organotin compounds are one of the organometallic compounds, since they have the characteristic metal-carbon bonds. However, they are very different from general organometallic compounds, especially, from organotransition metal compounds in their stability. For example, tetramethyltin is very stable, and it decomposes between 440 and 493 °C over a range between 5 and 185 mm Hg initial pressure [23].

Basic organotin compounds, their related compounds and their physical properties are shown in Table 2 [4, 16, 24]. They are liquid or solid at room temperature, and they are stable and easy to handle.

Tributyltin (TBT) has a low aqueous solubility and relatively high affinity for particulate matter, providing a direct and potentially persistent route of its entry into benthic sediment. The need to determine the fate of TBT in such environments is consequently a priority issue, as the sediment becomes the major reservoir for this pollutant. Initial tests are shown in Fig. 1 [25]. The sorption of TBT to the sediment was rapid so that the major proportion was taken up within 10 min and an equilibrium was achieved within 2 h, when 85.7% of the added TBT had bound with suspended particles in the dark at 15 °C [25]. The difference of affinities of the organotin compounds with water and sediment relates to the difference of the half-lives of organotin compounds in water and the sediment as described in the next section.

Table 2 Physical properties of organotin compounds [4, 16, 24][a]

Compound	Melting point (°C)	Boiling point (°C or °C mm Hg^{-1})	Refractive index (n_D^{20} (°C))	Solubility [b] (ppm)
Bu_3SnF	248–252			6
Bu_3SnCl		145–147/5	1.4903 (25)	16
Bu_3SnBr		163/12	1.5022 (25)	
Bu_3SnI		168/8	1.5345	
Ph_3SnF	357(dec.)			1.2
Ph_3SnCl	105.5–107	249/13.5		5.2
Ph_3SnBr	121–122			
Ph_3SnI	121			
Bu_2SnF_2	156–157			
Bu_2SnCl_2	43	153–156/5		4
Bu_2SnBr_2	20			
Bu_2SnI_2		145/6		
Ph_2SnF_2	>360			
Ph_2SnCl_2	42–44	333–337/760		
Ph_2SnBr_2	38			
Ph_2SnI_2	71–72	176–182/2		
$BuSnF_3$		93/10		
$BuSnCl_3$		93/10		
$BuSnBr_3$		77–90/0.2 (dec.)		
$BuSnI_3$		154/5		
$PhSnCl_3$		142–143	1.5844	
$PhSnBr_3$		182–183/25		
Me_4Sn	-54	78/760	1.4386 (25)	
Me_3SnBu		149–150/726	1.4560 (21.5)	
Me_2SnBu_2		70/4.4	1.4640 (25)	
$MeSnBu_3$		121/10		
Bu_4Sn		145/10	1.4727	
Ph_4Sn	224–225			
$(Bu_3)_2SnO$				8
Bu_3SnOAc	99–100	220–230/10	1.8472	16
Ph_3SnOAc	121–122			2.9

[a] Reference therein
[b] Solubility values refer to seawater at room temperature

2.3
Toxicities of Triorganotin Compounds

Tin is a ubiquitous element. Our bodies contain 0.2 ppm [26, 27]. We ingest 0.2 ppm based on kg body weight from daily foods [28]. Inorganic tin at trace level has never been implicated in plant and animal malfunctions and on the whole has been assumed to be an innocuous background material [29]. Most

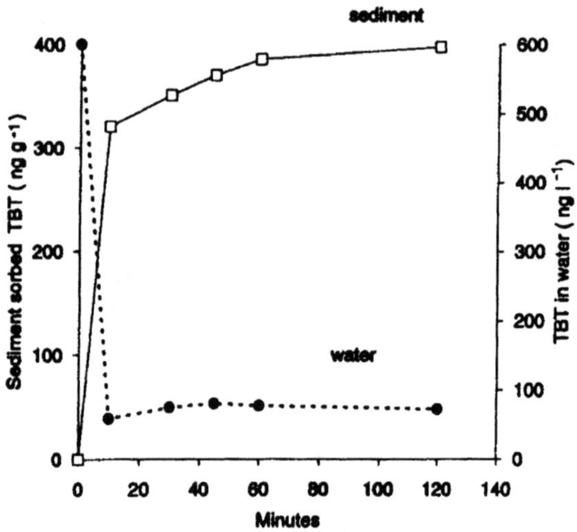

Fig. 1 The equilibration of TBT between particulates and water. The vessel initially containing 500 ml water, 1 g sediment and 300 ng ^{14}C-labeled TBT [25]

inorganic tin and organotin compounds entering mammals through food are excreted within a short period of time [29]. For example, the biological retention of $SnCl_2$ in rats is 2 days ($\geqq$ 98%) [30]. Inorganic tin has little or no effect on the animals, and common organotin compounds and inorganic tin compounds pass rapidly through the animals [29].

In the organotin series expressed by formula; $R_n SnX_{4-n}$ (R = carbon-bonded organic group; X = inorganic substituent; and n = 1–4), it is now well estab-

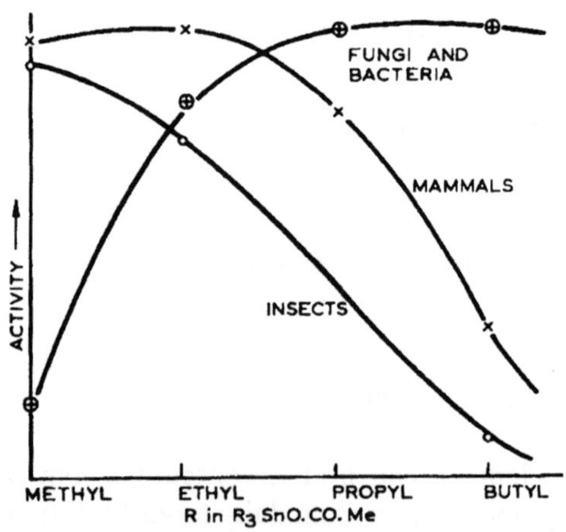

Fig. 2 Dependence of the biological activity of tri-*n*-alkyltin acetates on the nature of the alkyl group for different species [13, 14]

lished that maximum toxicity to all types of living species occurs with the triorganotin compounds R_3SnX. As the length of the carbon chain is progressively increased within a particular series of tri-*n*-alkyltin salt, R_3SnX, the mammalian toxicity reaches a maximum value when R is an ethyl group, and then falls off rapidly with further lengthening of the alkyl chain. The tri-*n*-

Table 3 Acute oral toxicity of organotin compounds [3, 14, 18, 24, 31]

Compound	LD_{50} (rats, mg kg^{-1})
R_4Sn	
Me_4Sn	195–331
Et_4Sn	9–16
Bu_4Sn	> 4000
Oct_4Sn	> 4000
R_3SnX	
Me_3SnCl	9–20
Me_3SnOAc	9.1
Et_3SnCl	10
Et_3SnOAc	4.0
$Pr_3SnOCOMe$	118
$Pr_3SnOSnPr_3$	120
Bu_3SnCl	122–349
Bu_3SnF	200
$Bu_3SnOCOMe$	125–136, 125–380
$Bu_3SnOCO(CH_2)_9Me$	205
$Bu_3SnOCO(CH_2)_7CH = CH(CH_2)_7Me$	195
$Bu_3SnOCOPh$	132
$Bu_3SnOSnBu_3$	112–234, 148–234
Oct_3SnCl	> 4000
$Oct_3SnOCOMe$	> 1000
Ph_3SnCl	125, 125–135
Ph_3SnF	1170, 160
Ph_3SnOH	500–600, 108–360
$Ph_3SnOCOMe$	125–491, 125–150
R_2SnX_2	
Me_2SnCl_2	74–237
Et_2SnCl_2	66–94
Bu_2SnCl_2	112–219
Bu_2SnO	487–520
Oct_2SnCl_2	> 4000
Oct_2SnO	2334–2350, > 4000
$RSnX_3$	
$MeSnCl_3$	575–1370
$BuSnCl_3$	2200–2300
$OctSnCl_3$	2400–3800

octyltin derivatives are essentially nontoxic as shown in Table 3 [13, 14, 18, 31] and in Fig. 2 [13, 14]. With regard to the toxicity of a TBT salt, (n-Bu_3SnCl) in mammals, acute oral LD_{50} values are in the range of 122–349 mg/kg (rats) as shown in Table 3. Their biological activity is almost entirely due to the Bu_3Sn moiety, since in common with other trialkyltin compounds, the variation of the inorganic substituent X does not usually have a significant effect on their toxicity. The "X" group tends to affect properties such as volatility and solubility.

The acute mammalian toxicities of the common triphenyltin (TPT) biocides are generally very similar to, and often lower than those of their tributyltin analogues. It is also interesting to note that anionic groups appear to have an effect on the toxicity in these compounds as shown in Table 3.

Although TBT and TPT compounds are not unduly hazardous to humans, as discussed above, they are nevertheless very effective biocides against a wide range of marine fouling organisms. Information on the activities of various TBT and TPT derivatives against marine algae, barnacles, shrimps and tubeworms has been published [13, 14].

Gadd [29] reported on some toxic effects of TBT compounds on microbial processes. The inhibition of microbial processes by TBT has been recorded for all major groups, with the main interactions occurring at cellular membranes and chloroplasts, or the mitochondria in eukaryotes as shown in Table 4 [32–36]. Maguire [37] and White et al. [22], also reported on toxicity of TBT to organisms in detail.

Table 4 Some toxic effects of TBT compounds on microbial processes [32] [a]

Process affected	Organism(s)/organelle(s)	Inhibitory concentration (IC) (μM)
Respiration	Bacteria	0.04–1.7
Photosynthesis	Cyanobacteira	~ 1 (IC_{50})
Nitrogen fixation	*Anabaena cylindrica*	< 1 (IC_{50})
Primary productivity	Microalgae	0.00055–0.0017
Growth	Microalgae	0.00017–0.0084
Energy-linked reactions	*Escherichia coli*	0.15–> 50 (IC_{50})
Growth	*Aureobasidium pullulans*	27 (IC_{50})
Growth/metabolism	Fungi	0.28–3.3
	Bacteria	0.33–16
Photophosphorylation and ATP synthesis	Chloroplasts	0.56–5
H^+-ATPase activity	*Neurospora crassa*	
plasma membrane		0.06 (IC_{50})
mitochondrial		0.01 (IC_{50})
vacuolar		0.6 (IC_{50})

[a] Gadd [29] compiled with data from original sources cited in publication from [30–33]

The chronic toxicity of the organotin compounds arises from their abilities as endocrine disruptors leading to disorders such as imposex, intersex and masculinization of abalone (ear shell, *Haliotis* sp.) The symptom of imposex (imposed sexual organ) is defined as a superimposition of the male characteristic, such as a penis and vas deferens, in female gastropods [38, 39].

Schweinfurth and Günzel [40] reported on the mammalian toxicity and risk evaluation for humans. Since TBT compounds can produce severe irritation of the skin and mucous membrane, their contact with the skin and eyes as well as the inhalation of their spray mist or dust must be prevented. TBT vapor is not considered hazardous. In repeated dose studies, lymphotoxicity and hepatobiliary toxicity were predominant findings. There was no evidence of damage to the central nervous system. From the available data, it is concluded that TBTs do not represent a mutagenic, teratogenic or carcinogenic hazard to humans.

3 Degradation of Organotin Compounds

3.1 Organotin Compounds as Ideal Antifoulants

Most organotin researchers' faith in organotin compounds as ideal antifoulants remained unshaken till early in the 1980s; this was because much experimental data actually showed that the above was true. In the mid 1960s, Professor Matsuda frequently said to us in his research laboratories "Organotin compounds are ideal compounds because they change to nontoxic inorganic tin compounds after we have used them".

Triorganotin compounds undergo degradation in the environment via progressively less toxic di-(R_2SnX_2) and mono-organotin ($RSnX_3$) derivatives, to form a harmless inorganic tin residue. This cleavage of the tin-carbon bonds may occur photolytically by UV light, microbiologically by fungi or bacteria, or by chemical attack, and a typical breakdown scheme for the TBT and TPT compounds is illustrated in Fig. 3 [41]. Therefore, organotin researchers did not expect the early environmental problems caused by the organotin compounds.

3.2 Degradation of Organotin Compounds by Hydrolysis, Ultraviolet Light and Microorganisms

The degradation of the organotin compound may be defined as the progressive removal of organic groups from the tin atom as shown in Eq. 2 (see Fig. 3).

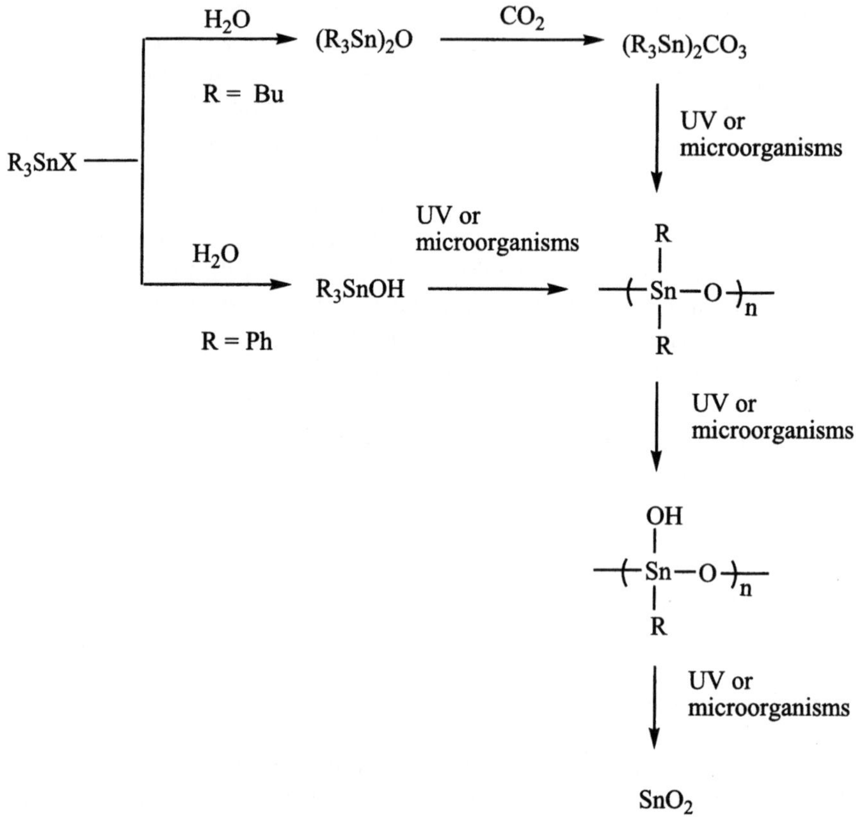

Fig. 3 Environmental degradation scheme for TBT and TPT compounds [41]

The degradation of the organotin compounds is caused by hydrolysis, ultraviolet light and microorganisms. The most dominant degradation is caused by microorganisms such as bacteria, algae and fungi [24].

Almost all of the sun's emitted radiation in the UV region with wavelengths shorter than 290 nm is absorbed by a thin band of ozone. However, the light of 290 nm wavelength possesses an energy of approximately 300 kJ mol^{-1}, and this is above the typical range [42] of tin–carbon bond dissociation energies (190–200 kJ/mol). Therefore, if absorption of light by the organotin takes place, degradation can occur [24].

Mailhot et al. [43] pointed out that photodegradation by a photoredox process by iron(III) occurs. Upon irradiation at $\lambda_{excitation} > 300$ nm, a photoredox process yielding iron(II) and OH radicals was observed. The disappearance of TBT was proved to involve only an attack by OH radicals in the presence of iron(III): the quantum yield of TBT disappearance was determined.

$$Fe^{3+} + H_2O \rightarrow Fe^{2+} + {}^{\cdot}OH + H^+ \quad (3)$$

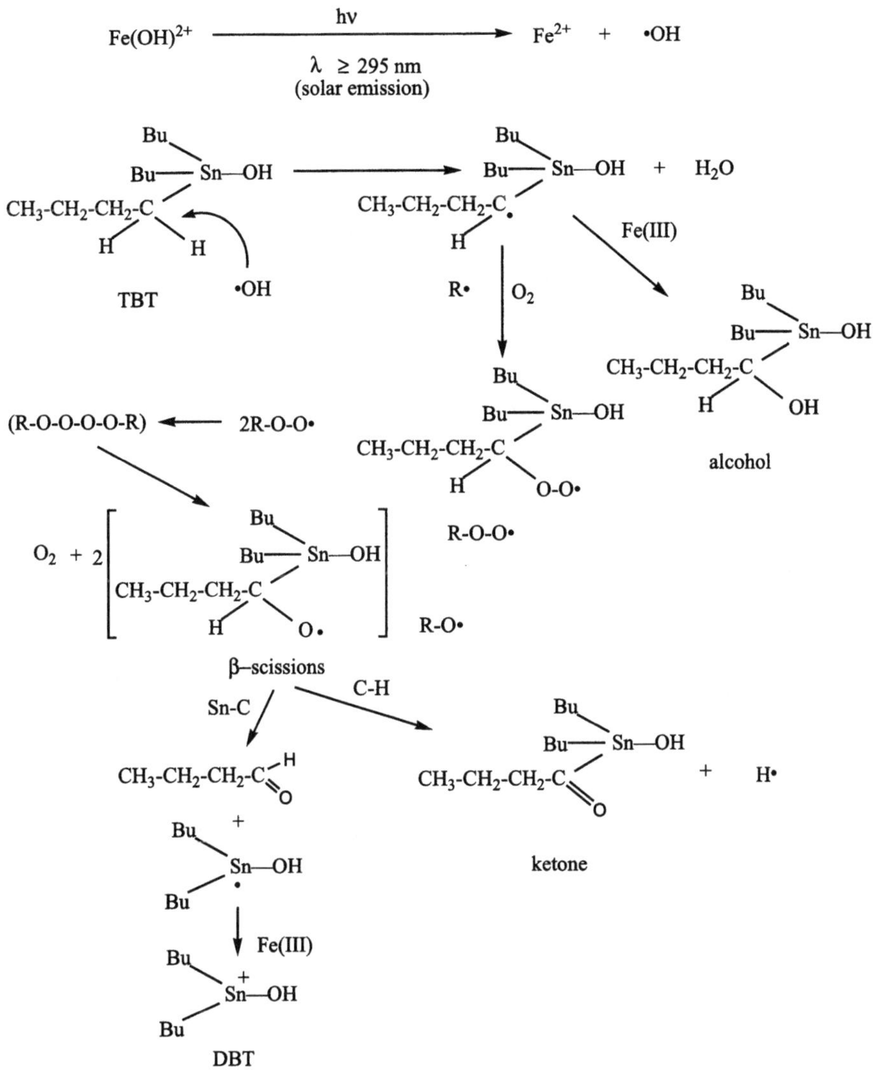

Scheme 1 Proposed mechanism for iron(III)-photoinduced degradation of TBT [43]

The complete mineralization of TBT was achieved with a long irradiation time, leading to innocuous inorganic tin. A mechanism for iron(III)-photoinduced degradation of TBT is proposed as shown in Scheme 1 [43].

Mailhot et al. [44] considered also that the degradation of TPT carried out by the photoredox process takes place in iron(III) aquacomplexes. TPT elimination was proved to come only from attack by hydroxyl radicals generated upon irradiation at 365 nm of $Fe(H_2O)_5OH^{2+}$. The first step is the formation of a hydroxyl radical adduct with the benzene ring. The main process is the stepwise dephenylation of the starting TPT as shown in Scheme 2. Hydroxylated phenyltin derivatives were also formed, but only as minor photoproducts. The process was shown to be efficient with artificial light as well as with solar light [44].

On a field of crops, the degradation of organotin compounds such as fungicides mainly proceeds photolytically. TPT is a fungicide used on a variety of crops throughout the world. In a study of pecan orchards in central Georgia in the USA, TPT concentrations in foliage and soil immediately after its application (0.5 days) were up to 33 and 1.6 $\mu g\,g^{-1}$ dry weight, respectively, and those after 10 or 15 days were decreased to 11 $\mu g\,g^{-1}$ dry weight (33%, 10 days) and 0.8 $\mu g\,g^{-1}$ dry weight (50%, 15 days), respectively. Besides TPT, diphenyltin (DPT) and monophenyltin (MPT) were present in the leaves and soil, with MPT generally the predominant compound. The proportion of MPT to TPT in pecan leaves increased with time after spraying, since MPT is indicative

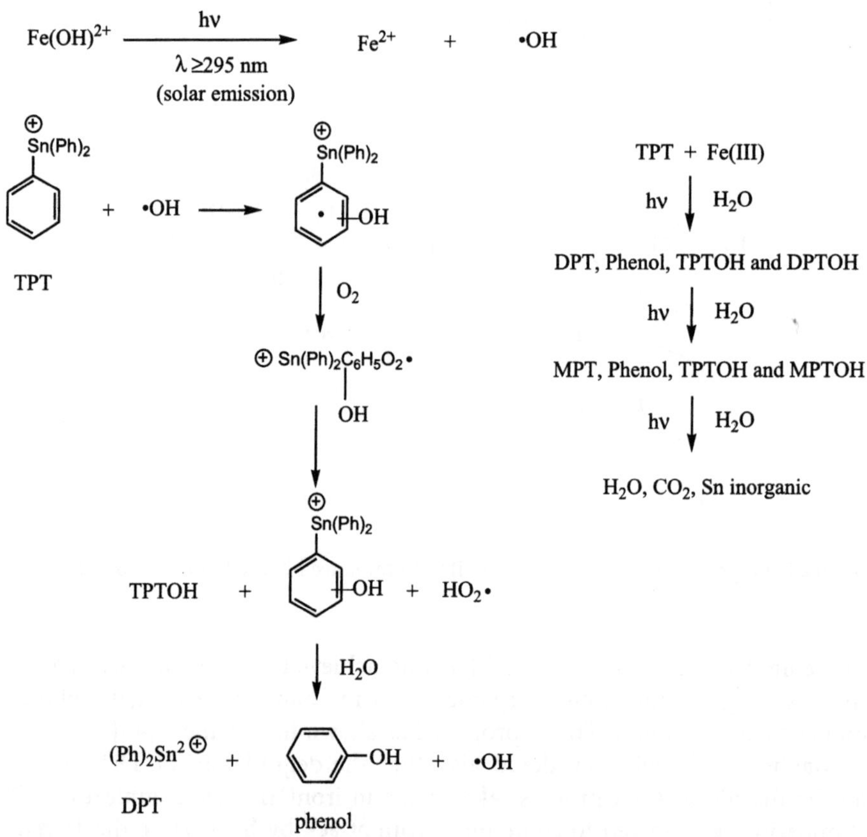

Scheme 2 Degradation of TPT by photoredox process taking place in iron(III) aquacomplexes [44]

of the photolytic degradation of TPT. The microbial degradation of a radiolabeled TPT in soil or sediment samples was slow, with only 5% degraded during a 14 day incubation period.

With regard to the organotins in aqueous systems, less is known about the patterns of degradation. The reason for this is that, in general, with the organotin compounds, their aqueous solubility decreases as the length of the alkyl chain increases, and tributyl- and triphenyltin compounds have solubility's typically of only a few ppm as shown in Table 2. Maquire et al. [45] showed that the UV degradation of tributyl species in water occurs in natural sunlight and the environmental half-life of tributyltin oxide (TBTO) in water was estimated to be approximately 89 days [45]. Hence, obviously, UV degradation is only important in situations where the organotins are exposed to sunlight and is important close to the surface of water because seawater contains a large amount of suspended solid and the organotins have a high affinity for the soil and sediment [24].

In Brest naval harbor water from September 1993 to December 1997, the levels of contamination were more or less the same (14–75 ng L^{-1}) during the four-year study period. The variations of the DBT concentrations showed a seasonal pattern and had a statistically significant relationship with those of water temperature [46]. This lends support to the hypothesis of a dominant effect of microbiological activity (favored by the increase in temperature) in the degradation of TBT.

A colder temperature slowed the degradation as shown in Table 5 [47–51]. This evidence also shows that the degradation of TBT takes place by microorganisms.

Table 5 Degradation of TBT in the marine environment [47–51]

Comments	Initial TBT concentration (μg L^{-1})	Reaction order	Half-life	Rate constant	Refs.
Natural water					
radiolabeled TBT	0.615	First	3.5 days	0.20/day	[48]
ambient marine water	0.57		6 days		[49]
natural water	0.8	First	60 days (at 5 °C)		[49]
filtered seawater	0.55	First	2.5 days	0.28/day	[50]
Sediment					
sediment cores from a marina	0.3	First	1.85 years	0.375/y	[51]
Biota					
eelgrass grown in TBT-spiked seawater	0.14	First	9.6 days	0.072/d	[50]

^{14}C-labeled triphenyltin acetate, TBTO, TBT fluoride and TPT chloride in soil could be broken down to inorganic tin; since carbon dioxide was evolved, and since as a similar breakdown did not occur in sterile soil, the degradation was ascribed to the ability of certain microorganisms to metabolize the organotin compounds [24, 52–55]. Gram-negative bacteria *Pseudomonas aeruginosa* and *Alcaligenes faecalis* and green alga, *Ankistrodesmus faecalis* are reported as deakylation organisms [56, 57].

Sheldon [55] reported the breakdown of ^{14}C-labeled TBTO in soil and showed that its degradation occurred faster under aerobic rather than anaerobic conditions.

Mussels are used as biomonitors to evaluate the concentrations of the organotin compounds in Dutch fresh water at two locations near potato crops that were sprayed with TPT at the end of the summer of 1992 [58]. The half-life times in the field were calculated to be 100 to 200 days for TPT as shown in Table 6 [58–62]. TBT is more easily depurated by zebra mussels than TPT [58].

The biodegradation of TBT and DBT in unfiltered seawater in summer is rather fast; their half-lives are about a week. But pretreatment with a glass fiber filter makes the half-life of TBT much longer (about 80 days) [63].

The organotin compounds are mainly degraded by microorganisms, and this is the depuration of organotin compounds by organisms. This depuration can be done not only by microorganisms but also by many other organisms such as mussels, fish and mammals under not so high concentrations of the organotins.

Table 6 Uptake and depuration rate constant of TBT and TPT for the zebra mussel (*Dreissena polymorpha*) and other organisms [58–62]

Compound	Organism	Study type	$T_{1/2}$ (day)	Refs.
TPT	*Dreissena polymorpha*	field	200±89	[58]
TPT	*Dreissena polymorpha*	field	105±49	[58]
TBT	*Dreissena polymorpha*	field	26 [a]	[59]
TBTO	*Mytilus edulis*	lab (18 °C)	2–14 [a]	[60]
TPT	*Mytilus edulis*	field	133 [a]	[61]
TPT	*Poecilia reticulate*	lab (22 °C)	49	[61, 62]

[a] $T_{1/2}$ was calculated after transplantation to a clean area by Becker van Slooten and Tarradellas [59]. In another study by Laughlin et al. [60], the animals were transplanted from the field to laboratory aquaria for elimination studies. In a study by Shiraishi and Soma [61], the mussels were not transplanted, but the decrease in body concentrations, after the legal restriction on TPT application, was investigated

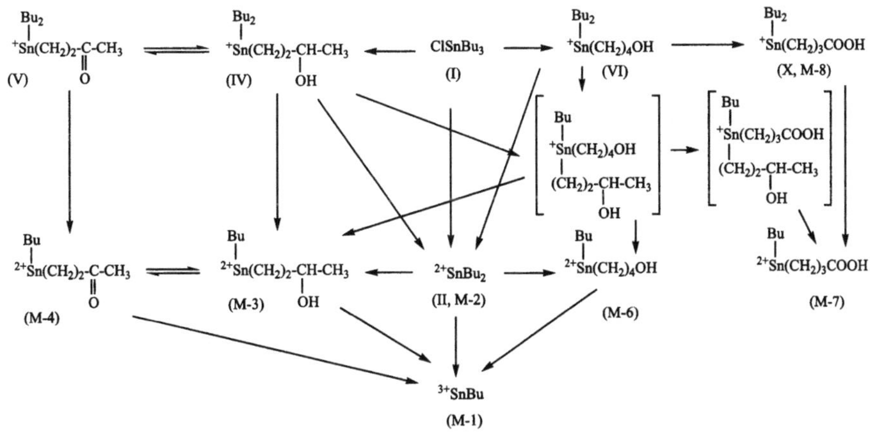

Scheme 3 Speculated pathway of TBT chloride metabolism [64]

The acute oral toxicity of TBT chloride in rats (LD_{50}) is 122–349 mg TBT kg^{-1} body weight (see Table 3). However, mammals such as rats are able to metabolize a small amount of TBT compounds. For example, Wister rats, which are orally administrated at a dose of 2 mg TBT kg^{-1} body weight after 12 h of fasting, were investigated for 7 days and *n*-butyl(3-carboxypropyl)tin compound in the liver and *n*-butyl(3-hydroxybutyl)tin compound in the kidney were found as the main metabolites, respectively. Matsuda et al. speculated the pathway for progressive removal of the organic groups from tin atoms as shown in Scheme 3 [64].

The depuration of TBT with mussel, *Mytilus edulis* was determined by the analysis of dissected organs. A transplant study showed that the depurations of TBT from gill tissue and digestive gland tissue are biphasic two-compartment processes involving a rapid TBT loss process and a concurrent slower TBT depuration process as shown in Table 7 [65]. Whole-body and gonadal tissue depuration followed a slower monophasic depuration process. Depuration half-life values range from 2.2–5.3 days for the fast depuration component and 28–69 days for the slow component [65].

Table 7 Half-life values for TBT and DBT depuration from various tissues in mussels [65]

Tissue	TBT rapid $T_{1/2}$ (day)	TBT slow $T_{1/2}$ (day)	DBT $T_{1/2}$ (day)
Whole animal	–	69	115
Gonad	–	28	58
Digestive gland	5.3	58	35
Gills	2.2	53	36

3.3 Half-lives of Organotin Compounds in Seawater and Sediment

After the environmental problems of organotin antifouling paints became evident, the stability of the organotin compounds such as TBT and TPT compounds in terms of the half-lives have been reported in many articles. These organotins are degraded by hydrolysis in water with sunlight and microorganisms at a room temperature. Hence, data for the half-lives in natural seawater, in fresh water or in pure water in the laboratory vary with the intensity of light, temperature, types or amounts of organisms, flow rates of seawater, types or amounts of ions in seawater, and amounts of the organotin compounds.

In 1983, the degradation of the TBTO as a representative organotin compound in water was reported. Its aqueous solubility at 20 °C is 0.7–7 mg L^{-1} at pH 5–7 and the vapor pressure is estimated to be 6.4×10^{-7} mm Hg. Dissolved in water, it neither volatilized nor lost butyl groups over a period of at least 2 months in the dark at 20 °C; in sunlight, however, it underwent

Table 8 Degradation of organotin compounds in water [67, 68]

Organotin, medium	Half life (days)	Comment (initial concentration)	Publication year	Refs.
TBT, seawater	4–19		1989	[69]
TBT, seawater	6–7	(0.5 μg/L)	1989	[70]
TBT, seawater	17	open seawater (28 °C)	1986	[71]
TBT, seawater	6–13	in sunlight, San Diego, USA	1996	[72]
TBT, seawater	7–19	in the dark, San Diego, USA	1986	[72]
TBT, seawater	6	in sunlight, San Diego, USA	1988	[73]
TBT, seawater	13	in one location, no degradation over 15 days in the dark, Osaka, Japan	1988	[74]
TBT, seawater	4–19	in sunlight, USA	1989	[75]
TBT, seawater	15	Italy	1990	[76]
TBT, seawater	6	in the light, ^{14}C TBT (2 μg/L)		[77]
	7	in the dark, USA, marina	1988	
TBT, seawater	17	open seawater	1996	[71]
TBT, seawater	60	at 5 °C	1987	[78]
TBT, estuarine	6–7	at 28 °C	1996	[71]
TBT, estuarine	7–14	at 22–28 °C in the dark	1988	[74]
TBT, estuarine	7.77	0–39 days	1998	[80]
	2.55	for first 10 days		
	13.4	after 11 to 40 days		
TBTO, fresh water	> 89	in sunlight	1983	[66]
TBT, fresh water	26	in winter, in lab window	1992	[81]

slow ($t_{1/2} > 89$ days) photolytic decomposition, by stepwise debutylations to inorganic tin [66].

The half-lives of the organotin compounds in seawater are several days to several months; however, those in sediment have been reported to be about 1 to 15 years. These data show that the stabilities of the organotin in natural seawater are surprisingly higher than those previously indicated. In particular, in sediment the stabilities of the organotins are very high. The data for the half-lives of organotin compounds including the above data, are shown in Tables 8, 9 and 10 [66–96].

Lee et al. [97] reported the TBT degradation in low concentrations ($< 5\,ng\,L^{-1}$, both radiolabeled and unlabeled) in estuarine waters along the Georgia coast (USA) with half-lives ranging from 3 to 6 days under light and from 7 to 13 days in dark conditions. Evidence suggests that microalgae play important roles in TBT degradation in sunlit coastal water. TBT degradation rates were high in light as compared with those of degradations in the dark. Further, after adding nitrate for growth of microalgae, for example *Skele-*

Table 9 Degradation of organotin compounds in sediments [67, 68]

Organotin, medium	Half life	Comment (initial concentration)	Publication year	Refs.
TBT, sediment	about 2.5 years	Aukland, New Zealand	1995	[82]
TBT, sediment	about 8.7 years	Vancouver, Canada	1997	[83]
TBT, sediment	2.1 years		1995	[84]
TBT, sediment	1.85 years, first order 0.375/year, estimated from sediment core data	Aukland, New Zealand (TBT: 0.759 μgSn/g)	1989	[85]
TBT, sediment	> 8–15 years, estimated from sediment core data,	Arcachon Bay, France	1990	[86]
TBT, sediment	6.9 years, estimated from sediment core data,	Chinhae Bay, Korea	1998	[87]
TBT, sediment	2.1 years, estimated from sediment core data	Oleron Island, France	1995	[88]
TBT, sediment	100–800 days, faster under aerobic conditions than anaerobic conditions	Japan	1995	[89]
TBT, sediment	460 days anaeroabic sediment	Toba Bay, Japan	1992	[90]
TBT, sediment	360–775 days surfacial sediment	UK	1993	[91]
DBT, sediment	1.9 years		1995	[84]
MBT, sediment	1.1 years		1995	[84]

Table 10 Degradation of TBT compounds in water/sediment mixtures [67, 68]

Organotin, Medium	Half life	Comment	Publication year	Refs.
Entuarine/sediment	3.8 years	Georges River, Australia estimated from sediment core data	1993	[92]
Entuarine/sediment	0.9–5.2 years	Southeast England estimated from sediment core data	1993	[93]
Fresh water, estuarine/sediment	1.2 years	Much longer in anaerobic sediments	1993	[94]
TBT, fresh water/ sediment	> 11 months	Sterile	1978	[95]
TBT, fresh water/ sediment	4 months	Toronto Harbor, Canada	1988	[96]

tonema costatum and *Skeletonema tropicum*, their half-lives become only approx. 2 days (see Fig. 7) [97].

Adelman et al. [98] studied radiolabeled TBT in a 13-m^3 marine enclosure with a near-natural water column and benthos. TBT and its degradation products were monitored for 278 days. TBT concentration in the water column (initially 590 ± 20 ng L^{-1}) decreased at a rate of 0.20 g^{-1} for 15 days and then slowed to 0.10 day^{-1} [98]. Lines were fit to the data and first-order re-

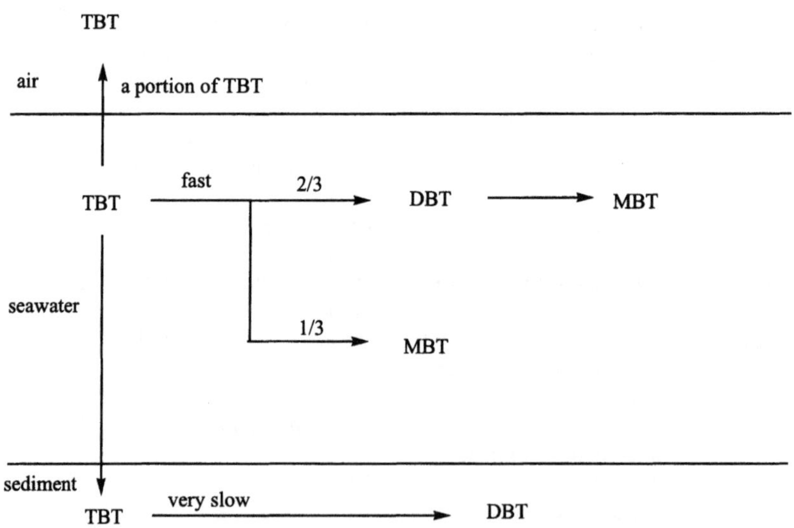

Fig. 4 Radiolabeled TBT into a 13-m^3 marine enclosure with near-natural water column and benthos [98]

moval rates were shown. Most of the TBT was lost from the water column through biodegradation which occurred at a rate of 0.08 day^{-1}. Two-thirds of the degradation proceeded through debutylation to DBT which in turn degraded to MBT at ~ 0.04 day^{-1}. One-third of the TBT was degraded directly to MBT in the water. Another portion of the TBT removed from the water column was apparently transported rapidly to the air–water interface to a measurable degree, and then lost from the tank as shown in Fig. 4 [98].

De Mora et al. [85] reported on sediment cores collected from Tamaki Estuary in Auckland, New Zealand in 1989. The data from three stations were used for the determination of a degradation rate for TBT in the sediment. The core profiles from these stations display a high degree of internal consistency when corrections are made for the differing sedimentation rates at each of the stations. A plot of the natural logarithm of concentrations against time yielded a straight line in all cases, thereby indicating that the degradation was a first order kinetic process. The average degradation rate was 37.5%/year giving a half-life of 1.85 year [85].

4 Bioaccumulation and Sediment Accumulation of Organotin Antifoulants

Many studies have shown that TBT is not persistent in most natural waters. However, the tendency of TBT to accumulate in sediment and biota implies that degradation processes operating in the sediment and biota will be of greater importance in determining the overall persistence of TBT in the

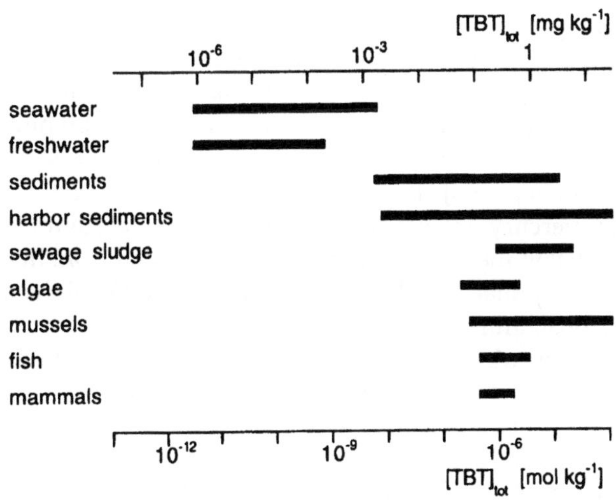

Fig. 5 Range of measured TBT concentration in different compartments of aquatic environments [99]

marine environment. Of particular concern is the stability of TBT in contaminated sediment where biodegradation processes may be inhibited [47]. Because of the long persistence of TBT in sediment, sediment may be a big reservoir for TBT in some locations [67].

The concentrations of TBT in seawater, sediment, fish, etc., are shown in Fig. 5 [99]. The concentrations are different (10^3-10^5 times) between the seawater or freshwater and sediments or fish. These data show that TBT is stable in the sediment, sludge and living organisms such as algae and fish. Bioaccumulation factors for TBT, range from 400 to 50×10^4 for fishes, 1500 to 2.2×10^4 for molluscs and were $> 3 \times 10^5$ for algae [67].

The main ingredient of the sediment is oxygen in the form of metal oxides, and those of fish and shellfish are proteins containing nitrogen elements as amino acids. The organotin compounds easily form a five- or six-coordination state by which the tin element is coordinated with electronegative atoms such as nitrogen, oxygen, sulfur or phosphorus [16]. In particular, the organotin compounds very easily form the five- or six-coordination state by the coordination of three sites with alkyl or phenyl groups and two or three sites with the oxygen or nitrogen atoms of metal oxides or proteins ($R_3SnL^1L^2$ or $R_3SnL^1L^2L^3$). Therefore, the organotin compounds show high affinities to those substances containing nitrogen or oxygen atoms. This is the first characteristic property of the organotin compounds described in Sect. 2.1.

5 Biomethylation of Organotin Compounds

Biomethylation proceeds through microorganisms such as bacteria having a methylating capability [100–105]. The principal naturally occurring methylating agents are: (a) methylcobalamin (vitamin B_{12}); (b) *S*-adenosylmethionine (methyl-group donor, active methionine); and (c) methyl iodide (probably formed by the methylation of the iodide ion by *S*-adenosylmethionine) [100]. Biomethylation is known in the methylation of metals such as mercury, arsenic, lead, chromium, tin, palladium and thallium. For example, methyl mercury compounds are produced from inorganic mercury in sediment by anaerobic bacteria through the action of the methlylating agent, methylcobalamine [100].

The main methylating agent for tin compounds is believed to be methylcobalamin (CH_3CoB_{12}), although the other methyl donors may be involved in extracellular reactions, e.g. methyl iodide, CH_3I [32].

The first laboratory studies on inorganic tin compounds were conducted in 1974 by Huey et al. [106] using pure cultures of tin-resistant *Pseudomonas* bacteria from Chesapeake Bay. Incubation with $SnCl_4 \cdot 5H_2O$ led to the production of what was believed to be a dimethyltin species [24]. The aquatic

Table 11 Aquatic environmental occurrence of methyltin compounds [24] [a]

Medium Location	$MeSn^{3+}$ (mol LSn^{-1} or mol kgSn^{-1})	Me_2Sn^{2+} (mol LSn^{-1} or mol kgSn^{-1})	Me_3Sn^{+} (mol LSn^{-1} or mol kgSn^{-1})	Refs.
Drinking water				
Florida	4.20×10^{-12} -6.81×10^{-11}	3.36×10^{-12} -1.85×10^{-11}	4.20×10^{-12} -2.02×10^{-11}	[107]
Fresh water				
Florida	$< 8.40 \times 10^{-14}$ -1.01×10^{-10}	8.40×10^{-14} -6.30×10^{-11}	$< 8.40 \times 10^{-14}$ -6.39×10^{-11}	[107]
Lake Michigan	2.49×10^{-11} -7.90×10^{-11}	n.d.–2.86×10^{-10}		[108]
Rivers in Southeast, USA	n.d.–1.18×10^{-11}	n.d.–2.61×10^{-11}	n.d.–1.43×10^{-11}	[109]
Rhine River, Germany	5.92×10^{-10}	1.77×10^{-9}	1.35×10^{-11}	[109]
Main River, Germany		3.4×10^{-12}	2.5×10^{-12}	[109]
Ontario	5.04×10^{-10} -1.01×10^{-8}	n.d.–3.36×10^{-9}		[110]
Estuarine water				
Florida	$< 8.40 \times 10^{-14}$ -7.14×10^{-11}	5.88×10^{-12} -3.86×10^{-11}	$< 8.40 \times 10^{-14}$ -3.36×10^{-11}	[107]
Baltimore Harbor		$< 4.20 \times 10^{-11}$ -4.12×10^{-9}	$< 4.20 \times 10^{-11}$ -1.68×10^{-10}	[114]
Seawater				
Florida	$< 8.40 \times 10^{-14}$ -1.26×10^{-10}	5.04×10^{-12} -5.88×10^{-11}	$< 8.40 \times 10^{-14}$ -8.23×10^{-12}	[107]
California	n.d.–3.33×10^{-11}	n.d.–2.04×10^{-10}		[108]
Rain water				
Florida	5.04×10^{-12} -1.85×10^{-11}	$< 8.40 \times 10^{-14}$ -6.22×10^{-11}	$< 8.40 \times 10^{-14}$ -9.24×10^{-12}	[107]
Sediment				
General	n.d.–8.91×10^{-8} (dry weight)	n.d.–1.13×10^{-7} (dry weight)	n.d.–1.63×10^{-7} (dry weight)	[111]
Fish				
(*Upenelus molluccensis* and *Mullus barbatus*)	2.27×10^{-7} (dry weight)	2.31×10^{-8} (dry weight)	1.05×10^{-7} (dry weight)	[111]
Seaweed	1.40×10^{-7} (dry weight)	3.11×10^{-7} (dry weight)	7.56×10^{-9} (dry weight)	[112]
Algae				
San Diego Bay	n.d.–4.99×10^{-9} (dry weight)	1.82×10^{-9} -1.00×10^{-8} (dry weight)	n.d.–1.50×10^{-9} (dry weight)	[113]

[a] This table is revised from the table which Wilson, Nicholson and Prosser presented in ITIR Publication No 669 in 1986 [24]

Table 12 Aquatic environmental occurrence of tetramethyltin and butylmethyltin compounds [24] [a]

Medium Location	Me_4Sn ($mol\,LSn^{-1}$ or $mol\,kgSn^{-1}$)	Bu_2SnMe_2 ($mol\,LSn^{-1}$ or $mol\,kgSn^{-1}$)	Bu_3SnMe ($mol\,LSn^{-1}$ or $mol\,kgSn^{-1}$)	Refs.
Estuarine water				
Baltimore Harbor	$< 8.40 \times 10^{-11}$ -2.52×10^{-9}			[114]
Invertebrates				
California coast	n.d.–3.35×10^{-8} (dry weight)			[115]
Algae				
San Diego Bay	n.d.–6.98×10^{-8} (dry weight)			[115]
Sediment				
Ontario	n.d.–4.68×10^{-8}	n.d.–4.68×10^{-8} (dry weight)	n.d.–7.47×10^{-8} (dry weight)	[57]
Port Dover Harbor Top 2 cm		10.9 ± 3.6 ($\mu g\,kg^{-1}$) (dry weight)	12.3 ± 3.1 ($\mu g\,kg^{-1}$) (dry weight)	[116]

[a] This table is revised from the table which Wilson, Nicholson and Prosser presented in ITIR Publication No 669 in 1986 [24]

environmental occurrence of methyltin compounds is shown in Tables 11 and 12 [24, 107–117].

Vella and Vassallo reported on TBT-contaminated sediment from Marsamxett Harbor, Malta, placed in 0.5 L chambers through which air was displaced by continuous pumping for 11 days, which released mainly methylbutyltins with concentrations (as for tin) reaching a maximum at 48 h showing mean values of 8.7 (Me_3SnBu), 22.1 (Me_2SnBu_2) and 93.0 $ng\,m^{-3}$ ($MeSnBu_3$) [118].

Marine sediment contaminated with TBT residues can host methylbutyltins, presumably formed by the environmental methylation of TBT and its debutylated metabolites. These substances are far more volatile than polar TBT, DBT and MBT, and would be released into the atmosphere from sediment that has been dredged up from the sea and allowed to dry on land. The preliminary results presented here suggested that air pollution from such sediment is probably minor; however, more work is necessary to establish the extent of this problem, especially in situations where the vapors emitted can be accumulated in some way, and also in regard to any ecological effect caused by any land uses to which a consideration similar to the sediment may be applied [118].

Recently, Amouroux et al. [119] reported on volatile organotin compounds in the Arcachon Bay (southwest France). These compounds, formed by the biomethylation of organotin compounds, were found both in silt and sand. The methylation products are mono-, di-, tri- and tetra-methyl derivatives. The tetraorganotins do not generally have highly physiologically active properties; therefore, they are used mainly as intermediates for the production of other organotin compounds. The amount of monomethytin compounds is about half of that of the total methyltin compounds in the silt sample and is about 1/10 000 of that of TBT. Hence the total amount of these products is very small compared with that of the TBT compounds. At present, the volatile methyl derivatives are considered to be of little concern in causing environmental problems.

Volatile butylmetyltin compounds formed by the biomethylation were reported in the 1980s by Maquire and coworkers [120, 121], Rapsomanikis and Harrison [101], recently by Yonezawa et al. [105], and Vella and coworkers [104, 122–124]. In 1994, Yonezawa et al. [105] reported that these biomethylations represent one path of degradation of the triorganotin compounds through microbial activities in the sediment. The degradation of the triorganotin compounds proceeds by two methylation reactions with a sulfate-reducing microbial activity and debutylation with a nitrate-reducing microbial activity.

6
Degradation of Organotin Compounds by Organisms and Their Metabolites

Yamaoka et al. [125–134] screened TPT-degrading bacteria with a simple technique using post-column high-performance liquid chromatography with 3,3′,4′,7-tetrahydro-fluavone as a post-column reagent for the determination of TPT and its metabolite, DPT. They found that TPT degradation was catalyzed by low-molecular-mass substances in the extract [126].

Yellow compound pyoverdins were isolated from *Pseudomonas chlororaphis*. The degradation of TPT by pyoverdin (20 mg) was carried out in distilled water (30 ml) containing 6 $\mu g\,L^{-1}$ concentration of TPT at 20 °C for 96 h in aerobic conditions. TPT and DPT in seawater were degraded to MPT with pyoverdins [127]. The degradation of TPT in seawater increased with elevating temperature between 4 and 37 °C. The optimum degradation of TPT in seawater was observed at pH 7–8.5. The degradation of TPT and DPT in distilled water can be faster than in seawater. Also, the degradation of TPT in both water and seawater was faster than that of DPT while TBT, DBT, MBT and MPT in water and seawater were not degraded by the pyoverdin [127].

The TPT-degrading bacterium *Pseudomonas chlororaphis* CNR15 produces extracellular yellow substances to degrade TPT. F-I is suc-pyoverdin from *P. chlororaphis* ATCC 9446, which is a peptide siderophore produced by fluorescent pseudomonads. F-IIa and F-IIb are also pyoverdins [130]. F-I and F-IIa degraded TPT to MPT via DPT and degraded DPT and DBT to MPT and MBT, respectively.

The TPT-degrading activity of F-I was remarkably inhibited by the addition of metal ions chelated with pyoverdin. On the other hand, the activity of the F-I was increased 13- and 8-fold by the addition of Cu^{2+} and Sn^{4+}, respectively. These results suggest that metal-chelating ligands common to pyoverdins may play important roles in the Sn-C cleavage of the organotin compounds in both the metal-free and metal-complexed states as described below (see Fig. 6) [130].

Using a screening technique based on pyrocatechol violet colorimetry for the detection of TPT metabolites, DPT, MPT, and inorganic tin were isolated. An isolated strain CNR22, *Burkholderia cepacia* was identified. The organotin degradation activity was observed in the culture supernant of the strain CNR22 grown without organotin; the reaction was catalyzed by low-molecular-mass iron chelators of a microbial origin [133]. These four substances are two azotochelins and two ornibactins [134]. The azotochelins

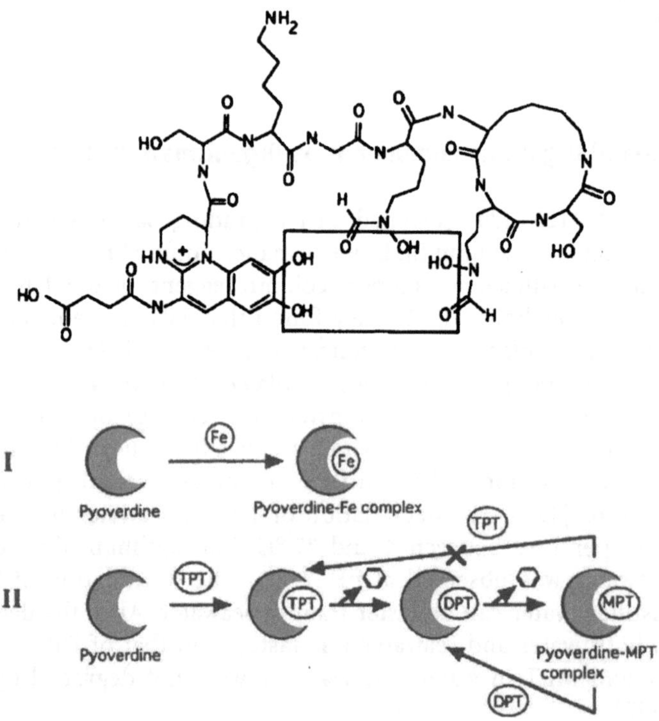

Fig. 6 The structure of pyoverdin from *Pseudomonas chlororaphis* CRN 15 possessing TPT degradation activity. Proposed TPT and DPT degradation pathway involving the complexation with pyoverdin. *Scheme I*, the physiological Fe-complexation of pyoverdin; *scheme II*, the potential organotin cometabolism pathway by pyoverdin [131]

show an activity for degradation of both TPT and TBT and the ornibactins show a weak activity for degradation of DPT.

They reported the adsorption and degradation of TPT by immobilized *Pseudomonas chlororaphis* CNR15 cells. The cells were entrapped in calcium alginate beads. Pyoverdin, which is a siderophore that participates in the degradation of TPT, was secreted from the immobilized cells, and the concentration of the pyoverdin increases for at least one month, and DPT and MPT are formed [132]. The degradation activity was influenced by temperature and pH, and showed a maximum at 30 °C and pH 8.8 [132].

The pyoverdins also degrade DBT to MBT in seawater. The optimum degradation of DBT in seawater was at pH 4.8–8.2, at a temperature of 25–30 °C [129]. The TPT degradation by pyoverdin was markedly inhibited by transition metal ions, Zn^{2+} and Sn^{4+}, while under the metal-free condition, its degradation seems to form a stable 1 : 1 complex with MPT without

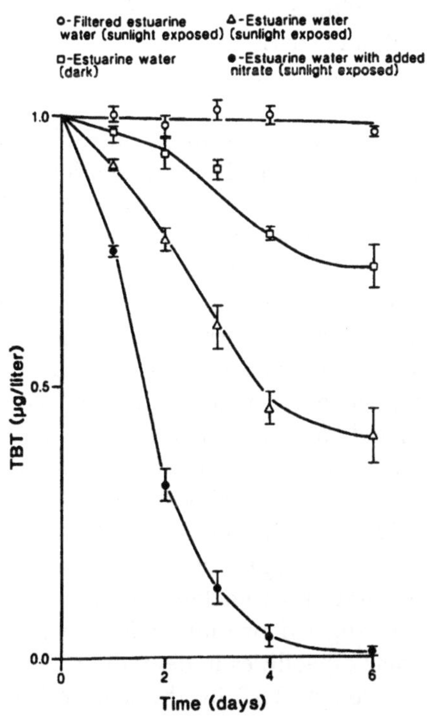

Fig. 7 Degradation of [^{14}C] TBT added to Skidaway River water under dark and natural sunlight conditions [97]. The initial concentration of [^{14}C] TBT was 1.0 µg L^{-1}. For filtered samples water was passed through a 0.2-µm filter. For the added nitrate study there was an addition of sodium nitrate (10 mg L^{-1}), which resulted in a chlorophyll concentration of 21 µg L^{-1}. Initial chlorophyll concentration in the estuarine water was 6 µg L^{-1}. *Vertical bars* represent one standard deviation with $n = 3$. TBT degradation was significantly higher ($P = 0.05$) in the light compared with dark degradation

releasing an intermediate, DPT [130]. The optimum pH for the activity was pH 7.5 to 8, and the relative activities were reduced to approximately 40–50% at pHs 5.5 and 9.5 [126]. These observations suggest that the chelating residue of pyoverdin may participate in a nucleophile attack at the tin atom and during the protonation of the phenyl group of TPT, to form more stable DPT- and MPT-pyoverdin as shown in Fig. 6 [131].

The half-lives of TBT in an estuarine river along the Georgia coast (at the low concentration of TBT ($< 5\ ng\ L^{-1}$) range from 3 to 13 days. Evidence suggesting that microalgae play an important role in TBT degradation in sunlit coastal water included the following:

1. TBT degradation rates were high in light compared to that in dark degradation and there was no evidence of TBT photolysis because the TBT degradation does not practically proceed in filtered estuarine water in sunlight as shown in Fig. 7.

2. (Hydroxybutyl)tins and dibutyltin were the major degradation products in the light and by the cultures of diatoms and dinoflagellates, while only dibutyltin was observed in the dark.

3. The TBT degradation increased in sunlight when nitrate was added as shown in Fig. 7 [97].

The degradation rates were always higher in sunlight-incubated samples than those in the dark (Fig. 7). The TBT degradation was considered to increase in the light because it is metabolized by the algae. Further support for this conclusion was that the TBT degradation rates increased significantly when nitrate was added to water before the samples were exposed to sunlight. The half-life of TBT was 1–2 days in nitrate-supplemented water (Fig. 7). The dominant algae in the normal and nitrate-supplemented estuarine water were the diatoms *Skeletonema costatum* or *Skeletonema tropicum* (summer months). Phytoplankton that can degrade TBT, included diatoms (*Skeletonema costatum* and *Chaetoceros curvisetus*), and a dinoflagellate (*Procentrum triestium*). Green algae (*Dunaliella tertiolecta*) and chrysophytes (*Isochrysis galbana* and *Cricosphaera ricoco*) showed a very limited ability to degrade TBT (Table 13 [97]). The stimulation of algal growth may be of use in enclosed aquatic areas with the organotins, since dense algal cultures may be able to rapidly degrade these organometallic compounds [97].

Tsang et al. [135] investigated two microalgal species, *Chlorella vulgaris* and *Chlorella* sp. for their capabilities in degrading TBT at its sublethal concentration. The biosorption of TBT by the algal cell wall was the major mechanism in reducing 40% of the initial TBT from the medium in the first 2 days. The half-life of TBT incubated with *C. vulgaris* was 60 h while that with *Chlorella* sp. was 80 h. At the end of the experimental period of 14 days, 27 and 41% of the original TBT were recovered as DBT and MBT in the cultures of *C. vulgaris*, respectively [135]. The capability of such debutylating processes therefore accounts for the higher tolerant ability of *C. vulgaris* than that of *Chlorella* sp. [135].

Table 13 Degradation of TBT by marine algae [97] [a]

Algae species	Concentration [b], μg L^{-1} TBT	DBT	Hydroxylated products	Polar products
Diatoms-Bacillariophyta				
Skeletonema costatum	0.1	0.2	0.1	0.04
Chaetoceros curvisetus	0.20	0.11	0.05	nd
Dinoflagellates-Pyrrhophyta				
Procentrum triestinum	0.30	0.05	0.01	nd
Golden algae-Chrysophyta				
Isochrysis galbana	0.40	nd	nd	nd
Cricosphaera ricoco	0.35	0.02	nd	0.01
Green algae-Chlorophyta				
Dunaliella tertiolecta	0.40	0.05	nd	nd

[a] Algae were cultured in seawater media. [^{14}C] TBT was added to cultured algae to give an initial concentration of 0.04 μg L^{-1}. After 2 days the algae culture containing TBT was extracted and analyzed for TBT and metabolites
[b] nd, not detected (< 0.01 μg L^{-1})

The fate of TBT with a subtropical seagrass bed including *Thalassia testudinum* and associated fauna was studied over 3 or 6 weeks by dosing weekly for 24 h with ^{14}C-labeled TBT. The TBT was rapidly removed from the water column (half-life of 10–20 h), primarily through absorption onto sediment and seagrass leaves. By contrast, 12 microcosms that received similar TBT doses but that contained only seawater had TBT removal half-life of 2–7 days. The accumulation of TBT in sediment and grasses was temporary, however, at harvest, the seagrass microcosms contained just 20–30% of the ^{14}C that had been adsorbed or assimilated during dose periods, and half of this label was in degradation products. The principal mechanism of TBT loss from solids was degradation followed by the desorption of degradation products (largely monobutyltin and CO_2 which are more soluble than TBT) [136].

7 Conclusions

The organotin compounds have a high affinity to sediment and organisms. Hence, the organotin compounds tend to be accumulated in the sediment and organisms, though they are degraded relatively rapidly in seawater or freshwater mainly by microorganisms.

Many countries started to ban the use of the organotin antifoulants for small vessels in the late 1980s. Consequently, in France, UK, USA, Canada, Australia, Japan, Spain, Norway, Switzerland, etc., decreases in TBT have been generally observed in the environment since the ban, particularly in areas dominated by the small craft affected by the ban. However, in a few areas frequented by large ships, the organotin compounds continue to be input to the surface of the sediment, especially, in some harbors in Mediterranean countries such as Spain, Italy and Greece, and in northern Europe, etc.

Therefore, recently, the number of articles investigating depuration of the organotin compounds by organisms have increased. Yamaoka et al. found two kinds of microorganisms, *Pseudomonas chlororaphis* and *Burkholderia cepacia* capable of degrading TPT. Dinoflagellates, diatoms, mussels, green algae and chrysophytes, etc. are able to depurate TBT in seawater. Biomethylation is also considered a depuration mechanism of the organotin compounds. The results of these researches may be utilized for depuration of high concentration areas of TBT and TPT such as areas frequented by the larger ships, seawater near a dry dock and aquaculture with fish nets using organotin antifouling paints. These may also be used for the treatment of waste water from dry docks and for the treatment of cooling water from power plants.

Generally, recently, the concentration of organotin compounds in seawater has tended to decrease in the world through restrictions for small boats and because of their fast degradation properties in seawater, though they have accumulating properties in sediment and organisms. Furthermore, with the help of these depuration technologies, we hope that the marine environment improves in the world up to the point in time when environmentally friendly best alternative antifouling paints are used routinely throughout the world.

Acknowledgements I would like to express my sincere appreciation to Dr. Sumio Chubachi for reading the full manuscript, which enhanced its accuracy and clarity, and for providing much valuable constructive criticism.

References

1. Champ MA (2000) Sci Total Environ 258:21
2. Bosselmann K (1996) Environmental law and tributyltin in the environment In: de More ST (ed) Tributyltin: Case Study of an Environmenal Contaminant. Cambridge Environmental Chemistry Series 8. Cambridge University Press, Cambridge, p 237
3. Stewart C (1996) The efficacy of legistration in controlling tributyltin in the marine environment In: de More ST (ed) Tributyltin: Case Study of an Environmenal Contaminant. Cambridge Environmental Chemistry Series 8. Cambridge University Press, Cambridge, p 264
4. Ingham RK, Rosenberg SD, Gilman H (1960) Chem Rev 60:459
5. Hauge RF, Hastie JW, Margrave JL (1973) J Mol Spectrosc 45:420
6. Harrison PG (1998) Compounds of tin: general trends. In: Smith PJ (ed) Chemistry of Tin, chap 2. Blackie Academic & Professional, London, p 18
7. Gillespie RJ, Nyholm RS (1957) Quart Revs (London) 11:339
8. Woodworth RC, Skell PS (1959) J Am Chem Soc 81:3383
9. Heeg MJ, Herber RH, Janiak C (1988) J Organomet Chem 346:321
10. Jensen KA, Clauson-Kass N (1943) Z Anorg U Allgem Chem 250:277

11. Harrison PG (1998) Compounds of tin: general trends. In: Smith PJ (ed) Chemistry of Tin, chap 2. Blackie Academic & Professional, London, p 10
12. Yngve V (1940) US Patent 2219463; (1941) Chem Abstr 35:1145
13. Evans CJ, Smith PJ (1975) Organotin-Based Anti-fouling Systems. ITRI publication No. 505. ITRI, St. Albans, UK
14. Evans CJ, Smith PJ (1975) J Oil Col Chem Assoc 58:160
15. (1987) Tin Chemicals, The Formula for Success. ITRI (International Tin Research Institute) publication No. 681. ITRI, St. Albans, UK
16. Omae I (1989) Organotin Chemistry. J Organomeallic Chemistry Library 21. Elsevier, Amsterdam
17. Evans CJ, Karpel S (1985) Organotin Compounds in Modern Technology. J Organometallic Chemistry Library 16. Elsevier, Amsterdam, p 135
18. Blunden SL, Cusack PA, Hill R (1985) The Industrial Uses of Tin Chemicals. The Royal Society of Chemistry, Burlington House, London, p 41
19. Evans CJ (1998) Industrial uses of tin chemicals. In: Smith PJ (ed) Chemistry of Tin, 2nd ed., Blackie Academic & Professional, London, p 442
20. Bluden SJ, Chapman AH (1986) Organotin Compounds in the Environment. ITRI Publication No. 665. ITRI, St. Albans, UK
21. Blunden SJ, Evans CJ (1990) ITRI Publication No. 711. ITRI, St. Albans, UK
22. White JS, Tobin JM, Cooney JJ (1999) Can J Microbiol 45:1999
23. Waring CE, Horton WS (1945) J Am Chem Soc 49:540
24. Wilson AD, Nicholson JW, Prosser HJ (1986) ITRI Publication No 669. ITRI, St. Albans, UK
25. Focardi S, Corsolini S, Aurigi S, Pecetti G, Sanchez-Hernandez JC (2000) Appl Organ Chem 14:48
26. Emsley J (2001) Nature's Building Blocks, An A-Z Guide to the Elements. Oxford University Press. Yamazaki A (translator) (2003) Japanease Ed. Maruzen, Tokyo, p 251
27. Schwarz K, Milne DB, Vinyard E (1970) Biochem Biophys Commun 40:22
28. Sakurai H (1997) Genso 111 no Shintishiki, Kodansha, Tokyo
29. Sherman LR (1986) The Effects of Inorganic Tin and Organotin Compounds upon Animals. Silicon, Germanium, Tin and Lead Compounds 9:323
30. Fritsch P, de Saint Blanquat G, Derache R (1977) Fd Cosmet Toxicol 15:147
31. Smith PJ (1978) Toxicological Data on Organotin Compounds. ITRI, Publication No. 538. ITRI, St. Albans, UK
32. Gadd GM (2000) Sci Total Environ 258:119
33. Cooney JJ, Wuertz S (1989) J Ind Microbiol 4:375
34. Cooney JJ, de Rome L, Laurenece OS (1989) J Microbiol 4:279
35. Gadd GM, Gray DJ, Newly PJ (1990) Appl Microbiol Biotechnol 34:116
36. Avert SV, Miller M, Gadd GM (1991) FEMS Microbiol Lett 84:205
37. Maguire RJ (1987) Appl Organomet Chem 1:475
38. Horiguchi T (1998) Kaiyo to Seibutsu 20:283
39. Horiguchi T (1998) Kaiyo to Seibutsu 20:283
40. Schweinfurth HA, Günzel P (1987) IEEE CH2498-4:1421
41. Sheldon AW (1975) J Paint Technol 54
42. Skinner HA (1964) Adv Organomet Chem 2:39
43. Mailhot G, Astruc M, Bolte M (1992) Appl Organomet Chem 13:53
44. Mailhot G, Brand N, Astruc M, Bolte M (2002) Appl Organomet Chem 16:27
45. Maguire RJ, Carey JH, Hale EJ (1983) J Agric Food Chem 31:1060
46. Chaumery CJ, Michel P (2001) Environ Technol 22:171
47. Stewart C, de Mora SJ (1990) Environ Technol 11:565

48. Hinga KR, Adelman D, Pilson MEQ (1987) Oceans '87. Washington, DC, p 1416
49. Thain JE, Waldock MJ, Waite ME (1987) Oceans '87. Washington, DC, p 1398
50. Francois R, Short FT, Weber JH (1989) Environ Sci Technol 23:191
51. de Mora SJ, King NG, Miller MC (1989) Environ Technol Lett 10:901
52. Barnes RD, Bull AT, Poller RC (1973) Pestic Sci 4:305
53. Suess A, Eben Ch (1973) Z Pflanzenkrankh Pflanzenschutz 80:288
54. Barug D, Vonk JW (1980) Pestic Sci 11:77
55. Sheldon AW (1978) Proc 18th Ann Marine Coatings Conf, Monterey, California
56. Barug D (1981) Chemosphere 10:1145
57. Maguire RJ, Wong PTS, Rhamey JS (1984) Can J Fish Aquat Sci 41:537
58. Stäb JA, Frenay M, Freriks IL, Brinkman UAT (1995) Environ Toxicol Chem 14:2023
59. Becker van Slooten K, Tarradellas J (1994) Environ Toxicol Chem 13:755
60. Laughlin RB, French W, Guard HE (1986) Environ Sci Technol 20:884
61. Shiraishi H, Soma M (1992) Chemosphere 24:1103
62. Tas JW, Opperhuizen A, Seinen W (1990) Toxicol Environ Chem 28:129
63. Maguire RJ (1992) Wat Sci Tech 25:125
64. Matsuda R, Suzuki T, Saito Y (1993) J Agric Food Chem 41:489
65. Page DS, Dassanayake TM, Gilfillan ES (1995) Mar Environ Res 40:409
66. Maguire RJ, Carey JH, Hale EJ (1983) J Agric Food Chem 31:1060
67. Maguire RJ (2000) Water Qual Res J Canada 35:633
68. Omae I (2003) Appl Organomet Chem 17:81
69. Seligman PF, Grovhoug JG, Valkirs AO, Stang PM, Fransham R, Stallard MO, Davidson B, Lee RF (1989) Appl Organomet Chem 3:31
70. Seligman PF, Valkirs AO, Lee RF (1986) Environ Sci Technol 20:1229
71. Batley G (1996) In: de More ST (ed) Tributyltin: Case Study of an Environmenal Contaminant, Cambridge Environmental Chemistry Series 8; Cambridge University Press, p 139
72. Seligman PE, Valkirs AO, Lee RF (1986) Environ Sci Technol 20:1229
73. Seligman PE, Valkirs AO, Stang PM, Lee RF (1988) Mar Pollut Bull 19:531
74. Hattori Y, Kobayashi A, Nonaka K, Sugimae A, Nakamoto M (1988) Water Sci Technol 20:71
75. Seligman PF, Grovhoug JG, Valkirs AO, Stang PM, Fransham R, Stallard MO, Davidson B, Lee RF (1989) Appl Organomet Chem 3:31
76. Bressa G, Cima L, Caravello GU, Turin P (1990) CHIMICA oggi 8:49
77. Seligman PF, Valkirs AO, Atang PM, Lee RF (1988) Mar Poll Bull 19:531
78. Train JE, Waldock MJ, Waite ME (1987) Oceans '87. Washington, DC, p 1398
79. Brickman FF, Olson GJ, Blair WR, Parks EJ (1988) In: Adams WJ, Chapman GA, Landis WG (eds) Aquatic Toxicology and Hazard Assessment. 10th ed. American Society for Testing and Materials, Philadelphia, ASTM STP 971, p 219
80. Dai S, Huang G, Chen C (1998) Appl Organomet Chem 12:585
81. Watanabe N, Sakai S, Takatsuki H (1992) Water Sci Technol 25:117
82. de Mora SJ, Stewart C, Phillips D (1995) Mar Pollut Bull 30:50
83. Stewart C, Thompson JA (1997) J Environ Technol 18:1195
84. Sarradin P-M, Lapaquellerie Y, Astruc A, Latouche C, Astruc M (1995) Sci Total Environ 170:59
85. de Mora SJ, King NG, Miller MC (1989) Environ Technol Lett 10:901
86. Astruc M, Lavigne R, Pinel R, Leguille F, Desauziers V, Quevauviller P, Donard OFX (1990) In: Patterson JW, Passino R (eds) Metals Speciation, Separation, and Recovery. Lewis Publ, Chelsea, p 263
87. Hwang HM, Oh JR, Kahng SH, Lee KW (1998) Mar Environ Res 47:61

88. Sarradin P-M, Lapaquellerie Y, Astruc A, Latouche C, Astruc M (1995) Sci Total Environ 170:59
89. Watanabe N, Sakai S, Takatsuki H (1995) Chemosphere 31:2809
90. Watanabe N, Sakai S, Takatuki H (1992) Mizu Kankyo Gakkaishi 15:672
91. Dowson PH, Bubb JM, Williams TP, Lester JN (1993) Wat Sci Tech 28:133
92. Kilby GW, Batley GE (1993) Austral J Mar Freshwater Res 44:635
93. Dowson PH, Bubb JM, Lester JN (1993) Environ Monitor Assess 28:145
94. Dowson PH, Bubb JM, Williams TP, Lester JN (1993) Water Sci Tech 28:133
95. Slesinger AE, Dressler I (1978) In: Good M (ed) Report of the Organotin Workshop. University of New Orleans, New Orleans, LA, 17–19 February, p 115
96. Maguire RJ, Tkacz RJ (1985) J Agric Food Chem 33:947
97. Lee RF, Valkirs AO, Seligman PF (1989) Environ Sci Technol 23:1515
98. Adelman D, Hinga KR, Pilson MEQ (1990) Environ Sci Technol 24:1027
99. Weidenhaupt A, Arnold C, Müller ST, Schwarzenbach TP (1997) Environ Sci Technol 31:2603
100. Krishnamurthy SJ (1992) Chem Educ 5:347
101. Rapsomanikis S, Harrison RM (1988) Appl Organomet Chem 2:151
102. Blunden SJ, Evans CJ (1985) ITRI Publication No. 711. ITRI, St. Albans, UK
103. Yonezawa Y, Nakata K, Miyakozawa Y, Ochi A, Kowata T, Fukawa H, Sato Y, Masunaga S, Urushigawa Y (1993) Environ Toxicol Chem 12:1175
104. Vella AJ, Adami JPT (2001) Appl Organomet Chem 15:901
105. Yonezawa Y, Fukui M, Yoshida T, Ochi A, Tanaka T, Noguti Y, Kowata T, Sato Y, Masunaga S, Urushigawa Y (1994) Chemosphere 29:1349
106. Huey C, Brinckman FE, Grim S, Iverson WP (1974) Proc Int Conf Transp Persist Chem Aquat Ecos, Ottawa, p 73
107. Braman RS, Tompkins MA (1979) Anal Chem 51:12
108. Hodge VF, Seidel SL, Goldberg ED (1979) Anal Chem 51:1256
109. Byrd JT, Andreae MO (1982) Science 218:565
110. Maguire RJ, Chau YK, Bengert GA, Hale EJ, Wong PTS, Kramar O (1982) Environ Sci Technol 16:698
111. Braman RS, Tompkins MA (1979) Anal Chem 51:12
112. Chau YK, Maguire RJ, Wong PTS, Glen BA, Bengert GA, Tkacz RJ (1984) Natl Water Res Inst Rept No 84-01. National Water Research Institute, Canada. Centre for Inland Waters Environment Canada, Burlington, Ontario
113. Ishi T (1982) Bull Jpn Soc Sci Fish 48:1609
114. Jackson JA, Blair WR, Brinckman FE, Iverson WP (1982) Environ Sci Technol 16:110
115. Seidel SL, Hodge VF, Goldberg ED (1980) Thalassia Jugoslavica 16:209
116. Maquire J (1984) Environ Sci Technol 18:291
117. Thompson JAJ, Sheffer MG, Pierce RC, Chau YK, Cooney JJ, Cullen WR, Maguire RJ (1985) Organotin Compounds in the Aquatic Environment: Scientific Criteria for Assessing Their Effects on Environmental Quality. Report No 22494 National Research Council Canada
118. Vella AJ, Vassallo R (2002) Appl Organomet Chem 16:239
119. Amouroux D, Tessier E, Donard OFX (2000) Environ Sci Technol 34:988
120. Maquire RJ (1984) Environ Sci Technol 18:291
121. Maquire RJ, Tkacz RJ, Chau YK, Bengert GA, Wong PTS (1986) Chemosphere 15:253
122. Vella AJ, Mintoff B, Axiak V, Agius D, Cassone R (1998) Toxicol Environ Chem 67:491
123. Vella AJ, Mintoff B, Axiak V (2000) Sci Total Environ 258:89
124. Vella AJ, Vassallo R (2002) Appl Organomet Chem 16:239

125. Yamamoto Y, Fuse H, Takimura O, Murakami K, Inoue H (2004) Kaiyo Sigen Kankyou Kenkyubumon Hokoku 5
126. Inoue H, Takimura O, Fuse H, Murakami K, Kamimura K, Yamamoto Y (2000) Appl Environ Microbiol 66:3492
127. Yamaoka Y, Inoue H, Takimura O, Oota S (2001) Appl Organomet Chem 15:757
128. Yamaoka Y, Inoue H, Takimura O, Oota S (2002) Appl Organomet Chem 16:277
129. Yamaoka Y, Inoue H, Takimura O (2002) Appl Organomet Chem 16:665
130. Inoue H, Takimura O, Kawaguchi K, Nitoda T, Fuse H, Murakami K, Yamaoka Y (2003) Appl Environ Microbiol 69:878
131. Inoue H, Takimura O, Yamaoka Y (2003) Recent Res Devel Microbiolgy 7:73
132. Takimua O, Inoue H, Fuse H, Murakami K, Yamaoka Y, Aihara M (2003) J Japan Soc Wat Environ 26:713
133. Inoue H, Takimura O, Yamaoka Y (2003) J Jpn Soc Water Environ 26:775
134. Inoue H, Takimura O, Fuse H, Murakami K, Yamaoka Y (2004) Nippon Mizu Kankyo Gakkai Koen Yokoshu, p 624
135. Tsang CK, Lau PS, Tam NFY, Wong YS (1999) Environ Poll 105:289
136. Levine SN, Rudnick DT, Kelly JR, Morton RD, Buttel LA (1990) Mar Environ Res 30:297

Hdb Env Chem Vol. 5, Part O (2006): 51–70
DOI 10.1007/698_5_049

Published online: 23 November 2005

New Trends in Sample Preparation Methods for the Determination of Organotin Compounds in Marine Matrices

C. Brunori[1] · I. Ipolyi[2] · P. Massanisso[1] · R. Morabito[3] (✉)

[1]ENEA/PROT-CHIM, Rome, Italy

[2]Corvinus University, Budapest, Hungary

[3]ENEA/PROT, Via Anguillarese 301, 00060 Rome, Italy
morabito@casaccia.enea.it

Abstract Tributyltin still remains a major issue for the marine environment and its determination is mandatory by legislation in several countries. As for speciation analysis in general, organotin analysis is a highly complex analysis with a long and laborious sample treatment procedure prone to various errors. It is widely recognized that the contribution of sample treatment to the total calculated uncertainty of a method is much larger than the contribution from the instrumental analysis. Furthermore, in contrary to instrumental analysis, sample treatment is far from being under control. Extraction, derivatization when needed, and preconcentration are usually carried out by exploiting a large variety of reagent combinations and techniques characterized by different efficiency and reproducibility.

In this chapter new approaches in sample treatment procedures for organotin speciation that have emerged within the last 5 years are presented schematically and discussed briefly underlining the main advantages and disadvantages of the individual solutions.

In the extraction and preconcentration steps the new tendencies are oriented towards the use of reduced-solvent or solvent-free systems, such as accelerated solvent extraction (ASE) or solid phase micro extraction (SPME). Derivatization has seen no significant developments within the last 15 years. The most significant development has probably been achieved in instrumental analysis with the application of isotope dilution techniques in order to trace the transformation/degradation of organotin species during the sample preparation steps. This technique will presumably become the most powerful tool for optimizing and validating extraction methods in the near future.

Keywords Derivatization · Extraction · Organotin species · Preconcentration · Sample treatment

Abbreviations

AAS	atomic absorption spectrometry
AED	atomic emission detection
APDC	ammonium pyrrolidine dithiocarbamate
ASE	accelerated solvent extraction
CRM	certified reference material
CT-GC-ICP-MS	cryogenic trapping gas chromatography inductively coupled plasma mass spectrometry
DBT	dibutyltin
DCM	dichloromethane
DDTC	diethyldithiocarbamate
DPhT	diphenyltin
ESI	electrospray ionization
ET-AAS	electrothermal atomic absorption spectrometry
GC-FPD	gas chromatography flame photometric detection
GC-ICP-MS	gas chromatography inductively coupled plasma mass spectrometry
GC-MS	gas chromatography mass spectrometry
GC-MS-MS	gas chromatography tandem mass spectrometry
GC-PFPD	gas chromatography pulsed flame photometric detection
HPLC-ICP-MS	high performance liquid chromatography inductively coupled plasma mass spectrometry
ICP-AED	inductively coupled plasma atomic emission detection
ID	isotope dilution
LLE	liquid liquid extraction
MBT	monobutyltin
MeSh	mechanical shaking
MIP-AED	microwave induced plasma atomic emission detection
MPhT	monophenyltin
MW	microwave
OT	organotin compounds
PLE	pressured liquid extraction
QF-AAS	quartz furnace atomic absorption spectrometry
SPE	solid phase extraction
SPMD	semipermeable membrane devices
SPME	solid phase micro extraction
TBT	tributyltin
TeBT	tetrabutyltin
TePhT	tetraphenyltin
TMAH	tetramethylammonium hydroxide
TOF-MS	time of flight mass spectrometry
TPhT	triphenyltin
US	ultrasound

1 Introduction

The release of triorganotin compounds [mainly tributyltin (TBT)] from immersed structures (mainly the hulls of boats, but also platforms) protected by organotin-based antifouling paints has been (and still is) responsible for one of the main problems of the direct, diffuse and continuous contamination of the marine environment.

Since their introduction as an active ingredient of antifouling paints, which dates back to the 1970s, there has been a great deal of evidence of its worldwide impact on non-target marine organisms such as bivalves and gastropods, and fish as well. Books, chapters of books and review papers have provided detailed descriptions of the history of organotin contamination and its effects on the marine environment [1–4].

Following the first legislations that limited the use but not the production of organotin-based antifouling paints, enacted in the 1980s in several countries following the example of France, several monitoring programmes were carried out for evaluation of organotin contamination in the different environmental compartments. However, reliability in the collected data was difficult to be achieved, at the beginning due to the lack of suitable and validated analytical methods for speciation analysis.

During the last 20 years outstanding progress has been made in the development of analytical methodologies and instrumentation for speciation analysis, which has given rise to an increased awareness of speciation-related issues in the international community. Since the turn of the millennium scientific publications on speciation still continue to proliferate, underlining the vital importance of information about bioavailability, toxicity and elements, and the long established fact that the determination of total element concentrations does not provide accurate information about the potential environmental and health impact of such elements. Frequently, papers report that speciation analysis is today routinely performed in many laboratories to control the quality of the environment, food and health, particularly in the case of tributyltin and methylmercury. For these species, in fact, instrumental analysis, including chromatographic separation and detection, can be considered today to be under control. However, there are still persistent difficulties in the presently applied procedures of elemental speciation which generally derive from the preceding steps—sample storage and sample preparation, and considerable efforts are still required to overcome them [5–10].

The main reason for these difficulties lies in the complexity of sample preparation required for the speciation of elements. It is indispensable to preserve the nature/integrity of the target species while separating the maximum possible amount from the matrix. In some cases the concentrations of target species may be extremely low even for the most sensitive techniques which even render the application of a pre-concentration step necessary. In addition, the sample preparation steps have to be harmonized with the instrumental set-up chosen for determination. It is not surprising therefore, that the experts of the field have recently made the overall conclusion that sample treatment is one of the most challenging steps in speciation analyses [3].

2 Sample Preparation Methods for Organotin Analysis

Sample preparation techniques for speciation analyses generally consist of several steps [11]. The necessary steps depend on the physico-chemical characteristics of the analytes to be determined and of the matrix to be analyzed. Furthermore, the suitability of the sample preparation steps with the chosen determination technique must also be assured.

Generally, the instrument combinations applied for speciation analysis are composed of a selective and sensitive detector coupled with some chromatographic separation step [high performance liquid chromatography (HPLC), gas chromatography (GC), gel electrophoresis (GE)] to provide time-resolved introduction of the analytes into the detector [5]. These separation techniques, especially GC, establish certain physico-chemical requirements for the analytes. This usually demands that the applied sample preparation procedure isolates the target species—by extraction—from the matrix and then converts them into suitable forms for separation. Derivatization is the most frequent approach to fulfil this task in the case where GC is chosen to carry out the separation. Besides producing volatile and thermally stable species, it reduces the occurrence of possible interferences during the subsequent analytical steps, particularly at the detection stage. These two steps must be considered crucial as low extraction efficiencies and/or low derivatization yields may lead to the underestimation of the actual analyte concentration in the investigated sample [8, 9]. Before the chromatographic separation, the phase containing the derivatized target analytes is often submitted to purification, by a so-called clean-up step. Furthermore, a preconcentration step is often included in order to improve the limit of detection of the overall method, mostly combined either with extraction or derivatization. These major steps are often further composed of a series of minor operations such as pH adjustment, agitation, change of solvent, phase separation, transfer from one vessel to another, dilution, evaporation of organic phase etc. [6].

Each single one of these sample preparation steps (both minor and major) is associated with possible sources of error. In general, all considerations that apply to trace organic analysis, also apply to speciation analysis, but in addition, due to the properties of the main target analytes of speciation analysis, the organometallic compounds, particular precautions have to be taken [7], specially for preserving the original form of the considered analytes.

Considering the different steps, including instrumental analysis, extraction generally accounts for a significant percentage of the total error of an analytical result, especially in speciation analysis and still much has to be done to improve extraction efficiencies while assuring species integrity. Several articles have been published so far with the aim of outlining an ideal extraction method for speciation purposes, however, many arrived to the conclusion that even in the case of two very similar sample types, slight modifications have to be carried out in the method conditions optimized on one material, to adjust it to the other. Such a thing as the ultimate extraction procedure has not been found yet, for none of the species of concern, from any matrix. Nevertheless, publications do propose methods for groups of samples, while underlining the sample used for the optimization of the method parameters.

Recently, the main emphasis of developments in the tin speciation of marine samples has no longer concentrated on the improvement of method detection limits, but on speeding up the lengthy sample preparation procedures and tracing the transformation/degradation of organotins that may occur during the sample treatment steps. In this context, recent developments in the field of sample preparation for speciation analysis have experienced a steady shift from traditional technologies. The novel approaches prefer (i) significant reductions in the use of organic solvents to decrease occupational hazards and problems of toxic chemical management; (ii) the application of isotopically labelled compounds to trace species-specific recoveries and alteration in the original speciation by degradation and/or possible occurrence of accidental species transformations such as transalkylation; (iii) miniaturization that allows easier in-situ and on-site manipulation; and (iv) the combination of previously well-separated sample preparation steps thus reducing the overall time required for sample preparation and simplifying the otherwise complex and time-consuming procedures [11–17].

The materials of marine origin can be divided roughly into three categories: water, sediment and biological materials. In this chapter the authors aim to point out the recent applications, and highlight the new developments, approaches and tendencies within the categories, especially concentrating on the most challenging step of speciation, the sample preparation methods.

2.1
Sample Preparation Methods for the Determination of Organotin Compounds in Seawater

Organotins are present in seawater at ng L^{-1} levels. Their quantification, therefore, requires highly sensitive techniques, and/or the collection of large sample volumes together with the application of preconcentration methods. Generally, the applied methods are (i) direct derivatization with organoborates or hydride generation in an acidic medium followed by liquid-liquid extraction (LLE), solid-phase extraction (SPE) or solid-phase microextraction (SPME) of the derivatizated compounds; and (ii) LLE with non-polar solvents [toluene, dichloromethane (DCM)] alone or in mixture with or without complexing agents and in the presence or not of acidic conditions and subsequent derivatization with Grignard reagents. After derivatization, gas chromatographic separation followed by different detection techniques such as mass spectrometry (GC-MS), tandem mass spectrometry (GC-MS-MS) or flame photometry (GC-FPD) for the determination of the species [18–21].

Martinez Vidal et al. [22] compared the use of LLE (using DCM, hexane, diethyl ether, acetone, acetonitrile and their different mixtures) and solid phase extraction (SPE) for the determination of eight organotin compounds (MBT, DBT, TBT, TeBT, MPhT, DPhT, TPhT and TePhT), with a spiking experiment, using DDTC (diethyldithiocarbamate) as the complexing agent. With LLE low recoveries were obtained for all the compounds except for MBT, but then the high values of MBT proved to be due to some interfering compound. With the SPE method, good recovery values were achieved, but no further details on recovery evaluation were reported in the paper. Tessier et al. studied the extraction of volatile tin compounds from marine water by cryogenic trapping-GC-ICP-MS [23]. The high sensitivity and selectivity of the ICP/MS detector coupled with cryofocusing assured detection limits below 1 fmol L^{-1} as tin. Følsvik et al. [24] studied whether semipermeable membrane devices (SPMDs) are applicable to monitor organotin compounds in the marine environment with time-integrated sampling. Triolein, one of the main components of neutral lipids in numerous aquatic organisms, suitable for concentrating organic contaminants, was used to fill the SPMD tubing

Table 1 Applied speciation steps for organotin determination in marine water reported in papers published since 2000

Extraction	Derivatization	Preconcentration	Analysis
• LLE: – DCM [21, 22] – hexane [18, 22] – toluene [20] – diethylether [22] – acetone [22] – acetonitrile [22] – mixtures [22, 24] • SPE [22] • SPMDs [24]	• hydride generation [19] • organoborate: – ethylation [18, 24] • Grignard: – pentylation [20, 22] – propylation [21]	• SPME • cryogenic trap [23]	• GC-MS [18] • GC-MS/MS [20, 22] • GC-FPD [19, 21] • GC-ICP-MS [23] • GC-AED [24]

for the sample treatment procedure that is based on imitating the neutral bioaccumulation processes. This SPMD setup was indeed successful in accumulating TBT and DBT from the water, but not MBT.

In respect of organotin analysis in seawater samples, it is worth stressing that the development of sample preparation methods has mainly been oriented to the reduction of time and to the use of solvent-free extraction systems or those with reduced solvent need such as SPME or SPMD after in situ derivatization. However, the complete validation of these methods is far from being achieved, mainly because of reproducibility problems [12]. The summary of the most applied extractants, derivatizing reagents, preconcentration systems and instrumental techniques in the determination of organotin species in seawater, derived from the literature published since 2000, is reported in Table 1.

2.2 Sample Preparation Methods for the Determination of Organotin Compounds in Sediment Samples

Speciation from solid samples—restricted to sediments and biota in the case of marine ecosystems—is obviously more challenging than from water. Although the high salt content of seawater may cause difficulties in the determination step (chromatographic separation and detection), the challenges of extraction, such as the process of isolating the target analytes from complex cell structures and biomolecules, and the number of possible errors are much higher.

Even if organotin compounds are not involved in mineralogical processes, they strongly interact with particulate matter in association with a variety of counterions, due to hydrophobic and/or ionic forces. The trisubstituted (TBT, TPhT) compounds are mainly bound by hydrophobic forces, while the monosubstituted ones (MBT, MPhT) by ionic forces. These forces influence the efficiency of the different extraction methods.

In recent literature, leaching mixtures for organotin extraction have been used in concert with various extraction devices, such as the Soxhlet, mechanical shaker, sonication probe or ultrasonic bath, microwave and more recently pressured liquid extraction (PLE). The most frequently adopted methods for organotin extraction from sediments are leachings with acids or acid-polar solvent mixtures. Acids with non-polar solvents and mixtures of different polarity grade solvents are seldom utilized.

The acidic conditions reduce the strength of mineral binding, break the bonds between the matrix and the target organotin analytes and provide positively charged ionic species. The extraction of di-substituted (DBT, DPhT) and, above all, mono-substituted organotin compounds (MBT, MPhT) from the aqueous to organic phase is enhanced by the addition of a complexing agent such as tropolone or diethyl dithiocarbamate. However, efficiency should always be critically evaluated, especially under strong acidic conditions [25], and it has to be kept in mind that high complexing agent concentrations might adversely affect chromatographic separation.

The choice of acid for organotin extraction mainly falls on acetic or hydrochloric acid whereas the polar solvent mostly chosen is methanol [14, 15, 26–36]. The extraction is mostly assisted by mechanical shaking, microwave energy or sonication [15, 18, 26, 29, 31–41]. Several authors have reported on the influence of, for example, sonication time, extraction temperature, extractant concentration and volume on the efficiency of organotin extractions from sediments [41]. The leached compounds are then derivatized in the aqueous medium, for example, buffered with AcOH/AcONa in the presence of a polar solvent and analyzed by GC separation. Results indicate that one of the most important factors within the extraction conditions is the acid concentration, too strong acid conditions leading to significant organotin degradation, mainly in the case of TPhT [42].

The influence of MW energy has been tested in several variations. From the closed-vessel microwave system to open focused microwave systems, integrating several steps of the sample preparation procedure or following the traditional step-by-step method. Yang and Lam [43] recently investigated the extraction of organotins from sediment in a closed-vessel microwave system with glacial acetic acid without a complexing agent. 6 min at 100 °C ensured quantitative recovery in the HPLC-ICP-MS determination of all the three butyltin species.

In extensive studies Ruiz Encinar et al. [14, 26] and Rodriguez-Gonzalez et al. [15] compared and evaluated the efficiency of all together four different solid-liquid extraction techniques—mechanical shaking, ultrasonic shaking, microwave, ASE—on sediment reference materials (PACS-2 and BCR 646), the extraction solution being a 3 : 1 (or 9 : 1 with ASE) mixture of methanol and acetic acid. The extraction time was higher than 4 h for mechanical shaking and in the order of minutes for the other three techniques. The best results in terms of butyltin recovery— quantitative for TBT and DBT—were provided by ultrasonication with a relatively short treatment time. Microwave extraction provided adequate data for DBT and TBT but only in a narrow range of MW conditions, while MBT seemed to be less prone to degradation. Ambient temperature leaching with mechanical shaking also provided data in agreement with certified values for all the three species, including MBT that until then seemed to require harsher extraction conditions and even longer extraction times (several hours) for quantitative recovery. A spike solution of ^{118}Sn and ^{119}Sn isotopes and the GC-ICP-MS technique, based on the determination of isotope ratios 120/119 and 118/119 for TBT, were used to check for transbutylation, and no evidence of the phenomenon was found under any of the tested extraction conditions.

In the last years, the application of ASE is becoming more and more popular for organotin speciation. With carefully optimized extraction conditions, it has been proven to give comparable results to ultrasonication and microwave-assisted extraction. The main advantage of ASE is that it allows a significant reduction of time and has increased potential for being used routinely. It is able to treat up to 24 sample at a time generating extracts in about 10 min, and, therefore carries considerable advantages for routine laboratories. Various extraction mixtures [1 M acetic acid in methanol, 0.5 M acetic acid in methanol with 0.2% (w/v) tropolone and 1 M acetate buffer in methanol] were tested with the same ASE settings (five static cycles of 5 min at 100 °C) and coupled to HPLC separation and ICP-MS detection [27]. 0.5 M acetic acid in methanol with 0.2% (w/v) tropolone proved to be the most efficient extractant. The work compares also the efficiency of the proposed ASE method with conventional acid extractions (acetic acid—stirring overnight) and medium polarity organic solvent extraction at acidic pH for the extraction of both butyl- and phenyl-tin compounds. The best overall extraction results were always assured by ASE, providing quantitative recovery for all the six compounds from spiked sediment samples. It has also been demonstrated that extraction time does not influence recoveries and neither does it induce degradation reactions, while static extraction temperatures above 110 °C do influence the decomposition factors [14, 15, 26]. Together with "isotopically labelled" spikes to check for possible degradation or transbutylation reactions, it is a promising powerful technique.

Isotope dilution with in-house synthesized ^{116}Sn-enriched organotin standards was applied to evaluate the efficiency of different extraction reagents in the GC-ICP-MS analysis of sediment CRMs (PACS-2 and BCR 646). Extractants of increasing ionic strength, from pure organic solvents to a mixture 1 : 1 with concentrated strong acids were applied in combination with ultrasonication and mechanical shaking [44, 45] in the presence as well as in the absence of tropolone. Extraction conditions like HBr : water (1 : 1) and a methanol : acetic acid (1 : 3) mixture gave results coinciding with the certified values for all the three butyltin compounds, while extraction with tropolone in diethyl ether or simply with HCl proved to be the most efficient for the phenyltin forms.

The use of complexing agents in organotin extraction have proved to be mandatory on several occasions [27, 28], especially in order to extract the mono-substituted compounds (MBT, MPhT), which tend to undergo strong ionic interaction and/or surface complexation in complex matrices [28], and the di-substituted compounds (DBT, DPhT) whereas TBT and TPhT are not affected at all by the presence of complexing agents like tropolone [46]. The presence of 1% oxine (8-quinolinol) complexing agent in the extraction of organotin compounds from spiked marine sediment under acidic conditions (various concentrations) with different polar solvents was found to be necessary to improve the recoveries of organotin compounds. Recoveries of 84–100% were achieved for both the phenylated and butylated compounds. Other methods with acidic conditions and non-polar solvent mixtures mostly foresee the utilization of tropolone [47] or ammonium pyrrolidine-dithiocarbamate (APDC) [48] as the complexing agent. A good

Table 2 Applied speciation steps for organotin determination in marine sediment reported in papers published since 2000

Extraction	Derivatization	Preconcentration	Analysis
• leaching with organic polar	• hydride generation [28]	• SPME [12, 29],	• GC-MIP-AES [41]
[14, 15, 26–36, 44]	• organoborate:	[31, 34, 36, 39, 49]	• GC-ICP-AES [39]
and non-polar solvents	– ethylation [14, 15],	• cryogenictrap [23]	• GC-ICP-MS [14, 15, 26, 33, 44]
[18, 21, 22, 44, 47]	[18, 26, 29, 31, 33],	• pervaporation [50]	• HPLC-ICP-MS [27, 43]
+ complexing agents using	[36–39, 41, 44, 47]		• GC-AAS [28, 35]
MW [15, 26, 43],	– propylation [34]		• GC-ICPMS [23]
US [15, 26, 29, 31–36, 40, 41, 44, 48],	• Grignard:		• GC-MS [18, 29, 30, 32–34, 40],
ASE [14, 27],	– ethylation [42]		• GC-PFPD [37, 48, 50]
MS [18, 26, 37–39, 41]	– propylation [21]		• GC-FPD [21, 38, 47]
• SPE [35, 40]	– pentylation [30, 32, 48]		• GC-MS/MS [22]
			• GC-FID [31, 36]

performance was obtained also by Martinez-Vidal et al. [22], who extracted sediment samples only with toluene in the presence of DDTC as the complexing agent in an ultrasonic bath, achieving recoveries of organotin species from spiked samples ranging between 96–108%.

The application of preconcentration methods is necessary to further increase the detection limit of analytical methods. Solid phase microextraction [12, 29, 31, 34, 36, 39, 49], mostly utilized after polar solvent extraction in acidic media, is becoming very popular also for tin speciation. The cryogenic trapping of a continuously stirred suspension of sediment and water [23] purged with He flow directly connected with a GC-ICP-MS system has also proved to have a good potential for the analysis of samples with very low organotin levels. For methyltin species, Gomez-Ariza et al. [50] utilized the advantages of the pervaporation module as a preconcentration that allows the direct determination of the species without preliminary treatment. After preconcentration, the species to be determined were ethylated, separated and determined by GC–PFPD. Recovery values in the range of 90–98% were achieved for the mono-, di- and also the trimethylated forms with this extremely simple and easily automatizable method.

In the sample preparation of sediment samples for tin speciation, the main developments within the last years concerned the extraction and preconcentration steps. With the application of alternative energies —microwaves, high pressure, ultrasounds—the efficiency of the extraction step has been increased and the required treatment time has been decreased, whereas the application of SPME has mainly improved the preconcentration step. On the other hand, the derivatization step has continued to be carried out essentially with the techniques developed twenty years ago. Furthermore, the possibility of using isotope dilution techniques for tracing possible degradation and interconversion of organotin compounds during sample treatment has become more accessible to a larger number of laboratories.

A summary of the most applied extractants, derivatizing reagents, preconcentration systems and instrumental techniques in the determination of organotin compounds in marine sediment, derived from the literature published after 2000, is reported in Table 2.

2.3 Sample Preparation Methods for the Determination of Organotin Compounds in Biological Samples

TBT present in the marine environment accumulates through the marine food chain, resulting in the occurrence of this compound as well as its breakdown products in marine biota organisms. Due to their lipophilic characteristics, organotin compounds may be incorporated in parts of organisms where lipids are present in a higher quantity such as the entrails, gonads and gills. These biological tissues offer the advantage of being highly soluble in polar solvents, which are used to leach/digest the biological matrices as much as possible to remove the analytes from the binding sites.

The most-reported extraction methods for organotin compounds from biota samples are also based on acidic digestion (mostly with HCl or acetic acid) in the presence of a polar solvent which is mainly methanol [16, 17, 32, 33, 37, 51–53]. The preference of methanol is due to its osmotic activity that induces rapid bursting of the cells and facilitates the homogenization of biological materials.

The extensive comparative study of Pellegrino et al. [6] proved that the use of a methanol/tropolone/hydrochloric acid extraction mixture assisted by ultrasound, out of 12 selected extraction methods (combination of organic solvents of various polarity—methanol, dichloromethane, toluene, pentane, hexane; the presence of a complexing agent (tropolone); the effect of different acidic conditions—HCl, HBr, acetic acid; and different extraction devices—ultrasonic bath, mechanical shaker, Soxhlet apparatus, microwave oven), provides the statistically most similar results to the certified butyl- and the indicative phenyltin values of BCR 477 CRM.

As it has become popular for organotin extraction from sediments, pressurized liquid extraction has also been gaining popularity for the leaching of marine biota samples. Wahlen and Catterick [51] utilized accelerated solvent extraction in the determination of various organotin species in CRM samples (BCR 477, candidate BCR 710) with the HPLC-ICP-MS technique achieving accurate and precise results. A program of five static cycles of 5 min at 100 °C and 1500 psi, with a solution of 0.5 M sodium acetate/1.0 M acetic acid in methanol was applied.

Wasik and Ciesielski [17] reached excellent extraction efficiencies without significant species degradation by applying ASE parameters as follows: four static cycles with a solution of 90% v/v methanol with acetic acid:sodium acetate containing 0.03% w/v tropolone, at 800 psi and 125 °C. The most important factor affecting extraction efficiency turned out to be solvent composition and, in particular, the presence of the complexing agent. It considerably improves DBT and above all, MBT recovery. The extraction efficiency of ionic organotins with polar or medium polarity solvents is controlled by two opposite effects. On the one hand, the complexation of the ionic forms, MBT in particular generates complexes that are less soluble in a medium polarity solvent, and thus extraction decreases. On the other hand, the formation of a tropolone–analyte complex that shields (owing to the steric effect) the analyte molecules from hydrogen bond forming functional groups of the protein matrix, increases extraction efficiency. According to the authors, this latter effect is prevalent and, consequently, the addition of tropolone does indeed increase extraction efficiency.

Looser et al. [52] proposed the extraction of biological samples with methanol/acetic acid by mechanical shaking at ambient temperature in the dark followed by aqueous ethylation, LLE and GC-MS separation/determina-

Table 3 Applied speciation steps for organotin determination in marine biota reported in papers published since 2000

Extraction	Derivatization	Preconcentration	Analysis
• acid & polar solvent – HCl, acetic acid/MeOH [16, 17, 32, 33, 37, 51–53] • acid & varying polarity solvent – HCl/toluene, diethyl ether, ethyl acetate [22, 30, 54–56] • basic digestion – TMAH [18, 38, 57–60, 68] • PLE (ASE) [17, 51, 52] – sodium acetate & acetic acid/MeOH	• organoborate – ethylation [17, 18, 33, 37, 38], [51, 52, 60, 68] • Grignard – pentylation [6, 20, 22, 30, 32], [54, 55] – propylation [53]	• SPME [59]	• GC-MS [6, 20, 22, 30, 32, 33, 52, 55, 60] • GC-FPD [17, 38, 53, 54, 57–59] • GC-PFPD [16, 37, 68] • GC-ICP-MS [51] • GC-MS/MS [22] • HPLC-ICP-MS [51] • HPLC-FD [56]

tion. The method allowed a fast (without a clean-up step for sample intakes smaller than 500 mg), reliable (as showed by BCR 477 analysis), and simple determination (no special equipment like ASE, MW, etc. is required) of organotin compounds in biological samples.

Even though the acid extraction in the presence of a polar solvent is the most widespread method, leaching with acid and low or non-polar solvents [22, 30, 54–56] and basic digestion with tetramethylammonium hydroxide [18, 38, 57–60] are also utilized with good recovery values and LOD at the ng g^{-1} level for each compound.

Alkaline conditions for organotin extraction are restricted to biotic samples and even then not applied very often and only with TMAH as the extractant [18, 38, 57–60]. The concentration of TMAH ranges from 20 to 25% and is always aided by sonication or agitation. The temperature of the extraction ranges from 50–60 °C and the time of extraction from 1 to 2 hours.

A summary of the most applied extractants, derivatizing reagents, preconcentration systems and instrumental techniques in the determination of organotins in marine biota, derived from literature published after 2000, is reported in Table 3.

3 The Commonly Applied Instrumental Techniques for the Determination of Organotins in Marine Samples

A significant number of various instrumental techniques are reviewed in the literature for the determination of organotin compounds [5, 11, 61–66].

The most commonly applied techniques over the last 20 years have been based on gas and liquid chromatographic separation followed by different types of detectors. In the case of gas chromatography, a derivatization step is necessary prior to the separation, due to the low volatility of the target compounds. The conversion of ionic alkyltins into gas chromatographable species can be divided into two categories: (a) those based on in-situ hydridization [with sodium borohydride ($NaBH_4$)] or alkylation (with most of all $NaBEt_4$ and $NaBPr_4$) in protic solvents [aqueous or (m)ethanolic environment]; and (b) those based on the liquid-liquid extraction (LLE) of the ionic organotin compounds and the following derivatization using Grignard reagents (methyl-, ethyl-, propyl-, pentyl-, hexylmagnesium chlorides/bromides) in non-protic solvents [9]. After derivatization a clean-up step is usually necessary unless the derivatized compounds are directly determined by purge-and-trap analysis or SPME is utilized to extract the derivatized compounds either from the aqueous or the gaseous phase prior to the gas chromatographic separation [67].

Methods for the direct analysis of volatile organotin compounds, for example, cryogenic trapping, gas chromatography, inductively coupled mass spectrometry (CT-GC-ICP/MS) systems [23] are seldom applied.

The high efficiency achieved by capillary GC allows the separation of the derivatized compounds with non-polar or semi-polar stationary phases, such as dimethyl-polysiloxane or 5% diphenyl-dimethyl-polysiloxane. Most of the capillary columns generally utilized for OT separation, have a length of 25–30 m, an i.d. of 0.2–0.25 mm and a film thickness of 0.1–0.3 μm [11, 25, 61, 63].

Flame photometric detection (FPD), atomic emission detection (AED), atomic emission spectrometry (AAS) and mass spectrometry (MS) are the mostly used methods for the detection of tin species and are characterized by detection limits down to pg levels for injected tin species [11, 25, 61, 63].

The FPD system is based on the emission of tin species in a hydrogen-rich flame; selectivity for tin is obtained at 610 nm using a cut-off or interference filter. However, the FPD signal may be disturbed by the coextracted sulfur species. Dual flame FPD had been introduced to improve the selectivity of the technique, but it provided lower sensitivity than the single-flame FPD. The new generation FPDs in which the continuous flame is replaced by a pulsed flame, the pulsed FPD (PFPD) detector, has recently been introduced. The main advantages are the increased sensitivity and lower matrix-dependency [68].

The application of the AED detector is generally used in combination with microwave-induced plasma (MIP-AED) rather than with inductively coupled plasma (ICP-AED) for its high sensitivity, even if the robustness of the ICP is a significant advantage when complex matrices are analyzed [67].

In the GC-AAS systems tin is quantified at 286.3 or 224.4 nm and the atomization of the compounds is mainly achieved by electrothermal (ET-AAS) or quartz furnace (QF-AAS) [25, 28].

Mass spectrometric (MS) detection with electron impact (EI) ionization for fragment production and a quadrupole or ion trap for ion detection, has distinct advantages over the previously mentioned detectors [6, 11, 65]. It is more sensitive, highly selective and is able to provide structural information about the target compounds. Especially, when a single fragment of a specific organotin compound is chosen for selected ion monitoring (SIM) analysis mode. Recently, ICP has been widely utilized as the ion source for fragment production coupled to AES, MS or time-of-flight (TOF)-MS detection. The application of MS in combination with isotope-dilution (ID) analysis is becoming very popular [14, 15, 26, 44, 45]. The sensitivity and selectivity of MS systems can be further improved by tandem mass spectrometry (MS/MS) [20, 22].

The absolutely essential derivatization step in gas chromatography is not necessary in high performance liquid chromatography (HPLC). This renders the HPLC methods significantly simpler and faster. The separation of the underivatized ionic organotin compounds may be performed with normal- or reversed-phase, ion-pair as well as ion-exchange columns; the latter being the most commonly applied [25, 61–64]. Implementations with HPLC, however, are limited for organotin separation compared to GC, because of the scarce sensitivity of the general detector for the organotin levels found in nature. Even though in the last years with the utilization of MS [ICP-MS in particular (with or without ID), but also MS/MS, ESI-MS/MS and ESI-TOF-MS] [51, 64] and fluorimetry (after post-column derivatization) [56] as instrumental detection techniques, LC methods satisfy the requirements of selectivity and sensitivity for the determination of organotin compounds in environmental samples [69], and are more largely applied.

4 Conclusions

Notwithstanding the ban of TBT in antifouling paints, TBT can still be encountered in the marine environment, because of its slow degradation and high sorption to suspended matter and sediment and high tendency to be bioaccumulated in marine organisms.

Its determination involves the use of multistep speciation analysis methods including extraction, derivatization, clean up, preconcentration, separation and detection. In some cases, some steps are simultaneously applied such as extraction/derivatization/preconcentration or even avoided such as derivatization in the case where high performance liquid chromatography is applied.

Each step presents specific problems that can be faced with different approaches presenting both advantages and disadvantages.

Extraction

In the case of water samples, the applied extractants can be solvents or solid phases. The efficiency of LLE mainly depends on the type of chosen extractant(s) and on the experimental parameters such as pH, sample to solvent ratio and extraction time. The main advantages of using LLE are the high efficiency, reproducibility in recovery and high enrichment factors. However, its disadvantages are the high cost, laborious handling and the difficulties of adapting them to field analysis and the potential hazard of organic solvents to the operators' health. The efficiency of SPE depends on the type of adsorbent chosen (both in terms of adsorption capacity and breakthrough) and on experimental conditions such as pH and sample flow rate. It presents a number of advantages over LLE as it is easy to handle, adaptable to field analysis, the extracts can be more easily stored and transported, it is of lower cost, allows high routine and presents a very low potential health hazard. In addition, no additional preconcentration is needed when it is coupled directly to a separation/detection system. However, the main disadvantages are the lower enrichment factor, efficiency and reproducibility when compared to LLE.

In the case of solid samples, the extraction efficiency depends, in addition to the choice of extractant(s) and the experimental parameters, on the techniques applied to support the extraction such as Soxhlet, ultrasound, microwave, supercritical fluid and pressurized solvent systems. The most commonly applied method is ultrasonication, even if microwave energy and pressurized solvent extraction have begun to be widely applied as well. Ultrasonication presents the advantages of being simple and relatively inexpensive, but on the other hand, requires long treatment times as ultrasonic extraction is generally performed with acid(s) and/or polar solvents, successively diluted with water and followed by LLE. The use of microwaves and particularly of focused microwaves allows a significant reduction of treatment time in the release of the organotin compounds from solid matrices, but the following steps are the same as in the case of ultrasonication (LLE of the extract diluted with water). The main advantages of the recently popularly used accelerated solvent extraction (ASE) are the high sample throughput, significantly lower solvent consumption and consequently, lower cost, and the high potential of being used routinely.

In all the cases, however, the presence of a complexing agent has proved to be mandatory for enhancing the extraction efficiency of the mono- and disubstituted tin compounds.

Derivatization

The derivatization of organotins is achieved by hydridization and/or alkylation with tetraalkylborates (mainly tetraethyl- and, to a lesser extent, tetrapropylborate) and/or a variety of Grignard reagents (from methyl- to hexylmagnesium bromide). Hydridization is strongly influenced by the type and concentrations of acid used and the sample matrix. Particularly, the presence of high metal concentration of the matrix, which leads to boride formation; humic substances and fat content leading to foam formation, strongly affect the efficiency and reproducibility of the derivatization.

The yield of derivatization in the case of alkylation with borates is influenced also by the reagent concentration, derivatization pH and reaction time as well as by the substitution degree of the organotin compounds. The main disadvantage of alkylation is the lack of stability of the alkylating reagents.

Both hydridization and alkylation with borates are particularly suitable for water samples, where, exploiting the reagents' compatibility with water, they can be used for in situ derivatization showing, particularly in the case of alkylation, high yields of derivatization. On the contrary, both techniques are less suitable in the case of solid samples.

Grignard derivatization is widely applied. The reaction yields are influenced by the concentration and type of reagent, however, regardless of the type, all Grignard reagents provide very high yields of derivatization. Pentylation and hexylation are preferable, as their derivatives are less prone to volatilization losses during the successive preconcentration steps. The main disadvantages of Grignard derivatization are the "violent" nature of the Grignard reagents (potentially leading to degradation or volatilization), the necessity of destroying the excess reagent carefully (thereby increasing the number of steps) and the high time consumption. Contrary to the other two hydridization and alkylation methods, Grignard derivatization is particularly suitable for solid samples and less so for water samples.

No significant improvements have been experienced recently in the derivatization of organotin compounds. The techniques and their application have been essentially the same for more than 15 years.

Preconcentration

Recently, several preconcentration techniques have been applied in speciation analysis which include biotrapping (for inorganic species), flow injection, purge and trap, cryogenic trapping and SPME. SPME—and to a smaller extent even cryogenic trapping—has been successfully applied to organotin analysis. It is easy to handle, not invasive which is crucial for preserving the original speciation equilibrium, allows a reduced number of steps and the simultaneous extraction/preconcentration of species of different elements and, furthermore, it is a free-of-solvent technique thus with reduced health hazard for operators. The main disadvantage is certainly its scarce reproducibility that still represents the major drawback in the diffusion of this technique.

Instrumental Analysis

The mostly applied instrumental techniques for the routine determination of organotin compounds in environmental samples are certainly GC-FPD and GC-MS. Recently, an increase has been noted in the use of ICP-MS as the detector for tin speciation coupled either with GC and, mainly, HPLC.

The most significant development in this field is the application of isotope dilution techniques. This is worth stressing not so much for the improvement of determination performances, but above all for their potential in sample treatment optimization. In fact, the transformation/degradation of organotins species during the sample preparation steps are monitored more and more often by the utilization of species-specific isotope-dilution, which presumably will become the most powerful tool for optimizing and validating extraction methods in the near future.

References

1. de Mora SJ (1996) Tributyltin: case study of an environmental contaminant. Cambridge environmental chemistry series 8. Cambridge University Press, Cambridge
2. Chiavarini S, Cremisini C, Morabito R (1996) In: Caroli S (ed) Elemental Speciation in Bioinorganic Chemistry. Wiley, New York, p 287

3. Donard OFX, Lespes G, Amouroux D, Morabito R (2001) Organotin Compunds in the Environment: Still a Critical Issue. In: Ebdon L, Pitts L, Cornelis R, Crews H, Donard OFX, Quevauviller P (eds) Trace Element Speciation for Environment, Food and Health. RSC, Cambridge, p 142
4. Hoch M (2001) Appl Geochemistry 16:719
5. Donard OFX, Ritsema R (1993) Hyphenated Techniques Applied to the Speciation of Organometallic Compounds in the Environment. In: Barceló (ed) Environmental Analysis (Techniques, Application and Quality Assurance). 2nd edn. Elsevier, Amsterdam, p 549
6. Pellegrino C, Massanisso P, Morabito R (2000) Trends Anal Chem 19:97
7. Rosenberg E, Ariese F (2001) Quality Control in Speciation Analysis. In: Ebdon L, Pitts L, Cornelis R, Crews H, Donard OFX, Quevauviller P (eds) Trace Element Speciation for Environment, Food and Health. RSC, Cambridge, p 17
8. Quevauviller Ph, Morabito R (2000) Trends Anal Chem 19:86
9. Morabito R, Massanisso P, Quevauviller Ph (2000) Trends Anal Chem 19:113
10. Quevauviller Ph, Astruc M, Morabito R, Ariese F, Ebdon L (2000) Trends Anal Chem 19:180
11. Morabito R (2003) In: Mester Z, Sturgeon R (eds) Sample Preparation for Trace Element Analysis, Comprehensive Analytical Chemistry, Vol XLI. Elsevier, Amsterdam, p 991
12. Mester Z, Sturgeon R (2003) In: Mester Z, Sturgeon R (eds) Sample Preparation for Trace Element Analysis, Comprehensive Analytical Chemistry, Vol XLI. Elsevier, Amsterdam, p 371
13. Pawliszyn J (2003) Anal Chem 75:2543
14. Ruiz Encinar J, Rodriguez-Gonzalez P, Fernandez JR, Alonso JI, Diez S, Bayona JM, Sanz-Medelt A (2002) Anal Chem 74:5237
15. Rodriguez-Gonzalez P, Ruiz Encinar J, Garcia Alonso JI, Sanz-Medel A (2003) Analyst 128:447
16. Le Gac M, Lespes G, Potin-Gautier M (2003) J Chromatogr A 999:123
17. Wasik A, Ciesielski T (2004) Anal Bioanal Chem 378:1357
18. Ikonomou MG, Fernandez MP, He T, Cullon D (2002) J Chromatogr A 975:319
19. Jiang Gb, Liu Jy, Yang Kw (2000) Anal Chim Acta 421:67
20. Tsunoi S, Matoba T, Shioji H, Huong Giang LT, Harino H, Tanaka M (2002) J Chromatogr A 962:197
21. Vella AJ, Mintoff B, Axiak V (2000) Sci Total Environ 258:81
22. Martinez Vidal JL, Vega AB, Arrebola FJ, Gonzalez-Rodriguez MJ, Sanchez MC, Frenich AG (2003) Rapid Commun Mass Spectrom 17:2099
23. Tessier E, Amouroux D, Donard OFX (2002) Biogeochemistry 59:161
24. Følsvik N, Brevik EM, Berge JA (2000) J Environ Monit 2:281
25. Smedes F, de Jong AS, Davies IM (2000) J Environ Monit 2:541
26. Ruiz Encinar J, Rodriguez Gonzalez P, Garcia Alonso JI, Sanz-Medel A (2002) Anal Chem 74:270
27. Chiron S, Roy S, Cottier R, Jeannot R (2000) J Chromatogr A 879:137
28. Feng YL, Narasaki H (2002) Anal Bioanal Chem 372:382
29. Cardellicchio N, Giandomenico S, Decataldo A, Di Leo A (2001) Fresenius J Anal Chem 369:510
30. Regoli L, Chan HM, de Lafontaine Y, Mikaelian I (2001) Aquat Toxicol 53:115–126
31. Arambarri I, Garcia R, Millán E (2003) Chemosphere 51:643
32. Chiavarini S, Massanisso P, Nicolai P, Nobili C, Morabito R (2003) Chemosphere 50:311

33. Centineo G, Rodríguez-Gonzáles P, Blanco Gonzáles E, García Alonso JI, Sanz-Medel A (2004) J Mass Spectrom 39:485
34. Muñoz J, Baena JR, Gallego M, Valcárcel M (2004) J Chromatogr A 1023:175
35. Puri BK, Muñoz-Olivas R, Cámara C (2004) Spectrochim Acta Part B 59:209
36. Millán E, Pawliszyn J (2000) J Chromatogr A 873:63
37. Bancon-Montigny CH, Lespes G, Potin-Gautier M (2000) J Chromatogr A 896:149
38. de Mora SJ, Fowler SW, Cassi R, Tolosa I (2003) Mar Pollut Bull 46:401
39. Aguerre S, Pécheyran C, Lespes G (2003) Anal Bioanal Chem 376:226
40. Nogueira JMF, Teixeira P, Florêncio MH (2001) J Microcolumn Sep 13:48
41. Carpinteiro J, Rodriguez I, Cela R (2001) Fresenius J Anal Chem 370:872
42. Morabito R, Soldati P, de la Calle MB, Quevauviller PH (1998) Appl Organomet Chem 12:621
43. Yang L, Lam JWH (2001) J Anal At Spectrom 16:724
44. Kumar SJ, Tesfalidet S, Snell J, Frech W (2003) J Anal At Spectrom 18:714
45. Kumar SJ, Tesfalidet S, Snell JP, Nguyen Van D, Frech W (2004) J Anal At Spectrom 19:368
46. Pannier F, Astruc A, Astruc M, Morabito R (1996) Appl Organomet Chem 10:471
47. Biselli S, Bester K, Hühnerfuss H, Fent K (2000) Mar Pollut Bull 40:233
48. Godoi AFL, Montone RC, Santiago-Silva M (2003) J Chromatogr A 985:205
49. Cardellicchio N, Giandomenico S, Decataldo A, Di Leo A (2001) Fresenius J Anal Chem 369:510
50. Gómez-Ariza JL, Mingorance F, Velasco-Arjona A, Giráldez I, Sánchez-Rodas D, Morales E (2002) Appl Organometal Chem 16:210
51. Wahlen R, Catterick T (2003) J Chromatogr B 783:221
52. Looser PW, Berg M, Fent K, Muhlemann J, Schwarzenbach RP (2000) Anal Chem 72:5136
53. Horiguchi T, Li Z, Uno S, Shimizu M, Shiraishi H, Morita M, Thompson JAJ, Levings CD (2003) Mar Environ Res 57:75
54. Hung TC, Hsu WK, Mang PJ, Chuang A (2001) Environ Pollut 112:145
55. Ten Hallers-Tjabbes CC, Wegener JW, Van Hattum B, Kemp JF, Ten Hallers E, Reitsema TJ, Boon JP (2003) Mar Environ Res 55:203
56. González-Toledo E, Benzi M, Compañó R, Granados M, Dolors Prat M (2001) Anal Chim Acta 443:183
57. Albalat A, Potrykus J, Pempkowiak J, Porte C (2002) Chemosphere 47:165
58. Morcillo Y, Porte C (2002) Environ Pollut 107:47
59. Cassi R, Tolosa I, Bartocci J, de Mora S (2002) Appl Organometal Chem 16:355
60. Gallina A, Magno F, Tallandini L, Passaler T, Caravello GU, Pastore P (2000) Rapid Commun Mass Spectrom 14:373
61. Attar KM (1996) Appl Organomet Chem 10:317
62. Harrington CF, Guenter KE, Cullen WR (1996) Appl Organomet Chem 10:339
63. Abalos M, Bayona JM, Compaño R, Granados M, Leal C, Prat MD (1997) J Chromatogr A 788:1
64. Ebdon L, Hill SJ, Rivas C (1998) Trends Anal Chem 17:277
65. Bayona JM (2000) Trends Anal Chem 19:107
66. Pereiro IR, Díaz AC (2002) Anal Bioanal Chem 372:74
67. Aguerre S, Lespes G, Potin-Gautier M (2003) J Chromatogr A 999:61
68. Leermakers M, Nuyttens J, Baeyens W (2005) Anal Bioanal Chem 381 1272:1280
69. González-Toledo E, Compañó R, Granados M, Prat MD (2003) Trends Anal Chem 22:26

Hdb Env Chem Vol. 5, Part O (2006): 71–100
DOI 10.1007/698_5_050

Published online: 20 December 2005

Worldwide Occurrence of Organotins from Antifouling Paints and Effects in the Aquatic Environment

Karl Fent[1,2]

[1]Institute of Environmental Technology, University of Applied Sciences Basel, St. Jakobs-Strasse 84, CH-4132 Muttenz, Switzerland
karl.fent@bluewin.ch, k.fent@fhbb.ch, kfent@ethz.ch

[2]Department of Environmental Sciences, Swiss Federal Institute of Technology (ETH), 8092 Zürich, Switzerland
karl.fent@bluewin.ch, k.fent@fhbb.ch, kfent@ethz.ch

Abstract Organotins belong to the most toxic pollutants for aquatic life known so far. Widespread contamination of harbors and areas with high shipping activities occurs globally due to the ongoing use of tributyltin (TBT) in antifouling paints on large vessels. In the last decade, organotin pollution has spread to developing countries. Contamination decreased only slowly several years after regulation of antifouling paints in small vessels and remains widespread in sediments and marine biota. TBT degradation in sediment is very slow under anaerobic conditions and remobilization occurs via mixing and dredging. Ecotoxicological effects occur globally in marine gastropods, which are affected by masculinization (imposex) at ng L^{-1} levels leading to population declines. Present TBT levels in harbors and adjacent areas are still in the range of effect levels. Recently, also androgenic activity of TBT in fish was reported. A large variety of organisms, in particular early life stages, are susceptible to low TBT concentrations of a few 100 ng L^{-1}. Bioaccumulation leads to significant residues in aquatic biota including marine mammals.

Despite regulations, ecotoxicologically relevant contamination of marine ecosystems is persisting, particularly in sediments. The recent spread into developing countries indicates that organotin contamination has ongoing ecotoxicological relevance globally.

Keywords Ecotoxicological effects · Environmental contamination · Organotins · Tributyltin · Triphenyltin

Abbreviations

BCF	bioconcentration factor
DBT	dibutyltin
DOM	dissolved organic matter
K_p	distribution ratio between particulate and dissolved fraction
K_{ow}	*n*-octanol–water distribution coefficient
pK_a	acidity constant
PVC	polyvinyl chloride
MBT	monobutyltin
TBT	tributyltin
TBT^+	tributyltin cation
TBTCl	tributyltin chloride
TBTOH	tributyltin hydroxide
TPT	triphenyltin
TPT^+	triphenyltin cation
TPTCl	triphenyltin chloride
TPTOH	triphenyltin hydroxide

1 Introduction

Organotins belong to the most important anthropogenic organometallic compounds. Trisubstituted organotins are of considerable environmental concern due to their extremely high toxicity to aquatic life, which has resulted in population declines in marine oysters [1] and snails [2]. Organotins are produced for different purposes. About 70% of the total annual world production is devoted to the stabilization of PVC (mainly dibutyl-, and dioctyltins, monosubstituted organotins), catalysts in the production of polyurethane foams, silicones and in other industrial processes. About 20–25% are used as biocides in agrochemicals and in a broad spectrum of applications such as antifouling paints, and fungicides in timber and wood protection. Tributyltins (TBT) are also used for protection of textiles, leather and other materials, as well as a disinfecting agent [3]. The annual world production of organotins may reach 50 000 t [4]. The agricultural and biocidal use of organotins, mainly in antifouling paints, gives rise to the largest direct environmental input. The current application of TBT lies mainly in its use as a preservative for timber and wood, and other materials, but it TBT is still used in antifouling paints, mainly in developing countries.

Historically, TBT gained widespread application as an effective antifouling paint biocide on pleasure boats, large ships and docks in the 1970s and 1980s. In the late 1970s organotin-containing antifouling paints were found to cause detrimental effects on marine and freshwater biota, however. As TBT leaches directly into water, relatively high contamination of pleasure boat and commercial harbors and their surroundings, areas with high shipping activities, and coastal areas result. Detrimental effects on oysters [5] shed light on the ecotoxicological impact of this toxicant. Subsequent studies confirmed the general hazard of TBT and triphenyltin (TPT) to aquatic ecosystems [6–8]. Reports on detrimental effects on marine neogastropods [2, 9, 10] played a major role in establishing the link between their sterilization due to masculinization, population decline and TBT contamination. Additional studies about the contamination of marine [11] and freshwater [12–14] environments contributed significantly to the adoption of regulatory standards for TBT in antifouling paints, their restrictions and banning in small boats. Triphenyltin (TPT) has been employed formerly as a co-toxicant in antifouling paints and found in freshwater [14] and marine environments [15–17], however, it is mainly used in agriculture as a fungicide in crop protection.

Originally, studies on organotins were devoted to the toxicity assessment in aquatic organisms and mammals including the search for modes of action. When environmental effects were observed in the 1980s, studies focused on the development of analytical techniques to determine trace levels of organotins in ecosystems. Surveys in aquatic ecosystems and effect studies revealed in combination with fate studies a clearer picture about the environmental behavior, hazards and risks of organotins. The use of TBT-containing antifouling paints has been controlled or banned in developed countries, resulting in a decrease of TBT contamination of coastal and harbor waters. France banned the use in 1982 on boats of < 25 m in length, and the UK followed in 1987. In addition, an environmental quality standard of 20 ng L^{-1} for freshwater, and 2 ng L^{-1} for seawater had been set [3]. Many other countries regulated TBT-containing antifouling paints for large ships, and prohibited the use on vessels < 25 m in length (European Community in 1990), or implemented a sales ban (e.g. 1990 in Switzerland and Germany). Today, restrictions for small boats are implemented in developed countries and the contamination declined [18, 19]. However, TBT-containing antifouling paints remain an important source. In developed countries, organotin levels are still often higher than effect concentrations for most susceptible organisms. An important move forward was made by the International Maritime Organization (IMO) by a convention adopted in 2001 prohibiting the use of harmful organotins in antifouling paints used on ships and establishing a mechanism to prevent the potential future use of other harmful substances therein (IMO, www.imo.org). By January 2003, all ships should not apply or reapply organotin-containing antifouling paints, and by January 1, 2008, all ships shall not bear such compound on their hulls. However, despite the European

Commission's moves, the IMO Convention has not been ratified by member countries to date, and therefore has not yet entered into force as of 2005.

In this review, recent data on the environmental contamination and ecotoxicology of organotins is given, focusing on the advancement in the last decade. Earlier insights given in a broader sense on many different aspects of organotins in the environment [4] are supported by recent data. In the last decade, studies mainly focused on monitoring of larger geographical areas using sediment and biological samples, particularly in Asian countries. Another focus lied in the monitoring and understanding of the mechanisms behind the masculinization of marine gastropods (imposex). Progress in the last decade has mainly been made in the global assessment of the organotin pollution using both chemical analysis and biomonitoring using imposex as a biomarker.

One of the most important conclusions lies in the recognition that the organotin contamination is global, reaching developing countries, which seems to be linked to economic growth. However, despite regulations, organotin contamination of harbors and marine environments persists.

2 Bioavailability

The aqueous chemical speciation of organotins is important for their environmental behavior, and consequently for bioavailability and toxicity. Organotins undergo pH-dependent hydrolysis in water and their speciation is also dependent on dissolved ions. Trisubstituted organotins form positively charged diaquo complexes $R_3Sn(H_2O)_2^+$ at pH < pK_a, and these monovalent organometallic cations behave as weak acids. Acidity constants (pK_a) of 6.25 and 5.20 were determined for TBT and TPT, respectively [20]. The dominant TBT and TPT species at pH < pK_a is the cation, whereas at pH > pK_a the neutral hydroxide species TBTOH and TPTOH, respectively, dominates. In the presence of chloride ions (seawater), R_3SnCl species occur. Hence, at environmental pH conditions, mono-, di-, and trisubstituted organotins exist predominantly as neutral hydroxides in aqueous solution.

The speciation of organotins is pH-dependent and has consequences for their bioavailability, a fundamental parameter in ecotoxicology. Significant partitioning into biota occurs via uptake through membranes by hydrophobic mechanisms via uncharged species (both TBTOH and TBTCl). Several octanol–water partition coefficients ($\log K_{ow}$) for organotins were given [4], including 4.10 for TBTOH and 3.53 for TPTOH [21]. TBTOH and TPTOH are relevant for uptake, bioaccumulation [22, 23] and toxicity toward aquatic organisms [22, 24, 25], including fish cells [26], where a positive correlation between the in vitro cytotoxicity and $\log K_{ow}$ was found. The influence of the pH and dissolved organic matter (DOM) was demonstrated in our studies with frogs [24], *Daphnia magna*, yolk sac larvae of fish *Thymallus thymallus* [22], and the midge *Chironomus riparius* [25]. Trisubstituted organotin cations are less toxic than the neutral hydroxides, and this goes along with their pH-related K_{OW}. Significantly higher bioconcentration occurred at pH 8 than pH 5 or 6, which is related to the fact that these compounds mainly, but not exclusively, accumulate as hydroxides [22]. Toxicity was also significantly increased at pH 8, where TBT occurs predominantly as TBTOH (95%) [21]. This is correlated with higher K_{OW} at pH 8 [22].

Figure 1 shows that bioconcentration of TBT in *C. riparius* larvae is significantly higher at pH 8 than pH 5 due to the prevalence of bioavailable TBTOH. TPT bioconcentration was significantly higher due to the lack of significant metabolism, occurring with TBT [25]. These pH dependent differences are mainly based on the aqueous phase speciation. This was also shown in yeast [27], where uptake and toxicity of TPT was reduced considerably at pH 3.5. TPT^+ uptake was restricted to cell wall sorption, while TPTOH caused an increase in the order in the cell phospholipid bilayer and diffuses into the cell. The neutral TBTCl and TPTCl were more toxic than the hydroxides. Similar to our studies [26], toxicity in yeast was correlated with lipophilicity.

The distribution of TBT and TPT between water and DOM is pH-dependent and highest distribution ratios with maxima occur close to the acidity constant [21]. Sorption is governed by complexation of the cation (TBT^+, TPT^+) by negatively charged ligands of DOM mainly at low pH, but also on hydrophobic partitioning of hydroxid species to nonpolar groups (hydrophobic interactions) at ambient pH. Figure 2 shows that DOM leads to a concentration-related reduction in the TBT bioaccumulation in *Chirono-*

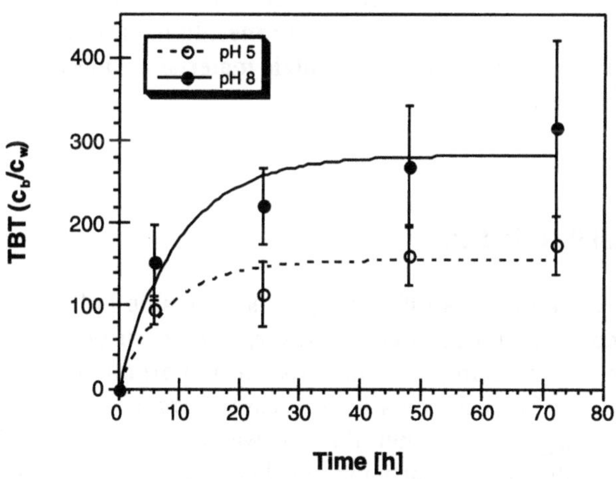

Fig. 1 Bioaccumulation of TBT and TPT in *Chrionomus riparius* at different pH. At pH 8, TBT occurs predominantly as TBTOH (98%), whereas at pH 5, TBT is present primarily as positively charged TBT^+ (fraction of 95%). After [23]

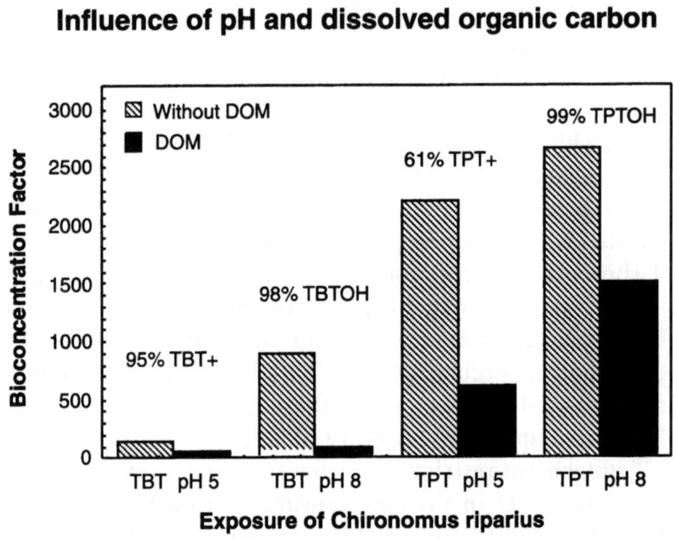

Fig. 2 Bioaccumulation of TBT and TPT in *Chrionomus riparius* at different pH and DOM concentrations. DOM (23 mg CL^{-1}) significantly reduced the bioaccumulation of TBT and TPT. The reduction was related to speciation depending on the pH (see Fig. 1). The fraction of TPTOH is more than 99% at pH 8, whereas TPT^+ and TPTOH are present in more similar fractions at pH 5 (61% and 39%, respectively). Undissociated TBTOH and TPTOH can penetrate biomembranes much more easily. After [25]

mus, but also in *Daphnia*, fish larvae *Thymallus thymallus* [22, 23]. The freely dissolved bioavailable fraction was reduced by complexation and hydrophobic interactions to DOM. This results in complexes that are too large or too polar to penetrate across biological membranes. In natural waters the portion of TBT and TPT sorbed to humic and fulvic material is unavailable for uptake via epithelial surfaces.

3
Environmental Distribution

Organotins leaching from antifouling paints from ship hulls are diluted into their surroundings. Biodegradation, adsorption to suspended particulates and subsequent scavenging to sediments are important removal processes. Bioaccumulation is also important. In mussels, organotin levels are of several orders of magnitude higher than in water [14, 28]. Because TBT and TPT associate with natural sorbents and because they are persistent under anoxic conditions [14, 19, 29], they accumulate in marine and freshwater sediments. TBT occurs in harbor waters to 95 to 99% in the dissolved phase. Particle-water ratios K_p range between 10^3–10^4 ($L\,kg^{-1}$) [4]. Particle-water partitioning of butyltins in the environment is controlled by the suspended solids concentration [30], pH, salinity and presence of natural DOM [31, 32], and therefore on sediment characteristics. Organic material in sorbents plays a pivotal role for sorption at ambient pH [21]. In muni cipal wastewater, organotins are predominantly associated with particulates, because of the much higher suspended solids concentrations [30] compared to surface water. TBT and TPT are very persistent in sediments and stored for years up to a decade, or longer [14, 33, 34]. Sediments act as important reservoirs, but ready desorption of organotins occurs [21, 31]. Resuspension by tide, storms or dredging will lead to increased organotin levels in the water column.

3.1 Degradation in Water and Persistence in Sediments

As abiotic transformation (photolysis and hydrolysis) is of minor importance, environmental degradation of TBT and TPT is principally biologically mediated [4, 35]. Decomposition occurs via microbial and algal degradation, and by a range of aquatic organisms including *Chironmids* [25]. TBT and TPT are degraded in water with half-lives in the order of 7–30 days found at high, but two months or longer at cold water temperatures [17]. Biodegradation is variable and dependent on environmental conditions and composition of micro-organisms [34, 36, 37]. Only a limited number of microbes are known to be able to degrade TBT [38]. Assuming zero-order kinetics, rapid half-lives from 3–19 days for the disappearance of TBT were recorded in warm water [37], where microalgae are important [39]. Complete mineralization proceeded only slowly with half-lives of 50–75 days. In deep sea and cold waters half-lives are in the order of several years [17]. Generally, low degradation rates of TBT and TPT were reported [33, 34, 40–42] with estimated half-lives in the order of several years, in particular under anaerobic conditions (0.9–5.2 years) [4]. The half-life in northern sediments was estimated to be more than 80 years [33]. However, fast degradation was reported in the marine sediments of San Diego Bay [43], and after a major organotin spill into a freshwater system [44], which must be regarded as exceptions. TBT degradation proceeds via successive dealkylation from TBT to DBT and MBT via hydroxylated metabolites, eventually leading to Sn^{4+}. Furthermore, methylation of organotins occurs in sediments, leading to remobilization to the atmosphere [45, 46]. Some bacteria develop resistance to high levels of organotins in sediments by an adaptation process. They express genes for a multidrug efflux pump and excrete toxic organotins [47].

Degradation is much more rapid in the water than in sediments, where TBT and TPT were persistent over long periods of time [14, 19, 48, 49]. Highest TBT and TPT levels occur at the top of sediment cores as shown in Fig. 3. High TBT concentrations reflect recent inputs, and the decrease below the

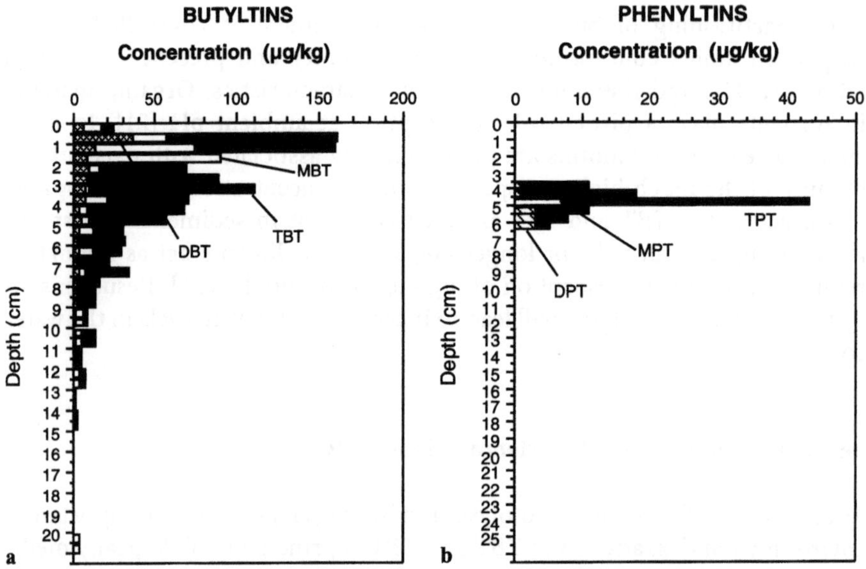

Fig. 3 **a** Butyltin and **b** phenyltin profiles in the sediment core of a pleasure boat harbor. Relatively high concentrations of TBT were found in the *upper* 6 cm. The decrease *below* mainly reflects the historical use of organotin-containing antifouling paints and input into the sediment. Degradation products DBT and MBT occur as well. Phenyltins occur only *below* the *upper* 3 cm demonstrating the recent lack of input. After [49]

historic build-up due to the increased use of TBT-containing paints in the late 1970s. Decreases with depth were paralleled by respective transformation products. Phenyltin sediment profiles indicate also slow degradation of TPT. The TPT input ceased as reflected by its lacking at the top of the core. Hence, sediment profiles likely reflect the historical use of the organotin antifouling paints rather than degradation alone. These profiles show that TBT is stored over periods of years or even decades in anoxic sediments, which to a lesser extent also holds for TPT [14]. Degradation of TBT is also slow in sewage sludge treatment [30, 50].

4 Contamination of Aquatic Environments

Only after the discovery of detrimental effects of TBT on oysters have efforts been directed to monitoring its environmental occurrence and assessing its effects. Widespread and significant pollution of marine [11, 41, 51–77], and freshwater ecosystems [4, 18, 34, 78–80] has subsequently been reported. TBT and degradation products DBT and MBT have generally been found in water, sediment and biota in developed countries, and recently in developing countries as well, indicating global occurrence in marinas and harbors, and waters with high shipping activities. Furthermore, TPT and its degradation products DPT and MTP have been detected, originating from older antifouling paints [14]. Phenyltins have still been detected in the last few years in sediments [54] (Figs. 3 and 4), biota [15, 81, 82] and in the deep-sea [16, 17].

Numerous monitoring studies indicate that TBT pollution originating from antifouling paints is global and has spread into developing countries [83–91], and even to the Arctic [92] and Antarctic [93]. Recent monitoring data using biota demonstrate global pollution of coastal areas. Studies mainly from Asian developing countries indicate a similar contamination pattern of harbors and estuaries as in developed countries. Moreover, monitoring data 7–10 years after regulations of TBT-antifouling paints demonstrate that the decrease is not substantial, as contamination remains at similar levels [94]. Only a few reports indicate a decrease in marine systems [95].

Besides leaching from antifouling paints, TBT reaches the aquatic system via wastewater due to its biocidal and fungicidal use. Leaching from PVC products leads to inputs of butyl- and methyltins into the aquatic environment via wastewater [30]. Municipal wastewater and sewage sludge are contaminated with organotins at low ng L^{-1} and 0.1–2.2 mg kg^{-1} (dry wt) [50], respectively. Moreover dialkylated organotins and even TBT are also released from containers into foodstuff and beverages [96] and ultimately to aquatic systems.

4.1
Seawater and Sediments

TBT contamination of marine and freshwater environments is well documented. In this review, emphasis will be placed on recent data after the regulation of TBT-containing antifouling paints in recreational boats, whereas data prior to antifouling paint regulations are compiled in [3, 4, 11, 34]. Recent reports document the global occurrence of organotins [74, 75, 87, 90]. Any decrease was slow or absent after legislation limiting TBT usage on pleasure boats. In France, the decline was found to level off after 10 years and relatively high TBT levels were detected even outside harbors [17]. In the UK 10–78 ng L^{-1} were detected in marinas and harbors in 1998, 8 years after legislation [97]. Similarly, coastal levels remained at 2–160 ng L^{-1} in 1999 in Japan [94]. The historic development of contamination using biota [98] documented no noticeable decrease. Elevated TBT levels in marine systems were, and still are, associated with increased pleasure and commercial boat activities, or vessel repair facilities and dry docks, but also with the use of antifouling paints on fish nets and cages in aquaculture [94]. Elevated levels of organotins were reported in estuaries including the Japanese coast with TBT levels up to 160 ng L^{-1} [94] or in ferry ports with 200 ng L^{-1} [99], but also in open waters [74], and areas with high shipping activities [83], and the deep sea [16]. Studies on the global occurrence of organotins in the last few years using biota and sediments as pollution markers demonstrate that the decrease is only marginal in many places [98], but occurred in others [60, 82, 100]. TBT water concentrations persist for years at levels considered as chronically toxic to most susceptible organisms [55, 97, 98].

Highest TBT levels were and are still found in and near marinas and ports due to the use of antifouling paints and painting and depainting operations in docks. Levels in harbor and port waters were in the range of 100–500 ng L^{-1} in the mid 1980s. They were significantly higher than in open surface waters, bays and estuaries where commonly values of up to 50 ng L^{-1} were observed. Recent surveys reported similar levels in developing countries, for instance 32 ng L^{-1} in South Korea [101], but also 7 ng L^{-1} in Japanese coastal waters [94]. In general TBT concentrations in seawater are variable. Much higher levels occur in the surface microlayer [102]. Contamination decreased slowly in countries, where TBT containing antifouling paints have been regulated [54, 63], but the decrease is not significant [97, 103], or even absent [94, 99]. Where no regulations have been implemented, levels are increasing to the same range as in other countries prior to legislation, as shown in Asia [70, 75, 84, 86, 87, 104], Oceanian countries, and South America [105]. It is a fact that TBT contamination occurs globally including Asian countries, Eastern Europe [57], South America or Bahrain, where extremely high levels of 14.7 μg L^{-1} were reported [64]. Larger vessels lead to contamination of commercial ports and shipping straits [72], or even the Arctic [92] and Antarctic [93]. Even after regulation of pleasure boats, large ships, icebrakers and dry-docks may be sources. In 1992, TBT ranged up to 17 ng L^{-1} in estuaries, but in harbors and dry-dock areas the concentrations are often high, exceeding 100 ng L^{-1}. Recent surveys indicate that seawater concentrations exceed the UK environmental quality standard value of 2 ng L^{-1} [106]. TBT at levels of 2–7 ng L^{-1} was also found in 15% of the offshore water samples in the North Sea, English Channel, Mediterranean, and off the coast of Japan.

The TBT decrease is not, or to a lesser extent, observable in sediments and biota. Temporal trends in a port indicate the decrease of TBT in water, sediment and biota, particularly for TPT, between 1989–1996 [63]. On the other hand, data on harbor and estuarine sediments in pleasure boat harbors indicate that TBT residues remain high (Fig. 4). Sediment levels are high in ports and boat harbors, depending on the boat frequencies, but lower in coastal regions and in some developing Asian countries (Fig. 5). However, this is not a general rule. Larger Bays, estuarine and harbor areas may be contaminated in the range of 0.1–2.71 μg g^{-1} TBT (dry wt) [42]. Recent TBT levels are also variable and range up to several μg g^{-1} (dry wt) in boat harbors, although considerable variation occurs between different sites (Fig. 5). Most contaminated areas are harbors with high vessel activity. A mean of 54 ng g^{-1} TBT was found in 35 sediment samples in a US coastal survey [76]. The contam-

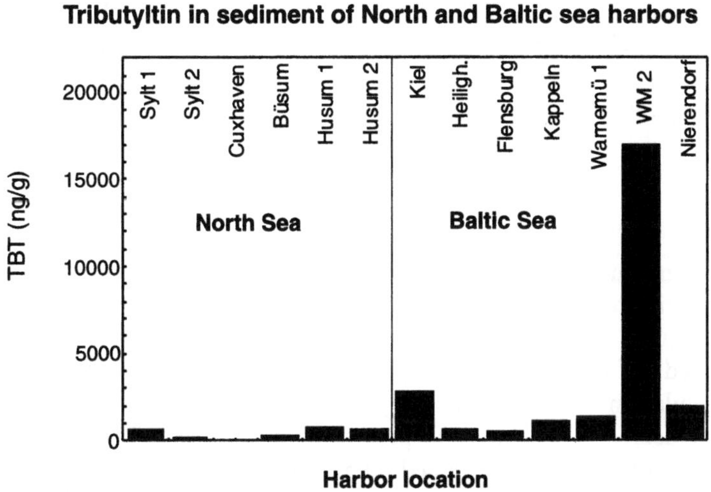

Fig. 4 TBT concentrations in sediments of boat harbors in the North Sea and Baltic Sea. Data after [54]

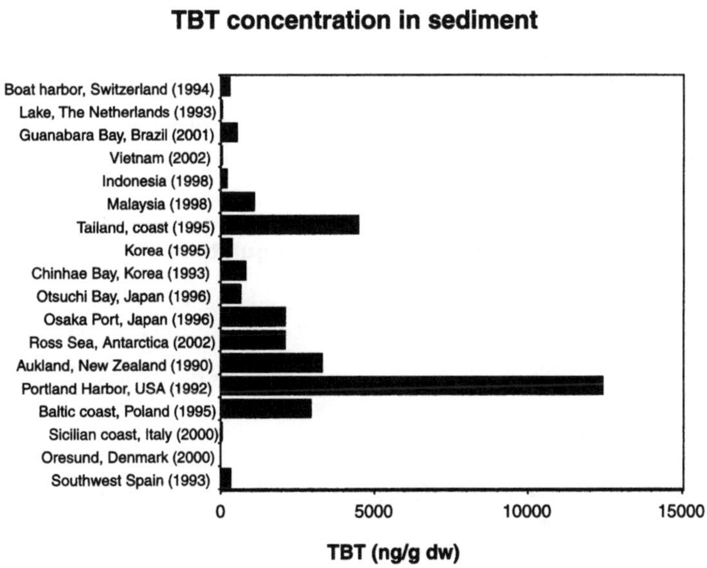

Fig. 5 Maximal TBT concentration in marine sediments, according to recent data (references see text)

ination of estuaries and offshore regions is still of concern including deeper areas offshore from Vancouver harbor (5.3 ng g^{-1} TBT) [41]. Sediments in Arcachon Bay, France, collected in 1990 are contaminated by organotins 8 years after regulation of antifouling paints, with levels of up to 1.48 $\mu g\,g^{-1}$, as in the years before regulation [107]. This and other studies show that considerable amounts of TBT were stored and conserved in sediments long after legislation [40, 54]. Furthermore, highly contaminated sediments occur in some developing countries (Fig. 5).

4.2 Freshwater

The contamination of freshwater harbors and lakes shows a similar pattern [4]. Prior to legislation, TBT has been detected in lake water up to levels of several $\mu g\,L^{-1}$ and degradation products DBT and MBT were in the range of hundreds of $ng\,L^{-1}$ [19]. Phenyltins were also present in harbor waters. After regulation of TBT-containing antifouling paints, levels in water and sediment decreased but are still present at $ng\,L^{-1}$ levels [19]. The decrease of TBT pollution in water and sediment after legislation has also been documented in many freshwater systems. Figure 6 shows the effect in a pleasure boat harbor. Even though TBT levels declined markedly, they persist in harbors and its surroundings, but also in river waters at $ng\,L^{-1}$ levels [18, 19]. In 1998–1999, TPT and DBT were found in some rivers at levels of up to 2.65 and 1.6 $ng\,L^{-1}$ in water and up to 221 and 389 $ng\,g^{-1}$ in fish in the USA [108].

In general, TBT and low proportions of DBT and MBT are still recorded in water, sediments and biota. Maximal TBT levels in harbor sediments were regularly in the range of 200–1000 $ng\,g^{-1}$, but very variable. TPT occurred at concentrations of up to several hundreds $ng\,g^{-1}$. In sediment cores taken in

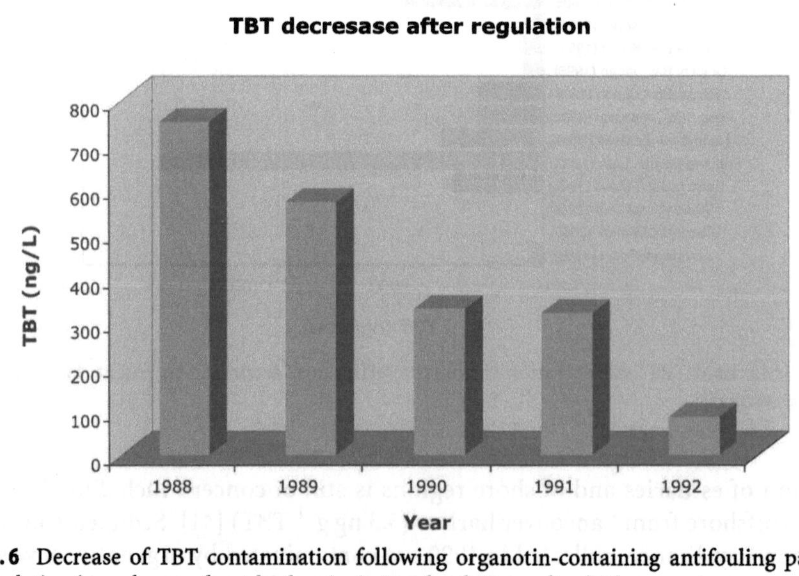

Fig. 6 Decrease of TBT contamination following organotin-containing antifouling paint regulation in a pleasure boat harbor in Switzerland. Data after [19]

boat harbors, high TBT and TPT concentrations were restricted to the top and decreased significantly below (Fig. 3). In 1994, TBT surface concentrations were up to 290 ng g^{-1}, which is about 5–10 times lower than in 1989 demonstrating the effect of regulation. The decrease continues with time. Similar TBT levels of 500–2500 μg kg^{-1} (dry weight) were found in harbor sediments of other lakes, and river systems. The data lead to the conclusion that legislation resulted in a marked reduction in boat harbor waters and sedimentary input, but trace levels persist in surface waters, probably also due to inputs via wastewater [19, 50].

4.3
Biota

Contamination of marine and freshwaters with butyltins has spread all over the world as evidenced by their detection in a wide range of oceans, environmental media and particularly biota [4, 73–75, 87, 109]. Organisms serve as important tools for assessing pollution. Biota tend to give a better indication of contamination then water samples, because they integrate the pollution situation over a longer period of time. Considerable butyltin and phenyltin, but also methyltin compounds were found in marine and freshwater organisms on a global scale. This holds true in particular for TBT and TPT showing bioaccumulation in biota including mussels [63, 82, 110, 111], fish [14, 15] and marine mammals [57, 90]. Because of the filter-feeding behavior and the generally high potential for bioaccumulation, bivalves have been widely used as sentinel organisms for monitoring [76, 87]. Common molluscs including *Littorina littorea* and *Mya arenaria*, *Mytilus edulis* [111], but also oysters may serve as bioindicators of TBT pollution in marine ecosystems, due to the limited ability for metabolism and elimination. In a study, sampling 13 species of squids at 77 stations in the world oceans, highest TBT and TPT levels of 279 and 519 ng g^{-1}, respectively, were found in squid livers in coastal areas. They were lower than in the open oceans. TBT concentrations were higher in the northern hemisphere than in the southern, where TPT was not detected [77]. Often, imposex in marine gastropods have been assessed in conjunction with residues analysis, demonstrating a correlation between TBT and TPT exposure and severity of imposex, for instance in the Lagoon of Venice, Italy [112].

Whereas previous data were obtained from Canada, USA and Europe, currently more data are reported from Japan and developing Asian countries. Although concentrations are generally low in mussels from coastal areas of developing countries, higher levels were found at locations with intensive maritime activities or aquaculture. Decreased levels of butyltins were found in fish [85] and marine mammals [90] collected from coastal waters in developing Asian rather than in industrialized countries such as Japan, Europe and the USA. Contamination was always related to intensive maritime activities (harbors and their vicinity, major shipping traffic), but also aquaculture areas. However, a survey of butyltin levels along the Japanese coastline from 1997–1999 shows that no decrease was recorded since the regulation of antifouling paints in 1990 [94]. Butyltins were detected in all samples of amphipods *Caprella* varying from 2.3–464 ng g^{-1} (wet wt). It can be expected that in developing countries contamination may increase with economic growth due to increase in boating, shipping and aquaculture activities [89].

Residues in fresh and seawater organisms are very variable. In marine biota, tissue TBT concentrations ranged from 0.2 μg g^{-1} in seaweed *Fucus*, to 36.8 μg g^{-1} (dry wt) in the clam *Mya arenaria* [113]. Typical organotin levels in low polluted areas in mussels are in the range of 50–200 and 20–100 ng g^{-1} (dry wt). In Pacific oysters, residues in digestive gland and gills of up to 7.0 and 17.4 μg g^{-1}, respectively, were recorded. Most oysters and mussels from US coasts were contaminated with TBT between < 0.005 to 1.56 μg g^{-1} (as Sn) [114]. Oysters used as monitors of TBT contamination of estuaries in England showed bioconcentration factors (BCFs) of around 10 000 [115]. Mean levels of TBT were higher in seed than adult oysters in the south of England, and reached 3.1 μg g^{-1}. Tissue concentrations in bay mussels (*Mytilus edulis*) of the US Pacific coast were in the range of 0.005–1.08 μg g^{-1} (wet wt), and were lower in spring than summer [110]. TBT was found in all oysters *C. virginica* in the Chesapeake Bay, USA, in concentrations from < 10 to 5600 μg kg^{-1} on the Atlantic Coast of Virginia after restrictions in the use of TBT-containing antifouling paints [60]. Butyltin concentrations in horseshoe crabs were high in the hepatopancreas, with TBT concentrations ranging from 0.12–2.2 μg g^{-1} (wet wt) in Japan [116]. Although the use of triphenyltin was restricted in Japan in 1989, high TPT levels were found in horseshoe crab hepatopancreas (1.1–12.0 μg g^{-1}) and crab eggs [81].

Recent monitoring data obtained from mussel watch programmes still indicate widespread organotin pollution in Canada [18], the USA [76], and recently in Asian countries [84, 87, 111] in coastal environments. Global pollution monitoring for open seas was performed using skipjack tuna [75], squids [77] and mammals [74]. These data are particularly important not only for assessing time trends [98], but also for direct comparison. Butyltins were found in tuna, demonstrating contamination in offshore waters and open seas on a rather global scale [75]. High concentrations were found in tuna liver offshore Japan (up to 400 ng g^{-1} wet wt), and comparable, but lower concentrations, around Asian developing countries (up to 270 ng g^{-1}). In green mussels, higher levels were found in Asian developed and industrialized areas than in developing countries. Table 1 summarizes selected recent data for different biota and locations.

Significant butyltin levels have also been found in different marine mammals including cetaceans and pinnipeds (dolphins, harbor porpoises, sea otters, whales) in Asia [67, 69, 90], the Mediterranean and US coastal waters [68], along the Polish coast of the Baltic Sea [57], and other locations.

Table 1 Global occurrence of organotins in biota (ng g^{-1} wet weight)

Species	Location (year)	Tissue	TBT (range) (ng g^{-1})	Total butyltins (ng g^{-1})	Refs.
Molluscs					
Mussel (*M. edulis*)	North/Baltic Sea (1985–1999)	body	17 (10–210)	15–480	[98]
	Korea (1994)	body	59–590		[197]
	Japan (1996)	body	24–1100		[198]
	Osaka Port, Japan (1996)	body	24–390		[63]
	Harbor, Canada (1995)	body	10–585		[18]
Oyster (*C. gigas*)	Korea (1994)	body	250–740 (dw)		[197]
Gastropods *H. trunculus*	Venice, Italy	body	n.a. [a]	237–725	[112]
Buccinum undatum	Denmark (1998)	body	8 (dw)		[72]
Nuculana pernula	Denmark (1998)	body	1316 (dw)		[72]
Green mussel (*P. viridis*)	Hong Kong (1999)	body	22–330	49–500	[87]
	South Korea (1998)	body	17–1200	50–2500	
	Malaysia (1998)	body	4–730	4–960	
	Thailand (1995)	body	3–680	4–800	
	China (1999)	body	16–86	30–110	
	Indonesia (1998)	body	2–38	4–64	
	Vietnam (1998)	body	2–84	2–100	
	India (1998)	body	1–570	1–760	
Clam (*Meretrix* spp.)	Vietnam (1998)	body	1–56	3–64	[86]
Amphipods (*Caprella*)	Japanese coast (1999)	body	n.a.	2.3–464	[94]
Zebra mussel *D. polymorpha*	Swiss harbors (1994)	body	820–2290	900–2460	[19]

[a] n.a., not analysed

Table 1 (continued)

Species	Location (year)	Tissue	TBT (range) ($ng\,g^{-1}$)	Total butyltins ($ng\,g^{-1}$)	Refs.
Fish					
Eelpout (*Z. viviparous*)	North/Baltic Sea (1985–1999)	muscle	10–22	n.a.	[98]
Skipjack tuna (*K. pelamis*)	Asia (1996–2001)	liver	4–220	4–400	[75]
Bluefin tuna	Italian coast	muscle	79	138	[109]
T.chalcogramma	Bering Sea, Alaska	muscle	1–6		[92]
	Otsuchi Bay, Japan (1996)	muscle		10–20	[198]
	Osaka Port, Japan (1996)			11–250	[63]
Different species	Mediterranean deep sea (1996)	liver	1–52	6–174	[16]
Different species	Asia, Oceania (1990–97)	muscle	0.1–13		[85]
Different species	Malaysia (1998)	body	2–190	6–210	[88]
Different species	Indonesia (1998)	body	2–52	3–84	[89]
Roach	The Netherlands, lake (1993)	muscle	120–6100		[80]
Mammals					
Bottlenoise dolphins	Italian coast (1996)	liver	610	1200	[109]
Harbor porpoises	Polish coast (1996–2003)	liver	160–472	785–2800	[57]
Striped dolphin		liver	177–1488	2220–7698	
Grey seal		liver	44–60	44–81	
Different porpoises	Asia, Pacific (1983–1996)	liver		22–5200	[90]

All animals contained butyltins with varying degrees in different tissues with butyltins ranging from 0.02 up to 10.2 mg kg^{-1} (wet wt). As with other biota, a decreasing contamination from coastal waters to the open seas was observed. Butyltins in the liver of Steller sea lion from Alaska were much lower (19 ng g^{-1}) than those from Japan (150–220 ng g^{-1}) [117]. Even in deep-sea fish [16], TBT levels of up to 175 ng g^{-1} (wet wt) were comparable to coastal fish. Moreover, phenyltins, particularly TPT, occurring up to 1430 ng/g, are much higher in shallow-water organisms. This confirms the long-range transport of TBT and TPT to the deep-sea.

Using archived biological samples, **time trends** of the pollution in the 1980s and 1990s were assessed. Decreasing levels of butyltins were found in water and mussels in Japan after regulation in 1990 [63]. This trend has, however, not been observed in open waters in tuna [75]. In Germany, concentration of TBT in North Sea and Baltic Sea organisms remained relatively constant during the period of 1985–1999, whereas levels of TPT decreased [98]. TBT concentrations remained at 17 ng g^{-1} for mussels, reflecting that TBT is still used despite the ban in 1990 for small boats in Germany. TBT levels in bivalve molluscs from the east and west coasts of the USA have been found to decrease systematically at some sites and remain static or fluctuating at others in the 1988–1990 national status and trends mussel watch project [118].

In **freshwaters**, zebra mussels (*Dreissena polymorpha*) show high bioaccumulation with little toxicity. We found that in lake harbors all organisms contained butyl- and phenyltins to variable levels, *D. polymorpha* showed highest values, which were related to water contamination [14, 28]. After regulation of TBT-antifouling paints a significant decrease was noted [19]. Transformation products of TBT occurred in only low proportions, which also holds for TPT indicating slow transformation. The ubiquitous presence of butyl- and phenyltins was also demonstrated in various biota of other lakes [80]. The concentrations of TBT and TPT ranged from 0.0015–28.2 and 0.081–7.28 μg g^{-1} (dry wt), respectively.

The occurrence of TBT and TPT is also widespread in **marine seafood** [91] and human exposure may take place via this route [91, 119]. In Japan, TBT and TPT levels were 0.02–1.3 and 0.03–1.3 μg g^{-1}, respectively [120]. Typical TBT and phenyltin residues in fish were in the range of 0.05–0.3 μg g^{-1} [81]. Edible mussels, clams and squids contained TBT in the range of 10–100 μg g^{-1}. Fish reared in nets treated with antifouling paints contained very high TBT concentrations of 1.5 μg g^{-1} [121]. TBT and TPT concentrations in fish muscle ranged between < 0.12–6.10 and 0.13–7.91 μg g^{-1} (dry wt), respectively. In freshwater fish 0.2 μg g^{-1} TBT in muscle tissue [14], and up to 0.49 μg g^{-1} in whole fish occurred. In Canada marine fish fillets contained low levels of TBT (< 1–2 ng g^{-1}) and molluscs between < 1–156 ng g^{-1}. Increased TBT levels were found in fish from aquaculture such as salmon (4–3.1 μg g^{-1}). Salmon from American public fish markets contained TBT of 0.08–0.2 μg g^{-1} [122]. The levels are, however, not considered to be harmful to humans, as they remained below the acceptable daily intake (ADI) value of 3.2, or 1.6 ng g^{-1} (Japan).

In summary, the contamination of marine waters is still widespread, although decreasing slowly in countries that implemented regulations. However, no significant decreases were noted in sediments and biota collected over a long period of time. Organotins are also widespread in marine systems in developing countries. Of particular importance is the contamination of biota, of which molluscs have the highest burdens, and sediments which may act as reservoirs.

4.4 Bioaccumulation

TBT is rapidly taken up from low ambient levels and accumulated to relatively high concentrations in organisms. BCFs range from 200 in some fish up to 50 000 in marine invertebrates, but in general, BCFs range from 1000–20 000 [4]. Molluscs show highest accumulation; the BCFs of TBT and TPT were in the range of $1.2–6.6 \times 10^4$. Bioaccumulation is dependent on ambient pH and DOM (Fig. 2). In *Chironomus* larvae, BCFs varied at pH 5 and pH 8 between 140 and 900 for TBT, and 2200 and 2680 for TPT [25, 123], confirming our previous results with *Daphnia* [22, 23].

TBT distributes differently into tissues, generally lipid-rich tissue are characterized by highest burdens. In fish highest TBT residues occur in the liver and peritoneal fat [4]. BCFs are in the range of 200–9400 in whole fish or muscle tissue, but higher in viscera (1000–10 000) [4, 124]. The bioaccumulation of TPT is in a similar range with BCF of 2000–3000 in liver, and 100–250 in muscle. The very high levels of butyltins in the range from 44–7700 ng g^{-1} (Sn) found in marine mammals in the Baltic Sea [57], Asia and the Pacific [74], USA [68], and the Mediterranean [109] indicate a high bioaccumulation.

Several studies focused on the question, whether or not biomagnification of TBT or TPT takes place in food webs. All data indicate that TBT shows no relevant biomagnification in aquatic ecosystems [4, 63, 75]. This holds true for TPT to a lesser extent, as some minor biomagnification was indicated in a freshwater lake foodweb [80]. If biomagnification takes place at all, it is only minor. However, it must be stated that organotin bioaccumulation in food webs is not well understood [74].

5 Ecotoxicological Effects

Organic derivatives of tin are by far more toxic than inorganic tin, and the toxicity of organotins increases with progressive introduction of organic groups at the tin atom, with maximum toxicity for trisubstituted compounds and decreasing toxicity with increased length of the organic moiety [3]. Of particular importance is the high toxicity of TBT, TPT and tricyclohexyltin derivatives, which have high fungicidal, bactericidal, algicidal, and acaricidal properties. Trisubstituted organotins are extremely toxic to aquatic life for a large variety of organisms as shown in vivo and in vitro [4]. The most sensitive species known so far are oysters and neogastropods, affected at $\mathrm{ng\,L^{-1}}$ levels of TBT. In both cases the impact of TBT has economical (oysters) and ecological consequences on the population level. Ecological effects on many snail species are found globally in contaminated coastal regions with shipping activity. Many ecotoxicological studies on organisms of different evolutionary level have been reported [1, 24, 125–130], and show rather large species sensitivity differences. Lowest effect concentrations are in the range of $1–10\,\mathrm{ng\,L^{-1}}$ [1, 2, 10, 130], which are among the lowest yet found for chemicals. Invertebrates and fish are affected as well, although at higher concentrations of $100–10\,000\,\mathrm{ng\,L^{-1}}$. The acute toxicity data basis is rather large [4, 131]. However, long-term or chronic effects on individual organisms are sparse and effects on the structure and function of aquatic ecosystems still unclear.

5.1 Modes of Action

TBT and TPT exert a number of important cellular, biochemical and molecular effects acting via different modes of action at μM concentrations. The inhibition of oxidative phosphorylation and ATP synthesis in mitochondria is a key process having been described decades ago [132–134]. Triethyltin acts at concentrations as low as 0.1 μM. Moreover, inhibition of photophosphorylation in chloroplasts, perturbation of calcium homeostasis [135], inhibition of ion transport through the cellular membrane by inhibition of ATPase [136], and inhibition of enzymes such as ATPases and cytochrome P450 monooxygenases (CYP) are among the biochemical processes affected by organotins [137]. Organotins act as ionophores, able to catalyze the ion exchange resulting in uncoupling of oxidative phosphorylation. Moreover, organotins have been known to act as potent inhibitors of the F_1F_0-ATP synthase. These enzymes in the mitochondrial membrane play a fundamental role in cellular energy metabolism and are constituents of nearly every living cell from bacteria to humans. The ATP synthases are rotary enzymes composed of two motors that are connected by a common shaft to exchange energy. TBT inhibits ATP hydrolysis by the Na^+-translocating or H^+-translocating ATP synthase in bacteria [138]. The subunit *a* ion channel is the specific target site for ATPase inhibition by organotins [138].

Triorganotins also interact with various intracellular enzymes that may result in toxicity, namely with the cytochrome P450 dependent monooxyenases (CYP) [139], and glutathion *S*-transferases [140]. TBT was shown to have a strong inhibitory effect in vitro [141, 142] and in vivo on hepatic microsomal CYP in fish [139, 143] and molluscs [144]. Similar effects have been shown for TPT [139]. Organotins act by binding to amino acids in the catalytic site of the CYP enzyme. The interaction with microsomal monooxygenase systems results in the loss of biotransformation capabilities. In addition, this may lead to inhibition of CYP19 (aromatase) and the formation of imposex in marine gastropods.

Perturbation of calcium homeostasis is involved in the cytotoxic action of organotins. Calcium ions play a critical role in chemical-induced toxic cell killing and programmed cell death (apoptosis). Elevated calcium concentrations appear to be responsible for cell death and stimulation of apoptosis by 1–10 μM TBT, in particular in thymocytes [145], resulting in immunotoxicity, or neurotoxicity in the case of trimethyltin. Organotins induce apoptosis by increase of intracellular Ca^{2+}, followed by generation of reactive oxygen species in mitochondria which causes oxidative stress, activation of caspases and finally results in DNA fragmentation and apoptosis [146].

Organotins act also as potent cell membrane toxicants leading to perturbations of plasma membranes and membrane proteins. Therefore, cells as a whole are affected by toxic organotins [147–149] and are found to be correlated with in vivo acute fish toxicity [26, 149], and the octanol–water partition coefficient [148]. In mussels induction of stress proteins (heat shock proteins) has also been shown [150]. Recently, a cDNA microarray technique was applied for the analysis of global gene expression profiles in TBT exposed ascidians [151]. More than 200 genes showed strong differential expression after 24 h exposure to 100 nM TBT. These genes encode proteins involved in stress response, detoxification, oxidoreduction reaction, biosynthesis and catabolism. This is a clear indication of pronounced molecular effects of TBT at the gene expression level.

5.2 Effects on Aquatic Life

TBT and TPT induce a number of physiological, morphological and toxicological effects in aquatic organisms. Numerous studies have shown that triorganotins are acutely and chronically toxic at low environmental levels. Aquatic life in highly polluted areas is significantly affected. The two most pronounced cases are the shell thickening and growth anomalies in oysters [1], and the occurrence of imposex in neogastropods, both occurring at low $ng\,L^{-1}$ [130]. The discovery of shell deformations in oysters resulted in regulation of TBT-containing antifouling paints. TBT act by inhibition of calcification in oysters at concentrations as low as $2\,ng\,L^{-1}$ TBT [152]. Anomalies in larvae and inhibition of reproduction was also observed [153]. Growth of oysters and molluscs was severely reduced by $240\,ng\,L^{-1}$ [154].

TBT and TPT are very toxic to aquatic life as demonstrated by numerous acute toxicity studies in a variety of marine and freshwater organisms. Data are reviewed in [3, 7, 131], and recent data support the high toxicities. For instance, growth was impaired in a polychaete exposed to TBT at concentrations often found in sediments (190 ng g^{-1} sediment) [155, 156] indicating that benthic organisms are affected similarly. Chronic toxicity has been observed in marine molluscs at low levels, but in general only a few species were analyzed. Other sensitive marine species including algae and zooplankton show acute and chronic effects at concentrations of a few 100 ng L^{-1} or less [4]. Acute effects are reported at levels above 420 ng L^{-1}, and chronic values for four invertebrate species ranged from 14–131 ng L^{-1} [95]. Chronic values for freshwater species (cladoceran and fish) ranged from 137–253 ng L^{-1}. Table 2 compiles some selected toxicity data.

Whereas most sensitive algae and invertebrates are affected at a few 100 ng L^{-1}, the acute toxicity of TBT in fish lies in the range of a few μg L^{-1}. The chronic toxicity is lower, however. Effects on the immune system and histological alterations in many tissues are found below 1 μg L^{-1} [157]. TPT has

Table 2 Toxicity of TBT to selected sensitive aquatic biota

Organism		Toxicity	TBT (μg L^{-1})	Refs.
Acute toxicity				
Algae	Diatoms	EC_{50}, 72 h	0.3–0.4	[199]
	N. oculata	EC_{50}, 24 h	0.258	[101]
Zooplankton	*Arcartia tonsa*	LC_{50}, 6 d	0.4	[200]
Amphipod	*Gammarus*	LC_{50}, 96 h	1.3	[157]
Molluscs	Oyster *C. virginica*	LC_{50}, 48 h	1.3	[201]
	Clam *M. mercenaria*	LC_{50}, 48 h	1.1	[201]
Crustaceae	Mysid shrimp	LC_{50}, 96 h	0.3	[202]
Fish	Minnows *P. phoxinus*	LC_{100}, 8 d	3.5	[127]
	Trout *O. mykiss* fry	LC_{100} 10–12 d	5	[203]
Chronic toxicity				
Algae	Diatoms	growth	0.1	[204]
Zooplankton	*Arcartia tonsa*	death, 144 h	0.3	[200]
Amphipod	Gammarus	24 d	0.53	[157]
Molluscs	Oyster *C. gigas*	LC_{100}, 12 d	0.18	[153]
	spat	growth	0.01	[205]
	Mytilus edulis larvae	LC_{50}, 15 d	0.04	[206]
Gastropods	*Nucella lapillus*	imposex	0.002	[9]
Fish	Minnows *P. phoxinus*	histology	0.7	[127]
	Guppy *P. reticulata*	immune system	0.4	[207]

a similar toxicity [158]. Effects to freshwater organisms are in the same range, but extremely susceptible species have not been identified. Data on the toxicity of TBT and TPT to fish are compiled [4]. Fish are more susceptible in their early life stages [127, 159], which also holds for tadpoles of frogs [24]. Target organs are the immune system (thymic atrophy, immunotoxicity), skin and eye (irritation), the CNS (neurotoxicity), and the developing skeletal, renal and neuronal tissues, in particular the eye [127]. Mesocosm and microcosm studies with TBT indicate that TBT levels less than 50 ng L^{-1} did not impact the structural and functional parameters of communities. Taken together, the experimental data indicate that chronic effects are probable at highly polluted sites such as harbors, areas with high shipping activity and their surroundings even presently.

5.3 Imposex and Masculinization

The occurrence of genital deformations occurring in the common dogwhelk (*Nucella lapillus*), a species of thick-shelled snails found on rocky shores on both sides of the North Atlantic, was very important with respect to the discovery of environmental effects induced by TBT from antifouling paints [2, 10]. These deformations are referred to as imposex, which is the development of male characteristics, notably a penis and vas deferens by females, is caused by extremely low TBT and TPT concentrations. The manifestation gradually becomes more severe at higher concentrations and prevents the release of egg capsules, rendering the female sterile. The occurrence of imposex correlated significantly with TBT levels in dogwhelks, and laboratory studies have confirmed that exposure to trace levels of 1–2 ng L^{-1} induce imposex [10]. After the discovery in the UK, imposex was found in Europe, North America and subsequently, in many species of neogastropods and mesogastropods around the world due to exposure to TBT originating from antifouling paints [65, 129, 160–171]. Imposex is known in more than 120 species belonging to more than 60 genera [172], and often associated with body burdens of organotins [112]. This masculinization may induce reproductive failure at severely affected stages, resulting in population declines. Throughout the widespread geographic range of *N. lapillus* imposex has been documented [10]. Imposex has been recognized as a very important biomarker of organotin contamination used for monitoring on a global scale [130, 172, 173]. TBT-induced imposex has also been shown in developing countries [56, 105, 164, 165, 174–177]. Although some studies showed a decrease of imposex due to antifouling paint regulations [65], others demonstrate its ongoing occurrence a decade later [178]. The presence of imposex is also related to TBT in the common whelk (*Buccinum undatum*) and the red whelk (*Neptunea antiqua*) in the open North Sea and Skagerrak [170]. Imposex was also described in a freshwater prosobranch snail after TBT and TPT exposure [179].

What are the modes of actions leading to imposex? First hypotheses were formulated by [166, 180]. Involvement of aromatase inhibition was proposed by [167] due to an increase in testosterone levels responsible for the masculinization of female marine gastropods by TBT and TPT. Among several hypotheses, inhibition of aromatase and associated disturbance of steroid metabolism with increase in testosterone and associated formation of male sex organs in female seems most evident [181, 182]. A strong reduction of estradiol levels was found in imposex-affected gastropods *Bolinus brandaris* [162]. CYP aromatase activity was higher in normal females than in imposex animals, evidently depressed aromatase activity was associated with imposex in affected females [183]. Evidence of masculinization with increasing testosterone and decreasing estradiol titers were also found in clams [184]. Testosterone metabolism was altered in the estuarine mysid *Neomysis integer* to which TBT was highly toxic (LC_{50} of $164\,ng\,L^{-1}$), and metabolic androgenization was induced even at $10\,ng\,L^{-1}$ [185].

Recently, additional mechanisms have been proposed and various lines of evidence provided. Androgen-dependent transcription in mammalian cells and cell proliferation in human prostate cancer cells were shown, an indication of the androgenic action of TBT and TPT [186]. Recently, TBT and TPT were demonstrated to bind to the human retinoid X receptors (hRXR) with high affinity [187]. They act as high affinity ligands for RXR similar to the natural ligand 9-cis retinoic acid. When this ligand was injected into females of the rock shell *Thais clavigera*, they developed imposex. This suggests that RXR plays an important role in the induction of imposex. TBT and TPT were also found being activators of peroxisome proliferator-activated receptor gamma besides binding to RXR [188]. Whether the activation of RXR is the key process for imposex development has yet to be shown. Besides, another hypothesis assumes that TBT acts as a neurotoxin to abnormally release the peptide hormone APGWamide, leading to imposex [189]. As of today, the exact mechanisms behind imposex formation are unclear, although most evidence points to the inhibition of aromatase as causative effect.

Inhibition of CYP by TBT and TPT has also been demonstrated in fish in vivo [143] and in vitro [139, 141]. The conjugation of testosterone was also significantly inhibited in fish [190]. Recently, inhibition of aromatase has been demonstrated in other species of molluscs [144], fish [191] and in mammals. This points to a possible androgenic activity of TBT resulting in additional masculinizing effects besides imposex in gastropods. TBT-induced masculinization has been seen in genetically female Japanese flounder with concomitant suppression of the aromatase gene expression [191]. Zebrafish early life stages are extremely sensitive to TBT. Nominal levels of 0.1, $1\,ng\,L^{-1}$ and higher were able to significantly bias the sex ratio towards males [192]. Moreover, TBT-induced dose dependent sperm abnormalities were seen. Fertilization success was reduced and sexual behavior altered in male medaka, exposed to TBT [193]. Recently, it was shown that concentrations of 100 ng L^{-1} TBT and TPT caused significant spermatogenesis in ovaries of exposed abalone *Haliotis gigantean*, which resembles gastropod imposex [194]. On the other hand, testosterone accumulation was suggested to be a reaction to the inhibition of androgen excretion due to a decrease in sulfur conjugation of androgens [195]. A two-generation reproductive study of TBT showed that the male reproductive system of rats is also affected. Testis weights, spermatid and sperm count were reduced at 125 mg kg^{-1} TBT [196]. The data in fish and mammals indicates that TBT has androgenic activity to different biota in addition to inducing imposex in gastropods.

6 Conclusions

The ongoing use of TBT-containing antifouling paints on large vessels leads to contamination of harbors, ports and marine environments with high shipping activities even today. Regulations for recreational boats resulted in a decrease of TBT levels in marinas, but it had a slow or marginal effect in coastal waters. Both decreases in TBT-containing antifouling paints and the lack of decreases have been reported in the literature. Organotins persist particularly in sediments and biota. Along coasts, levels had not significantly decreased even 10 years after regulations came into force. The widespread contamination found also in coastal regions of developing countries, in the open oceans and deep seas (particularly in sediments and biota) demonstrates the global occurrence of butyltins. The high toxicity, persistence and bioaccumulation of TBT are of particular environmental concern. Recently androgenic activity of TBT in molluscs, fish and mammals was reported, indicating important endocrine activity. The global occurrence of imposex in a large number of gastropod species and associated population declines in coastal areas emphasizes the hazards and risks associated with these antifouling paints. The use of TBT by commercial watercraft continues to contribute to the TBT load in aquatic systems. The problem can only be solved by a global regulation and ban of organotin-containing antifouling paints on ships and in aquaculture.

References

1. Alzieu C (2000) Ecotoxicology 9:71
2. Bryan GW, Gibbs PE, Hummerstone LG, Burt GR (1986) J Mar Biol Ass UK 66:611
3. WHO (1990) Tributyltin compounds. World Health Organization, Geneva, Switzerland
4. Fent K (1996) Crit Rev Toxicol 26:1
5. Alzieu C, Héral M, Thibaud Y, Dardignac MJ, Feuillet M (1982) Rev Trav Inst Peches Marit 45:101
6. Alzieu C, Sanjuan J, Deltreil JP, Borel M (1986) Mar Pollut Bull 17:494

7. Hall LW Jr, Pinkney AE (1985) CRC Crit Rev Toxicol 14:159
8. Thain JE (1986) Proceedings of the Organotin Symposium Oceans '86 Conference, vol 4. The Institute of Electrical and Electronics Engineers and Marine Technology Society, New York, Washington, D.C., p 1306
9. Gibbs PE, Bryan GW (1986) J Mar Biol Assoc UK 66:767
10. Gibbs PE, Bryan GW (1996) In: Champ MA, Seligman PF (eds) Organotin. Environmental fate and effects. Chapman & Hall, London, p 259
11. Champ MA, Seligman PF (1996) Organotin. Environmental fate and effects. Chapman & Hall, London
12. Maguire RJ, Tkacz RJ, Chau YK, Bengert GA, Wong PTS (1986) Chemosphere 15:253
13. Maguire RJ (1991) Wat Pollut Res J Canada 26:243
14. Fent K, Hunn J (1991) Environ Sci Technol 25:956
15. Tolosa I, Merlini L, De Bertrand N, Bayona JM, Albaiges J (1992) Environ Toxicol Chem 11:145
16. Borghi V, Porte C (2002) Environ Sci Technol 36:4224
17. Michel P, Averty B (1999) Environ Sci Technol 33:2524
18. Chau YK, Maguire RJ, Brown M, Yang F, Batchelor SP (1997) Water Qual Res J Canada 32:453
19. Fent K, Hunn J (1995) Environ Toxicol Chem 14:1123
20. Arnold CG, Weidenhaupt A, David MM, Müller SR, Haderlein SB, Schwarzenbach RP (1997) Environ Sci Technol 31:2596
21. Arnold CG (1998) PhD Thesis, Swiss Federal Institute of Technology (ETH) Zürich
22. Fent K, Looser PW (1995) Wat Res 29:1631
23. Looser PW, Bertschi S, Fent K (1998) Appl Organomet Chem 12:601
24. Fioramonti E, Semlitsch RD, Reyer H-U, Fent K (1997) Environ Toxicol Chem 16:1940
25. Looser PW, Fent K, Berg M, Goudsmit GH, Schwarzenbach RP (2000) Environ Sci Technol 34:5165
26. Fent K, Hunn J (1996) Mar Environ Res 42:377
27. White JS, Tobin JM (2004) Environ Sci Technol 38:3877
28. Becker van Slooten K, Tarradellas J (1994) Environ Toxicol Chem 13:755
29. Maguire RJ (1984) Environ Sci Technol 18:291
30. Fent K, Müller MD (1991) Environ Sci Technol 25:489
31. Unger MA, MacIntyre WG, Huggett RJ (1988) Environ Toxicol Chem 7:907
32. Randall L, Weber JH (1986) Sci Total Environ 57:191
33. Viglino L, Pelletier E, St Louis R (2004) Environ Toxicol Chem 23:2673
34. Maguire RJ (1987) Appl Organomet Chem 1:475
35. Maguire RJ, Carey JH, Hale EJ (1983) J Agric Food Chem 31:1060
36. Stewart C, de Mora SJ (1990) Environ Technol 11:565
37. Seligman PF, Valkirs AO, Lee RF (1986) Environ Sci Technol 20:1229
38. Blair WR, Olson GJ, Brinckman FE, Iverson WP (1982) Microbiol Ecol 8:241
39. Lee RF, Valkirs AO, Seligman PF (1989) Environ Sci Technol 23:1515
40. de Mora SJ, Stewart C, Phillips D (1995) Mar Pollut Bull 30:50
41. Stewart C, Thompson JAJ (1994) Mar Pollut Bull 28:601
42. Yoshitaka Y, Kisaburo N, Yoshikazu M, Akiko O, Takao K, Hiroaki F, Yuuji S, Shigeki M, Yoshikuni U (1993) Environ Toxicol Chem 12:1175
43. Stang PM, Lee RF, Seligman PF (1992) Environ Sci Technol 26:1382
44. Landmeyer JS, Tanner TL, Watt BE (2004) Environ Sci Technol 38:4106
45. Amouroux D, Tessier E, Donard OFX (2000) Environ Sci Technol 34:988
46. Donard OFX, Weber JH (1988) Nature 332:339

47. Jude F, Arpin C, Brachet-Castang C, Capdepuy M, Caumette P, Quentin C (2004) FEMS Microbiol Lett 232:7
48. Maguire RJ, Tkacz RJ (1985) J Agric Food Chem 33:947
49. Fent K (2004) Toxicology 205:223
50. Fent K (1996) Sci Total Environ 185:151
51. Alzieu C, Michel P, Tolosa I, Bacci E, Mee LD, Readman JW (1991) Mar Environ Res 32:261
52. Bacci E, Gaggi C (1989) Mar Pollut Bull 20:290
53. Seligman PF, Grovhoug JG, Valkirs AO, Stang PM, Fransham R, Stallard MO, Davidson B, Lee RF (1989) Appl Organomet Chem 3:31
54. Biselli S, Bester K, Hühnerfuss H, Fent K (2000) Mar Pollut Bull 40:233
55. Caricchia AM, Chiavarini S, Cremisini C, Fantini M, Morabito R (1992) Sci Total Environ 121:133
56. Chiavarini S, Massanisso P, Nicolai P, Nobili C, Morabito R (2003) Chemosphere 50:311
57. Ciesielski T, Wasik A, Kuklik I, Skora K, Namiesnik J, Szefer P (2004) Environ Sci Technol 38:1415
58. de Mora SJ, King NG, Miller MC (1989) Environ Technol Let 10:901
59. Dowson PH, Bubb JM, Lester JN (1993) Mar Pollut Bull 26:487
60. Espourteille FA, Greaves J, Huggett RJ (1993) Environ Toxicol Chem 12:305
61. Evans SM, Leksono T, McKinnell PD (1995) Mar Pollut Bull 30:14
62. Garcia Romero B, Wade TL, Salata GG, Brooks JM (1993) Environ Pollut 81:103
63. Harino H, Fukushima M, Kawai S (1999) Environ Pollut 105:1
64. Hasan MA, Juma HA (1992) Mar Pollut Bull 24:408
65. Horiguchi T, Li Z, Uno S, Shimizu M, Shiraishi S, Morita M, Thompson JAJ, Levings CD (2003) Mar Environ Res 57:75
66. O'Connor TP (2002) Mar Environ Res 53:117
67. Iwata H, Tanabe S, Mizuno T, Tatsukawa R (1995) Environ Sci Technol 29:2959
68. Kannan K, Senthilkumar K, Loganathan BG, Takahashi S, Odell DK, Tanabe S (1997) Environ Sci Technol 31:296
69. Kim GB, Tanabe S, Iwakiri R, Tatsukawa R, Amano M, Miyazaki N, Tanaka H (1996) Environ Sci Technol 30:2620
70. May-ming Lau M (1991) Arch Environ Contam Toxicol 20:299
71. Ritsema R (1994) Appl Organomet Chem 8:5
72. Strand J, Jacobsen JA, Pedersen B, Granmo A (2003) Environ Pollut 124:7
73. Takahashi S, Tanabe S, Takeuchi I, Miyazaki N (1999) Arch Environ Contam Toxicol 37:50
74. Tanabe S (1999) Mar Pollut Bull 39:62
75. Ueno D, Inoue S, Takahashi S, Ikeda K, Tanaka H, Subramanian AN, Fillmann G, Klam PKS, Zheng J, Muchtar M, Prudente M, Chung K, Tanabe S (2004) Environ Pollut 127:1
76. Wade TL, Garcia-Romero B, Brooks JM (1990) Chemosphere 20:647
77. Yamada H, Takayanagi K, Tateishi M, Tagata H, Ikeda K (1997) Environ Pollut 96:217
78. Dowson PH, Pershke D, Bubb JM, Lester JN (1992) Eviron Pollut 76:259
79. Schebek L, Andreae MO, Tobschall HJ (1991) Environ Sci Technol 25:871
80. Stäb JA, Traas TP, Stroomberg G, van Kesteren J, Leonards P, van Hattum B, Brinkman UAT, Cofino WP (1996) Arch Environ Contam Toxicol 31:319
81. Kannan K, Tanabe S, Tatsukawa R (1995) Chemosphere 30:925
82. Shiraishi H, Soma M (1992) Chemosphere 24:1103
83. Hashimoto S, Watanabe M, Noda Y, Hayashi T, Kurita Y, Takasu Y, Otsuki A (1998) Mar Environ Res 45:169

84. Hung TC, Lee TY, Liao TF (1998) Environ Pollut 102:197
85. Kannan K, Tanabe S, Iwata H, Tatsukawa R (1995) Environ Pollut 90:279
86. Midorikawa S, Arai T, Harino H, Ohji M, Duc Cu N, Miyazaki N (2004) Environ Pollut 131:401
87. Sudaryanto A, Takahashi S, Monirith I, Ismail A, Muchtar M, Zheng J, Richardson BJ, Subramanian AN, Prudente M, Duc Hue N, Tanabe S (2002) Environ Toxicol Chem 21:2119
88. Sudaryanto A, Takahashi S, Iwata H, Tanabe S, Ismail A (2004) Environ Pollut 130:347
89. Sudaryanto A, Takahashi S, Iwata H, Tanabe S, Muchtar M, Razak H (2005) Mar Pollut Bull 50:208
90. Tanabe S, Prudente M, Mizuno T, Hasegawa J, Iwata H, Miyazaki N (1998) Environ Sci Technol 32:193
91. Zhou Q, Jiang G, Liu J (2001) J Agric Food Chem 49:4287
92. de Brito APX, Ueno D, Takahashi S, Tanabe S (2002) Chemosphere 46:401
93. Negri AP, Hales LT, Battershill C, Wolff C, Webster NS (2004) Mar Pollut Bull 48:1142
94. Takeuchi I, Takahashi S, Tanabe S, Miyazaki N (2004) Mar Environ Res 57:397
95. Hall LW, Scott MC, Killen WD, Unger MA (2000) Human Ecol Risk Assess 6:141
96. Jiang GB, Liu JY, Zhou QF (2004) Environ Sci Technol 38:4349
97. Thomas KV, Fileman TW, Readman JW, Waldock MJ (2001) Mar Pollut Bull 42:677
98. Rüdel H, Lepper P, Steinhanses J (2003) Environ Sci Technol 37:1731
99. Michel P, Averty B, Andral B, Chiffoleau JF, Galgani F (2001) Mar Pollut Bull 42:1128
100. Huggett RJ, Unger RJ, Seligman PF, Valkirs AO (1992) Environ Sci Technol 26:232
101. Sidharthan M, Young KS, Woul LH, Soon PK, Shin HW (2002) Mar Pollut Bull 45:177
102. Cleary JJ, Stebbing ARD (1987) Mar Pollut Bull 18:238
103. Michel P, Averty B (1991) Appl Organomet Chem 5:393
104. Bhosle NB, Garg A, Jadhav S, Harjee R, Sawant SS, Venkat K, Anil AC (2004) Chemosphere 57:897
105. Fernandez MA, de Luca Rebello Wagener A, Limaverde AM, Scofield AL, Pinheiro FM, Rodriques E (2005) Mar Environ Res 59:435
106. Law RJ, Waldock MJ, Allchin CR, Laslett RE, Bailey KJ (1994) Mar Pollut Bull 28:668
107. Sarradin PM, Astruc A, Sabrier R, Astruc M (1994) Mar Pollut Bull 28:621
108. Jones-Lepp TL, Varner KE, Heggem D (2004) Arch Environ Contam Toxicol 46:90
109. Kannan K, Corsolini S, Focardi S, Tanabe S, Tatsukawa R (1996) Arch Environ Contam Toxicol 31:19
110. Short JW, Sharp JL (1989) Environ Sci Technol 23:740
111. Hong HK, Takahashi S, Min BY, Tanabe S (2002) Environ Pollut 117:475
112. Pellizzato F, Centanni E, Marin MG, Moschino V, Pavoni B (2004) Sci Total Environ 332:89
113. Kure LK, Depledge MH (1994) Environ Pollut 84:149
114. Wade TL, Garcia-Romero B, Brooks JM (1988) Environ Sci Technol 22:1488
115. Ebdon L, Evans K, Hill S (1988) Sci Total Environ 68:207
116. Kannan K, Yasunaga Y, Iwata H, Ichihashi H, Tanabe S, Tatsukawa R (1995) Arch Environ Contam Toxicol 28:40
117. Kim GB, Tanabe S, Tatsukawa R, Loughlin TR, Shimazaki K (1996) Environ Toxicol Chem 15:2043
118. Uhler AD, Durell GS, Steinhauer WG, Spellacy AM (1993) Environ Toxicol Chem 12:139
119. Belfroid AC, Purperhart M, Ariese F (2000) Mar Pollut Bull 40:226
120. Ishizaka T, Nemoto S, Sasaki K, Suzuki T, Saito Y (1989) J Agric Food Chem 37:1523

121. Forsyth DS, Cleroux C (1991) Talanta 9:951
122. Davies IM, McKie JC (1987) Mar Pollut Bull 18:405
123. Looser PW, Berg M, Fent K, Mühlemann J, S RP (2000) Anal Chem 72:5136
124. Yamada H, Takayanagi K (1992) Wat Res 26:1589
125. Alzieu C, Heral M (1984) Ecotoxicol Test Mar Environ 2:187
126. Bryan GW, Gibbs PE, Huggett RJ, Curtis LA, Bailey DS, Dauer DM (1989) Mar Pollut Bull 20:458
127. Fent K, Meier W (1992) Arch Environ Contam Toxicol 22:428
128. Hamasaki T, Sato T, Nagase H, Kito H (1993) Mutat Res 300:265
129. Horiguchi T, Shiraishi H, Shimizu M, Morita M (1997) Environ Pollut 95:85
130. Mathiessen P, Gibbs PE (1998) Environ Toxicol Chem 17:37
131. Hall LW, Bushong SJ (1996) In: Champ MA, Seligman PF (eds) Organotin. Environmental fate and effects. Chapman & Hall, London, p 157
132. Aldridge WN (1958) Biochem J 69:367
133. Aldridge WN, Casida JE, Fish RH, Kimmel EC, Street BW (1977) Biochem Pharmacol 26:1997
134. Aldridge WN, Rose MS (1969) FEBS Letters 4:61
135. Chow SC, Kass GEN, McCabe MJ, Orrenius S (1992) Arch Biochem Biophys 298: 143
136. Selwyn A, Dawson AP, Stockdale M, Gains N (1970) Eur J Biochem 14:120
137. Fent K (1998) In: Braunbeck T, Hinton DE, Streit B (eds) Fish Ecotoxicology. Birkhäuser Verlag, Basel/Switzerland, p 259
138. von Ballmoos C, Brunner J, Dimroth P (2004) Proc Nat Acad Sci 101:11239
139. Fent K, Woodin BR, Stegeman JJ (1998) Comp Biochem Physiol C 121:277
140. George SG, Buchanan G (1990) Fish Physiol Biochem 8:437
141. Fent K, Stegeman JJ (1991) Aquat Toxicol 20:159
142. Fent K, Bucheli TD (1994) Aquat Toxicol 28:107
143. Fent K, Stegeman JJ (1993) Aquat Toxicol 24:219
144. Morcillo Y, Porte C (1997) Aquat Toxicol 38:35
145. Aw TY, Nicotera P, Manzo L, Orrenius S (1990) Arch Biochem Biophys 283:46
146. Gennari A, Viviani B, Galli CL, Marinovich M, Pieters R, Corsini E (2000) Toxicol Appl Pharmacol 169:185
147. Babich H, Borenfreund E (1991) Toxicol in Vitro 5:91
148. Brüschweiler BJ, Würgler FE, Fent K (1995) Aquat Toxicol 32:143
149. Fent K (2001) Toxicol In Vitro 15:477
150. Clayton M, Steinmann R, Fent K (2000) Aquat Toxicol 47:213
151. Azumi K, Fujie M, Usami T, Miki Y, Satoh N (2004) Mar Environ Res 58:543
152. Chagot D, Alzieu C, Sanjuan J, Grizel H (1990) Aquat Living Resour 3:121
153. His E, Robert R (1985) Rev Trav Inst Peches Marit 47:63
154. Waldock MJ, Thain JE (1983) Mar Pollut Bull 14:411
155. Meador JP, Rice CA (2001) Mar Environ Res 51:113
156. Meador JP, Varanasi U, Krone CA (1993) Mar Biol 116:231
157. Bushong SJ, Hall LWJ, Hall WS, Johnson WE, Herman RL (1988) Wat Res 22: 1027
158. Schwaiger J, Fent K, Stecher H, Ferling H, Negele RD (1996) Arch Environ Contam Toxicol 30:327
159. Fent K (1992) Environ Pollut 76:187
160. Oehlmann J, Stroben E, Fioroni P (1992) Helgol Meeresunters 46:311
161. Hallers-Tjabbes CCT, Kemp JF, Boon JP (1994) Mar Pollut Bull 28:311
162. Morcillo Y, Porte C (1999) Environ Res A 81:349

163. Nias DJ, McKillup SC, Edyvane KS (1993) Mar Pollut Bull 26:380
164. Ramon M, Amor MJ (2001) Mar Environ Res 52:463
165. Shim WJ, Kahng SH, Huong SH, Kim NS, Kim SK, Shim JH (2000) Mar Environ Res 49:435
166. Smith BS (1981) J Appl Toxicol 1:22
167. Spooner N, Gibbs PE, Bryan GW, Goad LJ (1991) Mar Environ Res 32:37
168. Stroben E, Oehlmann J, Fiorini P (1992) Mar Biol 114:289
169. Stroben E, Oehlmann J, Fiorini P (1992) Mar Biol 113:625
170. Ten Hallers-Tjabbes CC, Wegener JW, Van Hattum AGM, Kemp JF, Ten Hallers E, Reitsema TJ, Boon JP (2003) Mar Environ Res 55:203
171. Wilson SP, Ahsanullah M, Thompson GB (1993) Mar Pollut Bull 26:44
172. Gibbs PE, Pascoe PL, Bryan GW (1991) Comp Biochem Physiol 100C:231
173. Oehlmann J, Stroben E, Schulte-Oehlmann U, Bauer B, Fiorini P, Markert B (1996) Fresenius J Anal Chem 354:540
174. Mensink BP, Tn Hallers-Tjabbes CC, Kralt J, Freriks L, Boon JP (1996) Mar Environ Res 41:315
175. Ten Hallers-Tjabbes CC, Kemp JF, Boon JP (1994) Mar Pollut Bull 28:311
176. Bauer B, Fiorini P, Schult-Oehlmann U, Oehlmann J, Kalbfus W (1997) Environ Pollut 96:299
177. Schulte-Oehlmann U, Oehlmann J, Fiorini P, Bauer B (1997) Mar Biol 128:257
178. Gibson GP, Wilson SP (2003) Mar Environ Res 55:101
179. Schulte-Oehlmann U, Bettin U, Fiorini P, Oehlmann J, Stroben E (1995) Ecotoxicology 4:372
180. Smith BS (1981) J Appl Toxicol 1:141
181. Bettin C, Oehlmann J, Stroben E (1996) Helgol Meeresunters 50:299
182. Oehlmann J, Bettin C (1994) Malacol Review Suppl 6:157
183. Santos MM, Ten Hallers-Tjabbes CC, Vieira N, Boon JP, Porte C (2002) Mar Environ Res 54:661
184. Morcillo Y, Porte C (2000) Environ Pollut 107:47
185. Verslycke T, Poelmans S, Wasch KD, Vercauteren J, Devos C, Moens L, Sandra P, De Brabander HF, Janssen CR (2003) Environ Toxicol Chem 22:2030
186. Yamabe Y, Hoshino A, Imura N, Suzuki T, Himeno S (2000) Toxicol Appl Pharmacol 169:177
187. Nishikawa JI, Mamiya S, Kanayama T, Nishikawa T, Shiraishi F, Horiguchi T (2004) Environ Sci Technol 38:6271
188. Kanayama T, Kobayashi N, Mamiya S, Nakanishi T, Nishikawa JI (2005) Mol Pharmacol (in press)
189. Oberdörster E, McClellan-Green P (2002) Mar Environ Res 54:715
190. Morcillo Y, Janer G, O'Hara SCM, L DR, Porte C (2004) Environ Toxicol Chem 23:990
191. Shimaski Y, Kitano T, Oshima Y, Inoue S, Imada N, Honjo T (2003) Environ Toxicol Chem 22:141
192. McAllister BG, Kime DE (2003) Aquat Toxicol 65:309
193. Nakayama K, Oshima Y, Yamaguchi T, Tsuruda Y, Kang IJ, Kobayashi M, Imada N, Honjo T (2004) Chemosphere 55:1331
194. Horiguchi T, Kojima M, Kaya M, Matsuo T, Shiraishi H, Morita M, Adachi Y (2002) Mar Environ Res 54:679
195. Ronis MJJ, Mason AZ (1996) Mar Environ Res 42:161
196. Omura M, Ogata R, Kubo K, Shimasaki Y, Aou S, Oshima Y, Tanaka A, Hirata M, Makita Y, Inoue N (2001) Toxicol Sci 64:224
197. Hwang HM, Oh JR, Kahng SH, Lee KW (1999) Mar Environ Res 47:61

198. Harino H, Fukushima M, Yamamoto Y, Kawai S, Miyazaki N (1998) Environ Pollut 101:209
199. Walsh GE, McLaughlan LL, Lores EM, Louie MK, Deans CH (1985) Chemosphere 14:383
200. U'ren SC (1983) Mar Pollut Bull 14:303
201. Roberts MH (1987) Bull Environ Contam Toxicol 39:1012
202. Valkirs AO, Davidson BM, Seligman PF (1987) Chemosphere 16:201
203. Seinen W, Helder T, Vernij H, Penninks A, Leeuwangh P (1981) Sci Total Environ 19:155
204. Beaumont AR, Newman PB (1986) Mar Pollut Bull 17:457
205. Lawler IF, Aldrich JC (1987) Mar Pollut Bull 18:274
206. Beaumont AR, Budd MD (1984) Mar Pollut Bull 15:402
207. Wester PW, Canton JH (1987) Aquat Toxicol 10:143

Hdb Env Chem Vol. 5, Part O (2006): 101–120
DOI 10.1007/698_5_051

Published online: 9 December 2005

Emission Estimation and Chemical Fate Modelling of Antifoulants

B. van Hattum[1] (✉) · A. Baart[2] · J. Boon[2]

[1]Institute for Environmental Studies, Vrije Universiteit, De Boelelaan 1087, 1081 HV Amsterdam, The Netherlands
bert.van.hattum@ivm.falw.vu.nl

[2]WL|Delft Hydraulics, P.O. Box 177, 2600 MH, Delft, The Netherlands
a.baart@wldelft.nl, j.boon@wdelft.nl

Abstract Triggered by the debates around entry of force of the antifouling convention of the International Maritime Organization and the upcoming ban of TBT, much progress has been made during the last ten years in the field of the environmental risk assessment of new and existing antifouling agents. In this chapter an overview is given of existing exposure assessment models and two important areas of uncertainties, such as the hydrodynamic exchange and emission estimation in estuarine and coastal harbors. Apart from tidal mixing, current and flow-induced horizontal mixing as well as density difference-driven exchange processes and sedimentation have a major impact on the environmental fate of antifouling compounds in coastal harbors. Existing generic screening models are not capable of accounting for the complex hydrodynamics and interaction with chemical fate processes. Considerable uncertainties are involved in the estimation of emissions, based on the biocide leaching rate extrapolated from laboratory studies and affected by temperature, salinity and pH, and shipping activity related parameters, i.e. number and dimensions of ships, speed, underwater surface area, and the time spent in- port. Two recent exposure assessment models for antifoulants in estuarine and marine environments, REMA and Mam-Pec are discussed. The environmental emission scenarios linked to these models were evaluated by a joint OECD-EU working group and formed the basis of the current emission scenario document (ESD-PT21) adopted for application in the frameworks of the Biocide Directive in Europe. The available models for antifoulants have been validated only for a limited number of compounds and no full quantitative sensitivity analysis studies have been performed. As it is expected that in the near future a large number of products needs to be evaluated, further refinement and validation of these models in realistic field studies is recommended.

Keywords Environmental fate · Models · Hydrodynamics · Leaching · Emissions

Abbreviations

2D, 3D	2 or 3 dimensional
DOC	dissolved organic carbon
EXAMS	exposure analysis modelling system, generic model developed by the Environmental Protection Agency, USA
HSE	Health and Safety Executive, London
DELWAQ	generic water quality model of WL|Delft Hydraulics, Delft
REMA	regulatory environmental modelling of antifoulants, exposure model developed by Water Research Centre, Swindon, UK
Mam-Pec	marine antifoulant model to predict environmental concentrations, exposure assessment model developed by Delft Hydraulics and Vrije Universiteit Amsterdam for CEPE
ESD-PT21	emission scenario document for product-type 21 (antifoulants)
TBT	tributyltin compound
ASTM	American Standards for Testing of Materials
CEPE	European Paintmakers Association, Brussels
IMO	International Maritime Organization, UN, London
EUSES	European Union system for the evaluation of substances
EQC	equilibrium criterion models, well-known generic screening models developed by the Canadian Environmental Modelling Centre, Peterborough, Canada
QWASI	quantitative water air sediment interaction models, developed by Canadian Environmental Modelling Centre, Peterborough, Canada
TOXFATE	contaminant fate model, developed by National Water Research Institute, Burlington, Canada
DELFT3D	integrated two or three-dimensional compound modelling system for inland and marine waters developed by WL|Delft Hydraulics, Delft
MIKE-3	3D modelling system for estuaries, coastal waters and oceans, developed by DHI Water & Environment, Hørsholm, Denmark
ECOS	estuarine simulation model, originally developed by Plymouth Marine Laboratory, Plymouth, UK
ECB	Environmental Chemical Bureau of Joint Research Centre of the European Commission, Ispra, Italy
SILTHAR	model to simulate siltation in harbor basins, developed by WL|Delft Hydraulics, Delft
K_{ow}	n-octanol water partitioning coefficient
K_d	sediment water distribution constant ($L\,kg^{-1}$)
K_{oc}	organic carbon adsorption coefficient ($L\,kg^{-1}$)
H	Henry's constant for air water partitioning ($Pa\,m^3\,mol^{-1}$)
PEC	predicted environmental concentration
CEFIC	European Chemical Industry Council, Brussels
TCMS	2,3,5,6-tetrachloro-4-methylsulfonyl (pyridine)

1 Introduction

As described in other chapters in this book the chemical fate of antifoulants is determined by many complex and interacting physical, chemical, and biological processes. Some of the major transport and transformation processes, which are commonly considered in environmental fate models, have been summarized in Fig. 1. The relative importance of each of the processes and pathways is highly compound- and habitat specific and may have a considerable spatial and temporal variation. For compounds with a high affinity to particulate matter or sediment, sediment transport phenomena will be of dominant importance. Stable dissolved compounds are likely to be affected most by river discharges or tidal currents. Biodegradation processes are highly temperature dependent and may be the dominant removal process in tropical water, while in temperate or polar zones this may be less. Photolysis may have a prominent role in the open sea even at greater depths in warm and transparent waters, while in turbid estuarine environments in temperate zones this may only be of importance in the upper water layers. Proper and realistic handling of photolysis in chemical fate models is complicated, as it varies with depth. This can only be done in adequate 3D models with proper relationships of effective rate constants with light penetration (transmittance or turbidity), temperature, and binding to particulate matter and DOC.

There is a broad distinction between organic and inorganic compounds both in the mechanisms and relative importance of different processes. For

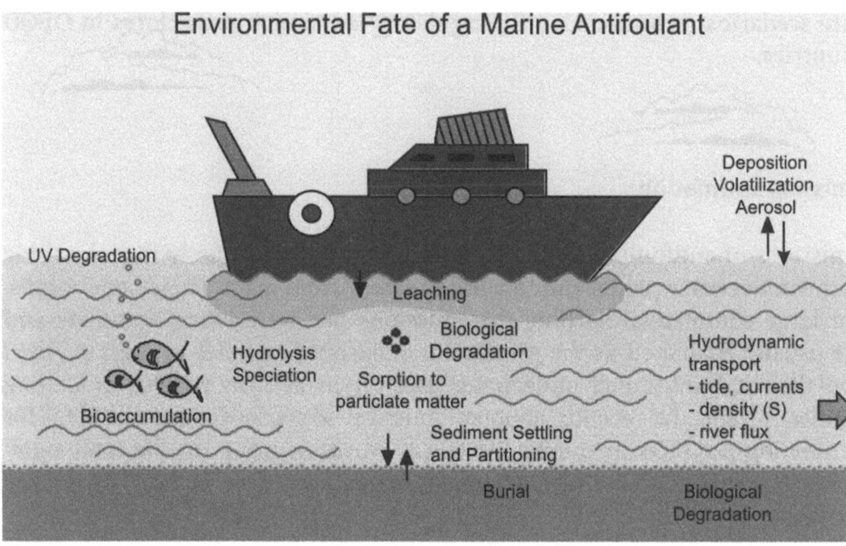

Fig. 1 Chemical fate processes of antifouling products in the marine environment. Source: [5]. Reproduced with permission

instance for copper, processes such as sorption, speciation, and redox reactions have a prominent role in the fraction of freely bioavailable and potentially toxic copper ions [1].

Among the trace element mechanisms, partitioning and rate constants are highly element specific, and may depend strongly on the physico-chemical speciation. The speciation is highly dependent on environmental factors, such as salinity, pH, pE, particulate matter, presence and nature of ligand-substrates. In contrast to the category of neutral hydrophobic compounds there are no simple and generic relationships with which partitioning and rate constants can be estimated. Only a few of the available generic chemical fate models, such as EXAMS [2] and DELWAQ [3] allow for the evaluation of speciation in the chemical fate of trace metals. In environments with low exchange rates or pseudo stagnant conditions the chemical and biological processes are likely to become more important. Especially in energy rich marine environments, such as the coastal areas where most of the main harbors are situated, the hydrodynamic transport and mixing processes of water masses and sedimentation tend to have a major impact on the environmental fate of chemicals. In this chapter some of the basic principles of hydrodynamic exchange processes in estuarine harbors have been described. As this phenomenon is not properly addressed in most regulatory screening-type models (Sect. 4), special models such as REMA [4] and Mam-Pec [5] have been developed for the prediction of the chemical fate of antifoulants. The environmental conditions and emission estimation procedures (Sect. 2) in these models have recently been evaluated by a joint EU and OECD working group [6], and were the basis of the recommended default emission and exposure scenarios recommended for regulatory admission procedures in OECD countries.

2 Emission Estimation

Emissions of antifouling products in harbors originate mainly from leaching of the underwater paints and to a lesser extent from maintenance operations. For large commercial harbors the emissions due to leaching dominate and are usually estimated as the product of a leaching rate (LR in $mg/cm^2/day$) and the total antifouled underwater area (A in m^2). In various modelling studies [5] and the recently adopted emission scenario document (ESD) for antifouling products by a joint EU-OECD working group [6] the total emissions (E_{tot} in $g\,d^{-1}$) in a harbor are estimated according to Eq. 1 or similar formulas.

$$E_{tot} = \sum_{i=1-n} (A_i{}^*N_{ib}{}^*F_i{}^*LR_b) + \sum_{i=1-n} (A_i{}^*N_{im}{}^*F_i{}^*LR_m) \quad (g\,d^{-1}) \qquad (1)$$

In which, A_i (m^2) represents the average underwater area of shipping category i, for example n length categories, N_{im} and N_{ib} represent the number of moving ships and ships moored at berth in the harbor at any time of the day, F_i is the application factor expressed as the fraction of ships in category i treated with a specific antifouling product, and LR_b and LR_m are the compound and paint specific leaching rates (g, m^2 day^{-1}) of ships at berth or moving. As will be described in the next section, large uncertainties may be involved in the proper estimation of each of the parameters in Eq. 1.

2.1 Leaching Rate

The leaching depends highly on the type of compound, characteristics and age of the paint matrix, the velocity of the ship, and environmental factors such as temperature and salinity [7]. Some typical leaching rates reported in the literature are indicated in Table 1.

Manufacturers usually conduct experimental determinations of leaching rates during the development of testing phases of new products with plates or rotating drums coated with the product and exposed to natural or semi-natural conditions. The results from these experimental studies cannot easily be translated to real-life leaching rates from ships to which the product is applied. Available ASTM (D5108-80) and ISO/DIS (15181-1,2) protocols have been criticized by various authors [8, 9]. For many of the new products established and certified analytical methods are hardly available [10]. Except for the study of Steen et al. [11], hardly any field studies on actual leaching rates from freshly painted ships have been published in the open literature. An interesting theoretical mass-balance approach was followed in a study by Boxall et al. [12], in which worst case leaching rates were estimated based on lifetime of the paint and national paint usage data. Their overall estimates were in line with the ranges reported in Table 1. In various reviews good descriptions can be found of the different classes of biocidal antifouling paints and the dependency of the biocide leaching rate on the physical and chemical processes at the paint–(sea)water interface [7, 13, 14]. The following main types of coatings were identified: soluble matrix; insoluble matrix, TBT self polishing copolymer, and TBT-free self polishing paints, each with their specific leaching patterns and changes of the leaching rate with the age of the paint matrix. For the soluble paint matrix, with typical in-service periods of 12–18 months, the biocide-release rate may decrease exponentially during the lifetime of the coating. The self-polishing copolymers and paints, which both may contain copper, have a different biocide release pattern with a rapid decrease in the first months to year, followed by a constant release rate during the remaining in-service period of up to 5 years.

As demonstrated in experimental and modelling studies [7] commonly occurring changes in speed (0–20 knots), temperature (10–30 °C), salinity

Table 1 Summary of leaching rate estimates reported from various experimental and modelling studies

Compound	Leaching rate µg/cm²/day	Type of study	Refs.
TBT	0.5–2.1	flume and rotary test system	[9]
	1.5–4	ASTM test system	[15]
	2.5	model Marina	[16]
	0.1–5	model Harbor	[17]
	1.3–3.0	model ships > 25 m	[18]
	4	model	[19]
Cu	18–21	flume and rotary test system	[9]
	25–40	ASTM test system	[16]
	4–6*	modified ASTM test	[8]
	1–20	not specified	[20]
	8–25	model ships > 12 m	[18]
	37–101	model ships > 25 m	[18]
Irgarol	2.6	flume test system	[21]
	5.0	ISO test system	
	2–16	model marina	[22]
	5	model marina	[23]
Sea-Nine 211	3.0	flume test system	[21]
	2.9	ISO test system	
	1 (0.1–5)	model harbor	[17]
	2.5	field and model study	[11]
Zinc pyrythione	3.3	ISO test system	[21]
Diuron	0.8	flume test system	[21]
	3.3	ISO test system	
Dichlofluanid	0.6	flume test system	[21]
	1.7	ISO test system	
TCMS pyridine	0.6	flume test system	[21]
	3.8	ISO test system	

* After 21 days. During the first 21 days leaching rates ranged between 7–61 µg/cm²/day

(15–35‰), and pH (7.5–8.5) have a significant impact on the actual leaching rate of Cu and TBT, with an increased leaching rate at higher speed, temperature, and salinity and at lower pH. In experimental studies conducted earlier by Thomas et al. [9] with the flume and rotary ISO test systems no such effects were found for Cu, and a lowering of TBT leaching at higher speeds. Due to the slow response of the leaching behavior to changes in these parameters [7, 9], actual leaching rates in coastal environments are probably highly variable and also dependent on previous time spent at sea or in port. Despite the large impact on the uncertainty of emission estimation, a surprisingly low number of experimental studies have been published in the open scientific literature. Much of the in-company information on experimental leaching rate studies used in regulatory procedures is confidential and unfortunately is not available for peer-review. Summarizing the available information from Table 1 it becomes evident, that for each of the compounds a broad range of leaching rate estimates has been observed. Copper leaching rates usually are higher than for other compounds. Leaching rates reported for TBT usually are below regulatory implied values of 4 $\mu g/cm^2/day$ in some countries (USA, Sweden). For the development of the Mam-Pec model [5] a set of default leaching rates was proposed based on expertise available within the CEPE Antifouling Working Group: 50 $\mu g/cm^2/day$ for copper, 4 $\mu g/cm^2/day$ for TBT (moving ships), and 2.5 $\mu g/cm^2/day$ for generic organic antifouling agents.

2.2 Underwater Surface Area

A second parameter in the emission estimation, the total antifouled underwater area of ships in a certain harbor area, depends on multiple factors, such as shipping intensities, dimensions and shape of the various categories of ships, cargo load, residence time in the harbor, and various other factors. Some paint suppliers [24] have published rule of thumb formulas for estimation of the underwater surface (A) and required amount of paint for recreational ships, from simple dimensions such as length (L) or length at the waterline (L_{WL}) width (W) or depth (D), as indicated in Table 2.

For commercial ships related formulas have been used to derive the estimated surface area, which surprisingly show limited differences in average estimated surface area. In the formula proposed by Willingham and Jacobson [17] and used in the emission estimation module in the Mam-Pec model [5] further simplifications were obtained by applying average ratios between L and W (W as 14–15% of L) and L and D (D as 5% of L). In the final ESD of the joint EU-OECD working group various approaches were com-

Table 2 Simple formulas for first estimation of average underwater area

Type of ship	Formula	Refs.
Recreational ships		
motor-launch (low draught)	$A = L_{WL}(W + D)$	[24]
sailing-yacht (intermediate draught)	$A = 0.75\, L_{WL}(W + D)$	[24]
sailing yacht (deep keel)	$A = 0.5\, L_{WL}(W + D)$	[24]
generic motor-boat	$A = 0.85\, L_{WL}(W + D)$	[45]
Commercial ships		
New York Harbor	$A = LW\, 1.3$	[17]
Port of Rotterdam	$A = L(W + D) + WD$	[5]
Finnish harbors	$A = 0.95\, L(0.8(D + W) + W)$	[45]

pared and a more elaborate formula (Eq. 2) was selected, which was derived from Finnish shipping data [25] and usually is referred to as the "Holtrop equation". The latter approach yielded slightly (8–16%) higher estimates of the surface area for corresponding length classes compared to the Mam-Pec model

$$A = L(2D + W)[C_m(0.53 + 0.63C_b - 0.36\,(C_m - 0.5) - 0.0013(L/D)]^{0.5} \quad (2)$$

In which: A = submersed ship area, L = length of ship, D = depth, W = width, C_M = empirical factor (ranging from 0.95–0.98, taken in the ESD taken as 0.975) on how full-bodied the main arch of the ship is, and C_B = empirical factor (ranging from 0.75–0.85, taken in the ESD as 0.8) on how full-bodied the underwater volume of the ship is. The factors C_M and C_B are calculated according to the formulas:

$$C_M = A_M/(BD) \quad (3)$$

$$C_B = V_d/(LBD) \quad (4)$$

In which: A_M is the area of the main arc of the ship, i.e. the area of the biggest cross-section of the ship, which is in general in the middle of the ship, and V_d is the underwater volume of the ship (displacement).

On the basis of a statistical survey of paint-usage data in relation to ship dimensions (DWT) from a large supplier for 300 ships and covering 9 of the 25 main Lloyds shipping categories, Van Hattum et al. [5] concluded that predicting submersed surfaces with simple generic regression formulas may result in deviations up to several 100% below or above actual measured surface areas. As it is evident that the submersed surface estimation introduces a large uncertainty in the emission estimation, it is recommended that in future studies a more refined approach, based on statistical surveys is followed.

2.3 Shipping Intensity

Information on the shipping density, a third parameter in Eq. 1, can be obtained from various sources, such as on-line port statistics from local port authorities or branch organizations such as the International Association of Ports and Harbours (IAPH), commercial suppliers (Lloyds Register Ltd), or trade oriented studies [26]. Although traffic intensities and port arrivals are monitored on a large scale in European waters, there still is no structured and aggregated reporting system and especially estimations of traffic intensities in open and coastal water may still have large uncertainties. Another problem is caused by the differences among harbors in reported dimensions and shipping types, such as length, depth, dead-weight tonnage (DWT), gross tonnage (GRT/GT), net tonnage (NRT/NT), compensated tonnage (CGT), cargo landed, number of containers, or economic parameters, such as revenue tons.

The average number of port arrivals in some main European ports [26] varied in 1996 between approximately 4400 ships per year for Helsinki (Finland) to more than 25 000 per year for Piraeus (Greece) and Rotterdam (Netherlands). In the first version of the Mam-Pec model [5] the shipping intensities and ship dimensions in the port of Rotterdam and the North Sea shipping lanes along the Dutch coast were used as a basis for the default emission scenarios for commercial sea going vessels. In the recent emission scenario document for the Biocide Directive [6] a similar approach was followed with different settings for the number of ships and submersed surface areas based on the Holtrop equation (Eq. 2). The recommended default values are summarized in Table 3.

Little information is available on the application factor of a product (F_i), which constitutes the fourth parameter in Eq. 1. In general, market share information of specific products is confidential, but it is clear that this type of information is crucial for a proper estimation of the emissions. Especially after implementation of the IMO treaty on antifoulants this will become more important, as TBT will be diminished or banned, and the market share of other biocide-based antifouling paints or biocide-free systems on sea going vessels will become more important. With currently existing differences in admission policies and regulation between countries, this may even vary on a local scale. Especially for the categories of smaller ships (< 25 m), where application of TBT is not allowed, a reliable estimation of the actual usage is not possible. In the EU-OECD emission scenario document (a worst case approach with values of 0.9–0.95 for F_i is recommended.

Table 3 Recommended default emission scenarios for antifoulants in Europe (Biocide Directive) and other OECD countries. Indicated are average underwater surface area and number of ships moving (N_m) or at berth (N_b) for the different length classes

Length class (m)	Surface area (m^2)	Shipping lane	Estuarine harbor		Marina
		N_m	N_b	N_m	N_b
< 50	31				500
50–100	1,163	3.9	11	1.8	
100–150	3,231	1.7	5	0.4	
150–200	6,333	1.6	5	0.4	
200–250	10,469	0.4	1	0.1	
250–300	15,640	0.5	2	0.1	
300–350	21,844	0.1			
Estimated emission* ($g\,d^{-1}$)		773	2303	191	345

* Of compound with a leaching rate 2.5 $\mu g/cm^2/day$ and 100% application of product. Adapted from: [5, 6]

3 Water Exchange Mechanisms in Coastal Environments

In estuaries many physical and chemical processes take place that have an impact on the concentration of a contaminant in the environment. The hydrological conditions in the estuary are, however, the major determining factor. Having a good grasp of the order of magnitude of the hydrological exchange of for example a harbor in an estuary with it's surroundings is essential in order to predict the concentration. In general, the exchange of water between a harbor basin and an estuary is caused by three phenomena [27, 28], that is by:

- filling and emptying by the tide;
- the horizontal eddy generated in the harbor entrance by the passing main flow;
- vertical circulation currents in the harbor generated by density differences between the water inside and outside the basin.

In some cases the above picture is complicated by the extra effects of a water discharge through the harbor basin to the estuary or sea. On the one hand such a discharge has a positive effect by flushing the basin, but on the other hand it has a negative effect by introducing or enhancing water exchange by density currents (as in e.g. some Dutch harbors, such as Delfzijl, IJmuiden, Harlingen, and Terneuzen). It even may introduce a serious sediment influx contributing to shoaling in the harbor basin. This mechanism is addressed as flushing with withdrawal of water (e.g. cooling water intake) defined as a negative flushing discharge rate.

Most quick assessment models only incorporate an empirical exchange (REMA, EUSES, Simplebox) or use only the tidal exchange. The Mam-Pec model is an exception as it incorporates all phenomena and allows for empirical exchange volumes as well. Most current true 3D models, such as Delft3D [29, 30], Mike-3 [31], or Telemac [32], incorporate all processes but require very experienced users with a high level of hydrological knowledge to use the models.

3.1 Tidal Exchange

The exchange by the first mechanism over a tidal period m, i.e. the tidal prism can easily be determined as:

$$V_t = 2\eta A_b \tag{5}$$

where: V_t = tidal prism of the harbor basin, η = tidal amplitude (Fig. 2), A_b = (storage) area of the basin.

The total water exchange volume is the sum of the tidal prism and the exchange volumes due to the horizontal eddy in the harbor entrance (V_h),

Table 4 Water exchange and siltation data of some harbors

Harbor	Region	Country	$V_t + V_h$ (10^6 m³/tide)	V_d (10^6 m³/tide)	V_e (10^6 m³/tide)
Braakman	North Sea	Netherlands	5.5	11.0	16.5
Paulina	North Sea	Netherlands	13.0	16.1	29.1
Um-Qasr	P. Gulf	Iraq	6.7	7.6	14.3
Port Gardner	Puget Sound	USA (WA)			
– present			1.1	6.1	7.2
– future design			2.4	19.0	21.4

Source: Eysink (1995)

due to density currents (V_d), and the extra water exchange due to flushing (V_{ef}):

$$V_e = V_t + V_h + V_d + V_{ef} \tag{6}$$

Some examples are given in Table 4. The quantities V_h, V_d and V_{ef} are less obvious than V_t and are dealt with in more detail in the next sections.

3.2 Water Exchange by a Horizontal Eddy

A current passing the entrance of a basin generates an eddy in this entrance (see Fig. 2). There, steep velocity gradients generate an exchange of water by turbulence. Through this mechanism silt laden water from outside penetrates the eddy and from there further into the harbor and to the center of the eddy. The rate of water exchange by this mechanism depends on the flow velocity in front of the harbor basin, the size of the entrance and the tidal prism. It will be obvious that the rate of exchange decreases with increasing tidal prism. The rate of "horizontal water exchange" can be approximated [33] by

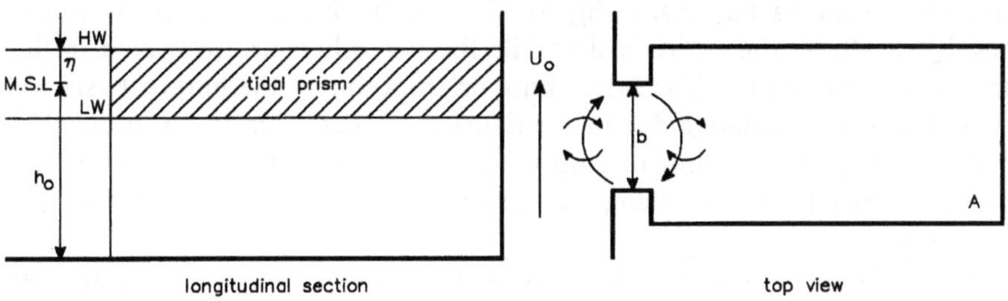

Fig. 2 Definition of tidal prism and horizontal exchange mechanism (HW = high water level, LW = low water level, MSL = mean sea level, u_o = flow velocity river, b = width at mouth of harbor)

the formula:

$$Q_h = f_1 h b u_o - f_2 Q_t \tag{7}$$

where: Q_h = rate of horizontal water exchange (m^3/s), f_1, f_2 = empirical coefficients depending on the geometry of the basin, h = depth of entrance (m), b = width of entrance (m), u_o = main flow velocity in front of the entrance (m/s), Q_t = filling discharge due to rising tide (= hbu_t) (m^3/s), u_t = tidal in- and outflow velocities in the entrance (m/s).

This formula is valid for rivers ($Q_t = 0$) and in tidal areas during flood; Q_h almost is negligible during ebb [33]. Hence, substitution of $h = h_o - \eta \cos \omega t$ and $u_o = u_{o,max} \sin \omega t$ (η tidal amplitude, ω tidal period) and integration over the flood period ($t = 0$ to $T/2$) yield the total volume per tide by horizontal exchange:

$$V_h = f_1 h_o b \frac{u_{o,max}}{\pi} T - f_2 V_t \tag{8}$$

where: V_h = total volume of water exchange per tidal cycle by horizontal exchange flow (m^3), h_o = depth in the entrance relative to MSL (m), T = tidal period (s), V_t = tidal prism of harbor basin (m^3).

The coefficients f_1 and f_2 generally are within the ranges 0.01–0.03 and 0.1–0.25 respectively and can be estimated based on existing knowledge from comparable situations. In some cases it may appear useful to determine more accurate values via hydraulic model investigations. Delft Hydraulics has determined these for several large harbors (Rotterdam, Antwerp). In case the equation yields a negative value for V_h it means that the horizontal exchange does not contribute to the total water exchange, in which case $V_h = 0$.

3.3 Water Exchange due to Density Currents

Water exchange is also caused by density differences between the water inside and outside the harbor basin (Fig. 3). This mechanism is very effective and, besides, it affects the entire basin while the two others are restricted to the area near the entrance. The tidal filling or emptying of the harbor basin reduces the water exchange due to the density currents. Bearing in mind that $2u_t h/2$ represents the filling or emptying discharge rate, Fig. 3 shows that the density-induced water exchange under all circumstances is reduced in the same way.

Hence the rate of exchange by density currents (no influence of horizontal exchange assumed) can be described by:

$$Q_d = (u_o - u_t) h b / 2 \tag{9}$$

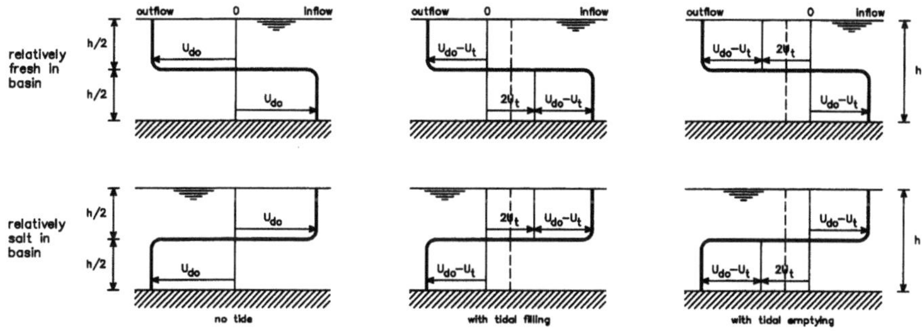

Fig. 3 Schematized flow profiles indicating reduction of density-induced exchange flow by tidal filling or emptying of basin. Source: [28]. Reproduced with permission

with

$$u_{\mathrm{do}} = f_3 \left(\frac{\Delta \rho}{\rho} g h \right)^{1/2} \tag{10}$$

where: Q_{d} = exchange rate due to density currents, u_{do} = exchange velocity without influence of tidal in- and outflow, ρ = density of water, $\Delta\rho$ = characteristic density difference, $(..)^{1/2}$ = characteristic parameter of density currents, f_3 = coefficient, g = gravitational constant (9.8 $\mathrm{m/s^2}$).

Figure 4 demonstrates the mutual relationships of a number of relevant parameters, such as the water level variation at the harbor entrance (h), the related river flow in front of it (u_{o}), the mean density variations outside (ρ_0) and inside the harbor (ρ_{ha}) and their difference, as also the tidal in- and outflow currents (u_{t}) and the undisturbed and effective density currents (u_{do} and u_{t} respectively) in case of harmonic relations. The lower figure distinctly indicates the reduction of the water by density currents due to the tidal flow velocities through the harbor entrance and the relevance of the phase lag φ_{t} and the ratio $\hat{u}_{\mathrm{do}}/\hat{u}_{\mathrm{t}}$ in this respect.

Assuming linear harmonic relationships, the density induced exchange flow rate can be integrated over a tidal cycle, which yields:

$$V_{\mathrm{d}} = f_4 h_{\mathrm{o}} b \left(\frac{\Delta \rho_{\mathrm{max}}}{\rho} g h_{\mathrm{o}} \right)^{1/2} T - f_5 V_{\mathrm{t}} \tag{11}$$

with for practical reasons:

$$\Delta \rho_{\mathrm{max}} = 0.5 \left(\overline{\rho}_{\mathrm{o,max}} - \overline{\rho}_{\mathrm{o,min}} \right) \tag{12}$$

and

$$V_{\mathrm{do}} = f_{4,\mathrm{max}} h_{\mathrm{o}} b \left(\frac{\Delta \rho_{\mathrm{max}}}{\rho} g h_{\mathrm{o}} \right)^{1/2} T \tag{13}$$

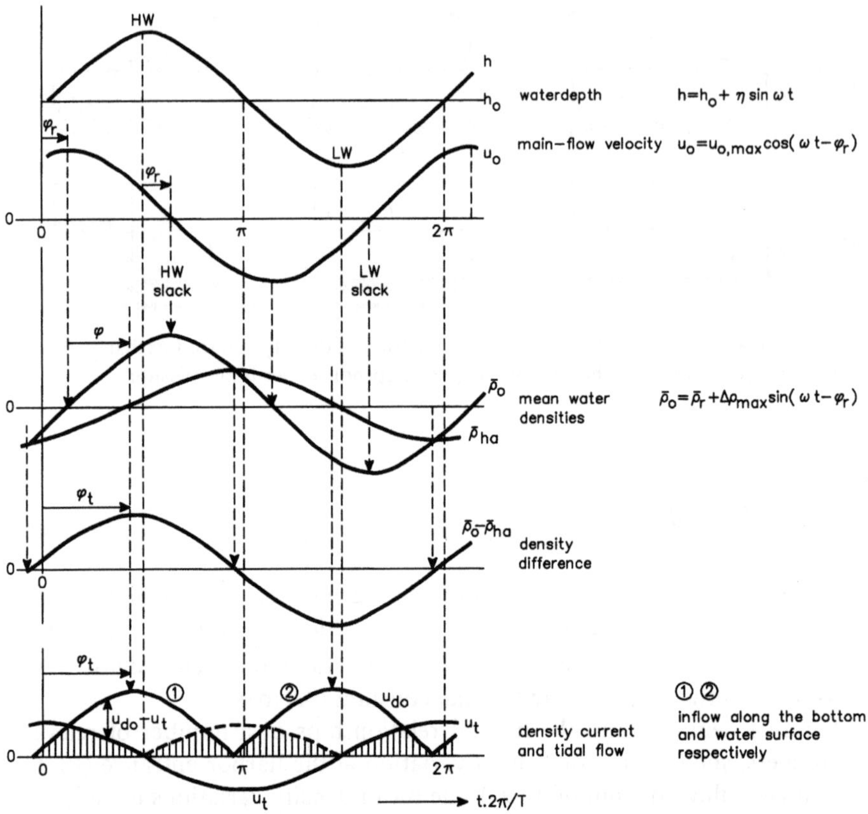

Fig. 4 Effect of phase lag between tidal and density currents through the harbor entrance on the density-induced exchange flow. Source: [28]. Reproduced with permission

where: V_d = exchange volume per tide due to density currents, f_4 = coefficient depending on V_{do}/V_{ha} with V_{ha} being the volume of the basin below MSL, f_5 = coefficient depending on V_{do}/V_t and phase lag φ_t between u_{do} and u_t.

It is obvious that in a large harbor (low values of V_{do}/V_{ha}) the average water density will hardly follow the density fluctuation of the water in front of the harbor. In the case of a small harbor basin and/or strong density currents (higher values of V_{do}/V_{ha}), however, the density of the water inside the harbor may follow the density fluctuations outside. This results in a reduction of the characteristic density difference inducing the density currents. This effect has been included in the coefficient f_4, which consequently reduces with increasing V_{do}/V_{ha}. This effect has been estimated theoretically on the basis of linear harmonic theory assuming a constant water level and neglecting tidal flow (total water exchange $V_e = V_{do}$).

Under specific conditions of low tide, low flow, and low density differences other exchange processes become more important in the exchange [34]:

- Non-tidal water level changes. In the absence of tide, water level differences still occur based on large-scale water movement or wind-related water setup.
- Wind induced exchange. When wind blows over a water surface, interaction of wind and water consists of shear stress at the surface and sometimes a normal pressure component on a wavy surface. Internal friction exists both in the airflow as in the water flow, as well as friction between water and bed-layer and walls. These interactions cause vertical wind and water velocity profiles, which could be relevant phenomena for the marina-sea exchange.

4 Environmental Fate Models

Table 5 summarizes a number of currently available environmental fate models, ranging from simple generic screening-type models to complex and detailed 3-dimensional location-specific models. The frequently used generic risk-assessment models, such as EUSES, Simplebox and the Mackay-type EQC models [35–37], usually include only a limited number of emission, transport, and chemical fate pathways required for a reliable assessment of the fate of antifouling products. Only the more sophisticated chemical equilibrium models, with both steady state or dynamic calculation options, such as EXAMS, ECOS, DELWAQ, QWASI, TOXFATE [2, 3, 38–40] are capable of a more comprehensive treatment of the subtle physico-chemical and biological processes and interactions. Only the complex 3D models Delft3D, Mike-3, Telemac-3D [29–32] and a few of the generic models (EXAMS, ECOS, DELWAQ [2, 3, 38]) can handle complex hydrodynamics and provide a 2D or 3D spatial resolution of the final results, which is of importance for estuarine and coastal environments with strong gradients. From the generic 2D models, both ECOS [38] and DELWAQ [3] can easily be adapted to complex estuarine hydraulic processes. For the EXAMS [2] model this can only be achieved after linking with tailor-made and location specific calculation modules, such as in the study on the chemical fate of TBT and Seanine in New York harbor [17]. Several models, such as REMA [4] and Mam-Pec [5] were specially designed for chemical fate prediction of antifoulants in estuarine and coastal environments and provide the combination of a realistic treatment of hydrodynamic and chemical fate processes and user-friendly entry and handling of model parameters and model runs.

The REMA model [4] is a special model developed for British estuaries for the regulatory evaluation of antifoulants by the Health and Safety Executive in the UK. It is based on the QWASI model [39] and contains three sections of an estuary with three adjacent marinas. Hydrodynamic exchange rates between marinas and estuarine sections need to be entered manually. The model does

Table 5 Characteristics of models available for chemical fate assessment of antifoulants

	C/f	Nr. of Media	Hydro-dynamics	Emiss. from ships	Skills required	Effect of S, T, pH	Generic	Spatial resolution	Steady state/ Dynamic	Ref.
Screening type models										
EUSES	c	4	–	–	low	–	g	1D	s	[35, 46]
EQC-based models	f	> 4	–	–	low	–	g	1D	s	[37]
Simplebox	c	> 4	+	+	high	(+)	g	2D	s/d	[36]
2D/3D type models										
QWASI	f	> 4	+	+	high	(+)	l	2D	s/d	[39]
TOXFATE	f	> 4	+	+	high	(+)	l	2D	s/d	[40]
EXAMS	c	> 4	+	+	high	(+)	g	2/3D	s/d	[2]
ECOS 2.1	c	> 4	+	+	high	+	g	3D	s/d	[38]
Delft3D/DELWAQ	c	> 3	++	+	high	+	g	3D	s/d	[29]
Mike-3.	c	> 4	++	+	high	+	g	3D	s/d	[31]
Telemac-3D	c	> 4	++	+	high	+	g	3D	s/d	[32]
Models developped for antifoulants										
Simple spreadsheet models	c	3	–	+	low	–	g	1D	s	[16, 47]
REMA	f	4	+	+	low	(+)	l	2D	s	[4]
Mam-Pec	c	4	++	++	low	(+)	g	2D	s	[5]

Explanation: C/f: concentration or fugacity based model; Nr. of Media: number of abiotic main and sub-compartments included (water column, sediment, air, soil, particulate matter); Hydrodynamics: ability to cope with more complex marine hydrodynamic features; Emiss. from Ships: ability to include (+) or estimate (++) typical emission patterns of antifoulants. Generic: generic model (g) or location specific (l). Other categories: –: missing; +: available as user editable option; ++: integrated in model; (+): option is partly available not allow an estimation of the hydrodynamic exchange processes. The treatment of the chemical fate processes is in many respects similar to the first version of the Mam-Pec model. Several default combinations of emissions and estuary types are included in the model: a small estuary that dries out, well-mixed estuaries with narrow and wide mouth, and a large complex estuary. The model has been validated successfully for these estuaries. The treatment of emissions is not flexible, fixed surface areas are included in the model for pleasure crafts and medium-sized ships and changing emission settings can only be done indirectly. The REMA model is limited to mainly emissions from pleasure crafts in typical estuaries in the UK and cannot be adapted to marine exposure scenarios for large commercial harbors or open sea shipping lanes. Model software and documentation can be obtained via the website of the HSE (www.hse.gov.uk) and the Environmental Chemical Bureau (ECB; www.ecb.jrc.it/biocides) of the European Commission.

The Mam-Pec model [5] is an integrated 2D (10×10 grid) hydrodynamical and chemical fate model based on DELWAQ and SILTHAR of Delft Hydraulics. The DELWAQ model has previously been linked to hydraulic models for the North Sea [41]. The SILTHAR model [28] has been used for the description of hydrodynamic exchange and sedimentation in several large international harbors (Port of Rotterdam, Hong Kong). Models runs are executed after entry or editing of input data in three different panels for (1) environment and hydrology, (2) compound properties, and (3) emission scenarios. The environmental and hydrodynamical parameters included in the model are: currents, tides, salinity, DOC, suspended matter load, sedimentation, port dimensions. The model allows estimation of hydrodynamical mixing in five generalized "typical" marine environments (open sea, shipping lane, estuary, commercial harbor, yachting marina). The default environments can easily be adapted to local situations. Several of the recently adopted environmental emission scenarios for the Biocide Directive [6] were based on slightly modified Mam-Pec scenarios (commercial harbors, shipping lane, marina) and the estimation and calculation routines in the model. Compound properties/process included in model are: K_{ow}, K_{d}, K_{oc}, H, volatilization, speciation, hydrolysis, photolysis, and biodegradation. The emission estimation is based on leaching rates, shipping intensities, residence times, ship hull underwater surface areas according to the principles described in Sect. 2 (Eq. 1). The model contains default emission settings for the matching environments. The EU-OECD environmental emission scenarios [6] are included in an updated version of the model, which will be available end of 2005. The model allows easy comparison of predicted environmental concentrations of PECs of different compounds. The model has been validated based on the results of a few recent monitoring studies. Model-predicted concentrations of TBT, Irgarol, Diuron, Seanine and other compounds matched within order of magnitude with measured concentrations [11, 42–44]. A full quantitative sensitivity analysis has not yet been conducted. On the basis of preliminary results, the following processes were identified as major determinants of the uncertainty: leaching rate, underwater surface area, shipping characteristics, hydrodynamic exchange, partitioning, sedimentation, and (bio)degradation. The first version of Mam-Pec (v1.0) was released in 1999. Since then updates have been released in 2002 (v1.4) and early 2005 (v1.6) compatible with common operating systems (Win9x-NT-2000-XP). The development of the model was sponsored by the Antifouling Working Group of the European Paint Makers Association (CEPE/CEFIC), with contributions of the European Commission—DG XI (1997–1999). Regulatory agencies in Finland, Suisse, Netherlands, USA, and other countries in biocide admission procedures have applied Mam-Pec recently. The model and documentation and are freely available via websites of CEPE (www.cepe.org) and other organizations of paint producers (www.antifoulingpaint.com).

5 Conclusions and Outlook

Most of the currently available screening models used in environmental risk assessment are not equipped to handle the complex hydrodynamic exchange and sedimentation processes in estuarine and coastal environments. As the largest emissions of biocides in antifouling paints occur in the marine and estuarine commercial harbors special models are required for a realistic risk assessment of antifoulants. Various user-friendly models are currently available, which allow a more or less realistic prediction of environmental concentrations in the marine environment. Some of these models and underlying emission scenarios have been incorporated in recent environmental emission scenario documents (ESD-PT21) for the Biocide Directive the European Community and other OECD countries. Large uncertainties are involved in the estimation of emissions based on leaching rates extrapolated from laboratory experiments and estimated underwater surface areas. Available models have been validated for a limited number of compounds in a few field studies. No full quantitative sensitivity analysis data are available. With the expected implementation of the antifouling treaty of the International Maritime Organization (IMO) in the coming years and the ban on TBT a large number of new products need to be evaluated. Against this background significant efforts are required to reduce the inherent uncertainty of model predictions, as well as realistic field studies for a proper validation of existing models.

References

1. Kramer KJM, Jak RG, van Hattum B, Hooftman RN, Zwolsman JJG (2004) Environ Tox Chem 23:2971

2. Burns LA (2000) Exposure Analysis Modelling System (EXAMS): User Manual and System Documentation. National Exposure Research Laboratory, U.S. Environmental Protection Agency, Athens, GA
3. Boderie PMA (1994) DELWAQ 4.0: Technical Reference Manual. WL|Delft Hydraulics, Delft
4. HSE (1999) REMA—regulatory environmental modelling of antifoulants. Biocides and Pesticides Assessment Unit, Health and Safety Executive, London
5. Van Hattum B, Baart AC, Boon JG (2002) Computer model to generate predicted environmental concentrations (PECs) for antifouling products in the marine environment, 2nd edn. Accompanying the release of Mam-Pec version 1.4, IVM Report (E-02/04), Institute for Environmental Studies, Vrije Universiteit, Amsterdam
6. Van der Plassche E, Van der Aa E (2004) Harmonisation of environmental emission scenarios: an emission scenario document for antifouling products in OECD countries (EDD-PT21). Royal Haskoning, Nijmegen
7. Kiil S, Dam-Johansen K, Weinell CE, Pedersen MS, Codolar SA (2003) Biofouling 19:37
8. Berg EA (1995) Eur Coat J 7:534
9. Thomas KV, Raymond K, Chadwick J, Waldock M (1999) Appl Organomet Chem 13:453–460
10. Thomas KV (1998) J Chromatogr A 825:29
11. Steen RJCA, Ariese F, Van Hattum B, Jacobsen J, Jacobson A (2004) Chemosphere 57:513
12. Boxall ABA, Comber SD, Conrad AU, Howcroft J, Zaman N (2000) Mar Pollut Bull 40:898
13. CEPE AWG (1998) Classes of biocidal antifouling paint commercially available in the European Union. CEPE, Brussels
14. Omae I (2003) Appl Organomet Chem 17:81
15. Fisher AS, Hill SJ, Waldock M, Thomas KV (1997) Health and Safety Executive Contract Report 3283/R51.82, Cited in [9]
16. Johnson A, Luttik R (1994) Risk assessment of antifoulants—position paper. Paper nr. 1994-05-03 presented at the 7th meeting of the Ad Hoc Group of Experts of Non-Agricultural Pesticides, 16–18 May 1994. National Chemicals Inspectorate, Sweden, National Institute for Public Health and the Environment, Netherlands
17. Willingham GL, Jacobson AH (1996) ACS Sym Ser 640:225
18. Lindgren P, Olsson B, Unger C (1998) Antifoulingsprodukter FARTYG PM-beslut 1998-10-20, KEMI, Stockholm (in Swedish)
19. Stronkhorst J (1996) TBT Contamination and toxicity of sediments: a persistent problem. In: Stewen U (ed) The present status of TBT-copolyment antifouling paints. Proceeding of an International One Day Symposium on antifouling paints for ocean-going vessels, 21st Febr. 1996, The Hague. Ministry of Transport, Public Works and Water Management, Rijkswaterstaat, DGSM; ORTEP, Association, The Hague
20. Hare CH (1993) J Prot Coat Lin 10:83
21. Thomas KV (2001) Biofouling 17:73
22. Ciba (1995) Summary on ecological and health effects of Irgarol 1051, Information Brochure Ciba Geigy, Basel
23. Scarlett A, Donkin ME, Fileman TW, Donkin P (1997) Mar Pollut Bull 34:645
24. International Paint Ltd (1999) Commercial brochure issued by International Paint Ltd, Gateshead (UK)
25. Holtrop J (1977) Internat Shipbuild Progr 1 24:270

26. ISL (1997) Shipping Statistical Yearbook 1997. Institute Of Shipping, Economics and Logistics, Bremen
27. Eysink WD, Vermaas H (1983) Computational methods to estimate the sedimentation in dredged channels and harbour basins in estuarine environments. Proceedings Int Conf on Coastal and Port Eng in Dev Countries, Colombo, Sri Lanka
28. Eysink WD (1995) Silthar, a mathematical program for the computation of siltation in harbour basins, WL|Delft Hydraulics, Delft
29. Lesser GR, Roelvink JA, van Kester JATM (2004) Coast Eng 51:883
30. Gerritsen H, Goede ED, Genseberger M, Uittenbogaard RE (2003) Validation document Delft3D-FLOW, Report nr. X0289. WL|Delft Hydraulics, Delft
31. McClimans TA, Pietrzak JD, Huess V et al. (2000) Cont Shelf Res 20:941
32. Hervouet JM (2000) Hydrol Proces 13:2209
33. Van der Graaf J, Reinalda R (1977) Horizontal exchange in distorted scale models, Report nr. S61, WL|Delft Hydraulics, Delft
34. Baart AC (2003) Mam-Pec, application at low exchange conditions. Report Z3662. WL|Delft Hydraulics, Delft
35. ECB (1997) EUSES—The European Union system for the evaluation of substances. Joint Research Centre European Commission, Environment Institute, European Chemicals Bureau, Ispra
36. Brandes LJ, Den Hollander H, Van de Meent D (1996) SimpleBox 2.0: a nested multimedia fate model for evaluating the environmental fate of chemicals. Report no. 719101029. National Institute of Public Health and the Environment (RIVM), Bilthoven
37. Mackay D, Di Guardo A, Paterson S, Kicsi G, Cowan CE, Kane DM (1996) Environ Toxicol Chem 15:1638
38. Harris JRW, Gorley RN, Bartlett CA (1993) ECOS version 2 user manual—an estuarine simulation shell. Plymouth Marine Laboratory, Plymouth
39. Ling H, Diamond M, Mackay D (1993) J Great Lakes Res 19:582
40. Halfon E, Allan RJ (1995) Environ Int 21:557
41. Baart A, Boon J (1998) Scremotox: A screening model for comparative ecotoxicological risk-assessment of substances entering the North Sea. WL|Delft Hydraulics, Delft
42. Van Hattum B (2001) TBT—a compound currently in the lime light. In: Gandrass J, Salomons W (eds) Dredged Material in the Port of Rotterdam—Interface between Rhine Catchment Area and North Sea. GKSS Research Centre, Geesthacht, p 277
43. Lamoree MH, Swart CP, Van der Horst A, Van Hattum B (2002) J Chromatogr A 970:183
44. Readman J (ed) (2002) Assessment of antifouling agents in coastal environments (ACE)—Final scientific and technical report (MAS3-CT98-0178). Plymouth Marine Laboratory, Plymouth, UK
45. Kovisto S (2003) Proposal for Finnish exposure scenarios for antifouling products. Finnish Environment Institute (SYKE), Helsinki
46. TGD (1996) Technical guidance document in support of commission directive 93/67/EEC on risk assessment for new notified substances and commission regulation (EC) No 1488/94 on risk assessment for existing substances—Part-III. Office for Official Publications of the European Communities, Luxembourg
47. Baur D, Jacobson A (1996) Modelling of marine antifoulants. Rohm and Haas Company, European Laboratories, Valbonne (France) and Research Laboratories, Spring House (PA, USA)

Hdb Env Chem Vol. 5, Part O (2006): 121–153
DOI 10.1007/698_5_052

Published online: 23 November 2005

Evaluation of Antifouling Booster Biocides in Marine Water and Sediments Based on Mass Spectrometric Techniques

M. D. Hernando[1] (✉) · M. Mezcua[1] · D. Barceló[2] · A. R. Fernández-Alba[1]

[1]Department of Analytical Chemistry, University of Almería, 04120 Almería, Spain
amadeo@ual.es

[2]Department of Environmental Chemistry, CID-CSIC, 08034 Barcelona, Spain

Abstract This chapter outlines the stages involved in chemical analysis of antifouling booster biocides, reviews the sample preparation strategies and the analytical methods for the determination of these compounds in marine water and sediment samples. Antifouling booster biocides which are reviewed in this chapter include: Irgarol 1051, Diuron, Sea-nine 211, TCMTB (2-thiocyanomethylthiobenzothiazole), Dichlofluanid and Chlorothalonil. Gas chromatography-mass spectrometry (GC-MS), liquid chromatography-mass spectrometry (LC-MS) as well as future prospects related to current mass spectrometric techniques are outlined.

Keywords Antifouling booster biocides · GC-MS · LC-MS · Sample preparation · Seawater · Sediments

Abbreviations

AFID	alkali flame ionization
APCI	atmospheric pressure chemical ionization
CI	chemical ionization
CID	collision-induced dissociation
CPDU	[1-(3-chlorophenyl)-3,1-dimethylurea]
CW-DVB	carbowax-divinylbenzene
DAD	diode-array detection
DCA	[3,4-dichloroaniline]
DCM	dichloromethane
DCPMU	[1-(3,4-dichlorophenyl)-3-dimethylurea]
DCPU	[1-(3,4-dichlorophenyl) urea]
DMSA	(*N′*-dimethyl-*N*-phenyl-sulphamide)
ECD	electron capture detector
EI	electron impact ionization
FTD	flame thermionic detector
GC-MS	gas chromatography-mass spectrometry
HLB	hydrophilic-lypophilic balance
HS-SPME	headspace solid-phase microextraction
IT-MS	ion trap-mass spectrometry
LC-MS	liquid chromatography-mass spectrometry
LC-MS/MS	liquid chromatography-tandem mass spectrometry
LLE	liquid-liquid extraction
LOD	limit of detection
MAE	microwave assisted extraction
MRM	multiple reaction monitoring
MS/MS	tandem mass spectrometry
NCI	negative chemical ionization
PA	polyacrylate
PCI	positive chemical ionization
PDMS	polydimethylsiloxane
PDMS-DVB	polydimethylsiloxane-divinylbenzene
PTV	programmable temperature vaporizing
Q-MS	quadrupole mass spectrometry
QqQ	triple quadrupole
RPLC	reversed-phase liquid chromatography
SFE	supercritical fluid extraction
SIM	selected ion monitoring
SME	solvent microextraction
SPE	solid-phase extraction
SPME	solid-phase microextraction
SRM	selected reaction monitoring
TBT	tributyltin
TCMTB	2-thiocyanomethylthiobenzothiazole
TFA	trifluoroacetic acid

1 Introduction

Extensive survey data for antifouling compounds have been reported for marine coastal, docks and harbour waters [1–4]. In addition, the environmental impact on the aquatic environment and fate of organotin compounds, in particular tributyltin (TBT) and copper is well documented [3–7]. As an alternative to organotin compounds, the use of which is now regulated and limited to vessels greater than 25 m in length, organic booster biocides were introduced. However, the potential of booster biocides to cause pollution has also been of concern for several research studies carried out during the last few years [8–11]. Maximum concentration levels found up to approx. 1700 ng L^{-1} have been reported in areas of high boating activity or enclosed marinas [9, 12–14], while in marinas with higher water exchange rates, lower concentration levels have been encountered (low ng L^{-1}) [15–17]. However, those concentrations can be sufficiently relevant to pose a risk to aquatic life.

Because of the potentially dangerous consequences of the presence of booster biocides in the environment, several analytical methodologies have been described in the literature. Most of the methods involve a preconcentration step, a clean-up of the extracts and a chromatographic separation by gas and liquid chromatography.

In general, because of their low concentration levels, trace-level aqueous samples have to be extracted and enriched prior to their analytical determination. Several methods have been developed to extract aqueous samples containing antifouling compounds. Liquid-liquid extraction (LLE) [10, 18] and solid-phase extraction (SPE) [10, 11, 19, 20] were the dominant approaches for sample preparation. Among these, the LLE method has well-known operating disadvantages (e.g. time-consuming and requires large amounts of organic solvents). In the past few years, the application of new techniques reflects the current analytical trends (reduction of organic solvents, automation of the sample preparation step). In addition to the common use of SPE, examples of current techniques applied are solid-phase microextraction (SPME), which is a simple procedure, inexpensive, efficient and no solvent for sample preparation is necessary, [21, 22] as well as solvent microextraction (SME) [23] and headspace solid-phase microextraction (HS-SPME) [24] for the analysis of booster biocides. Similar goals have led to the application of new extraction techniques for sediment samples, including microwave-assisted extraction (MAE) [25], supercritical fluid extraction (SFE), or SPME [26], as an alternative to traditional extraction methods such as ultrasonication.

Although FTD or ECD selective detection systems have been included in the development of analytical methods [21, 23], the reported methodology in the literature for booster biocides have been to a great extent based on the application of gas chromatography-mass spectrometry (GC-MS) systems, and currently also liquid chromatography-mass spectrometry (LC-MS). Among developed GC-MS methods, the modes of ionization such as chemical ionization (CI) or electron-impact ionization (EI) [27] as well as modes of operation, such as selected ion monitoring (SIM) or tandem MS-MS, have been applied to determine booster biocides in environmental water and sediment samples [27–29]. For thermolabile compounds, LC has been applied with diode-array-detection (DAD) systems [30] and with MS, which has been increasingly used because of its selectivity and sensitivity. The atmospheric pressure chemical ionization (APCI) technique has been a widely used interface for coupling LC-MS [19–31]. LC-MS/MS methods have also experienced progress in terms of application in the analysis of booster biocides, which is especially suitable for environmental sample analysis for routine confirmation of identity as well as structural elucidation or ultra trace analysis [32].

This chapter reviews the sample preparation and analytical strategies applied in the analysis of booster biocides in marine water and sediment samples. The review of booster biocides includes Irgarol 1051, Diuron, Sea-nine 211, TCMTB (2-thiocyanomethylthiobenzothiazole), Dichlofluanid

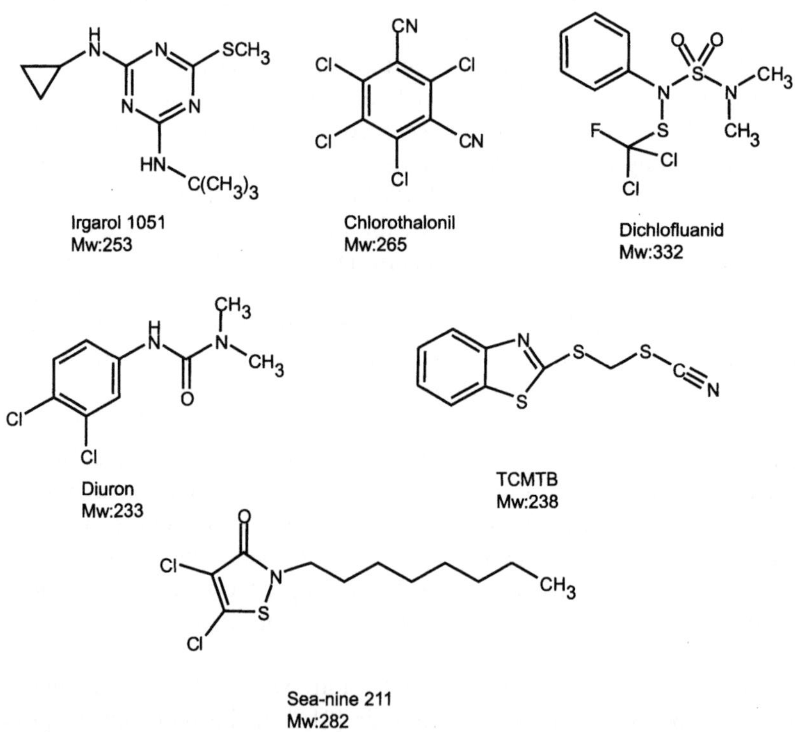

Fig. 1 Chemical structures of antifouling booster biocides: Irgarol 1051, Chlorothalonil, Dichlofluanid, Diuron, TCMTB and Sea-nine 211

and Chlorothalonil (Fig. 1 shows the chemical structures). Current extraction and mass spectrometric techniques as well as future prospects are outlined.

2
Sample Preparation

2.1
Marine Waters

Given that antifouling booster biocides are usually detected at trace levels in marine water samples, pre-concentration techniques have been applied previously to the analytical determination. Extraction of biocide residues has been carried out with different techniques. Tables 1 and 2 show the recovery results obtained with different extraction methods. LLE is a traditional technique that, in spite of its known disadvantages, is still applied [2, 10, 13, 14, 16, 33]. As an example, this technique has been used for the extraction of Irgarol 1051, Chlorothalonil and Dichlofluanid [2, 14, 16]. Organic solvents with apolar to medium polarity such as *n*-hexane, dichloromethane (DCM), acetone and methylene chloride have been used, obtaining recoveries higher than 75% in particular for Irgarol 1051. Information concerning recoveries for the other two compounds has not been reported.

SPE has become a common and effective technique to extract and enrich the analytes from environmental samples. SPE has been applied in the majority of reported studies [9, 19, 20, 27, 28, 37, 47, 51]. Characteristics that make suitable its use, are the low quantities of solvent needed, or that the procedures can be readily automated. Automatic off-line SPE is generally performed by passing a water sample volume of 0.1–1 L through cartridges or disks [27, 28]. C18-bonded silicas [9, 30, 31, 37, 42, 44, 48, 51], polymeric material [19, 20, 27, 28, 32, 34, 47, 51] and graphitized carbon black [47] have been selected as sorbents. Another approach is on-line SPE for the automated preparation of samples prior to the analysis [20]. Advantages of on-line systems include minimized adsorptive losses that can occur with off-line sample transfers and sample-handling procedures. The contamination process from external sources can be more difficult than in off-line SPE procedures. Limitations can be related to the reduced sample through-put and stability problems for the sample caused by extended storage times in the auto-sampler. The efficiency with which biocides are extracted from marine water by SPE procedures is satisfactory. Recoveries higher than 80% are generally obtained in the experiments performed to extract Irgarol 1051, TCMTB and Diuron [19, 30–32, 44, 47, 51]. Good efficiency in the extraction method was also observed for Chlorothalonil (approx. 78%) [19, 27, 28]. Lower recoveries have been reported for the results obtained for Dichlofluanid

Table 1 Summary of developed GC-MS methods for the determination of antifouling booster biocides in marine water and sediment samples. GC conditions and MS operation modes. Extraction methods and sensitivity (limits of detection, LOD)

Antifouling booster biocides	Matrix	Extraction method (sorbent)	Recovery (%)	GC Capillary column	Ionization mode	Detection system	Mode operation	LOD (LOD, ng L^{-1} in seawater and ng g^{-1} in sediment)	Refs.
Irgarol 1051	seawater	LLE	76	HP-5	EI	Q-MS	SIM	0.5	[33]
	seawater	LLE (dichloro-methane)	92	HP-5, DB-5	EI	MS	SIM	not reported	[10, 13]
	seawater	LLE (methylene chloride)	93.5	DB-5	EI	Q-MS	SIM	1	[16]
	seawater	LLE (*n*-hexane)	75 [a]	DB-5	EI	Q-MS	SIM	2	[2]
	seawater	LLE (dichloro-methane)			EI	MS	SIM	200	[14] *
	seawater	SPE (polymeric)	95	HP-5	PCI	IT-MS	MS/MS	50	[27] *
	seawater	SPE (polymeric)	95	HP-5	NCI	Q-MS	SIM	20.000	[28] *
	seawater	SPE (polymeric)	95	HP-5	EI	Q-MS	SIM	1.2	[28] *
	seawater	SPE (SDB disks)		DB-5	EI	IT-MS	MS/MS	0.1–1	[29]
	seawater, ports	SPE (PLRP-S)	84	HP-5	EI	Q-MS	SIM	10	[20] *
	seawater, harbour	SPE (C18)	93	SE-54		AFID		4	[37]
	seawater	SPE (C18)	81	HP-5	EI	Q-MS	SIM	1.5	[9]

* simultaneous determination: ([22]: I, D; [21]: I, S; [24]: I, S; [20]: I, D; [14]: I, C, D; [35]: I, C, D; [23]: C, D, S; [27]: I, C, D, S, T; [28]: I, D, S, T; [26]: I, C, D, S). I: Irgarol 1051, C: Chlorothalonil, D: Dichlofluanid, T: TCMTB, S: Sea-nine 211.

[a] reported recovery of internal standard TPT (tripropyl tin chloride)

Table 1 (continued)

Antifouling booster biocides	Matrix	Extraction method (sorbent)	Recovery (%)	GC Capillary column	Ionization mode	Detection system	Mode operation	LOD (LOD, $ng\,L^{-1}$ in seawater and $ng\,g^{-1}$ in sediment)	Refs.
	seawater, harbour	SPME (polyacrylate)		HP-1	EI	Q-MS	SIM	50	[22] *
	seawater	SPME (PDMS-DVB)	90	DB-5		ECD		2	[21] *
	seawater	SPME (PDMS-DVB)	90	DB-1		FTD		20	[21] *
	seawater	HS-SPME	118	DB-5	EI	Q-MS	SIM	30	[24] *
	sediments	LLE (dichloro-methane/acetone)			EI	MS	SIM	500	[14]
	sediments	sonication/LLE	82	DB-1	EI	Q-MS	SIM	0.7	[35] *
	sediments	soxhlet/acetone	61	SE-54		AFID		0.05	[37]
	sediments	water extraction/ SPME (PDMS)	66	DB-5	EI	Q-MS	SIM	8.0	[26] *
	sediments	acetone extraction/ SPME (PDMS)	91	DB-5	EI	Q-MS	SIM	0.5	[26] *
	sediments	MAE	85.4	ZB-5	EI	IT-MS	SIM	1.7	[25]
	sediments	SFE	87	DB-225		MS	SIM	3.0	[38]
Chlorothalonil	seawater	LLE (dichloro-methane)			EI	MS	SIM	200	[14] *
	seawater	SPE (polymeric)	78	HP-5	NCI	IT-MS	MS/MS	0.05	[27] *

* simultaneous determination: ([22]: I, D; [21]: I, S; [24]: I, S; [20]: I, D; [14]: I, C, D; [35]: I, C, D; [23]: C, D, S; [27]: I, C, D, S, T; [28]: I, D, S, T; [26]: I, C, D, S). I: Irgarol 1051, C: Chlorothalonil, D: Dichlofluanid, T: TCMTB, S: Sea-nine 211.

[a] reported recovery of internal standard TPT (tripropyl tin chloride)

Table 1 (continued)

Antifouling booster biocides	Matrix	Extraction method (sorbent)	Recovery (%)	GC Capillary column	Ionization mode	Detection system	Mode operation	LOD (LOD, ng L^{-1} in seawater and ng g^{-1} in sediment)	Refs.
	seawater	SPE (polymeric)	78	HP-5	NCI	Q-MS	SIM	0.5	[28] *
	seawater	SPE (polymeric)	78	HP-5	EI	Q-MS	SIM	2.5	[28] *
	seawater	SME	94	DB-5		ECD		2.0	[23] *
	sediments	LLE (dichloro-methane/acetone)			EI	MS	SIM	500	[14] *
	sediments	sonication/LLE	92	DB-1	EI	Q-MS	SIM	2.5	[35] *
	sediments	water extraction/ SPME (PDMS)	37	DB-5	EI	Q-MS	SIM	25	[26] *
	sediments	acetone extraction/ SPME (PDMS)	74	DB-5	EI	Q-MS	SIM	6	[26] *
Dichlofluanid	seawater	LLE (dichloro-methane)			EI	MS	SIM	200	[14] *
	seawater	SPE (polymeric)	53	HP-5	NCI	IT-MS	MS/MS	5.0	[27] *
	seawater	SPE (polymeric)	53	HP-5	NCI	Q-MS	SIM	1.5	[28] *
	seawater	SPE (polymeric)	53	HP-5	EI	Q-MS	SIM	2.5	[28] *
	seawater, ports	SPE (PLRP-S)	67	HP-5	EI	Q-MS	SIM	20	[20] *
	seawater	SPE	72	HP-5	EI	IT-MS	SIM	3	[34]

* simultaneous determination: ([22]: I, D; [21]: I, S; [24]: I, S; [20]: I, D; [14]: I, C, D; [35]: I, C, D; [23]: C, D, S; [27]: I, C, D, S, T; [28]: I, D, S, T; [26]: I, C, D, S). I: Irgarol 1051, C: Chlorothalonil, D: Dichlofluanid, T: TCMTB, S: Sea-nine 211.

[a] reported recovery of internal standard TPT (tripropyl tin chloride)

Table 1 (continued)

Antifouling booster biocides	Matrix	Extraction method (sorbent)	Recovery (%)	GC Capillary column	Ionization mode	Detection system	Mode operation	LOD (LOD, $ng\,L^{-1}$ in seawater and $ng\,g^{-1}$ in sediment)	Refs.
	seawater, harbour	SPME (polyacrylate)		HP-1	EI	Q-MS	SIM	80	[22] *
	seawater	SME	88	DB-5		ECD		3.0	[23] *
	sediments	LLE (dichloro-methane/acetone)			EI	MS	SIM	500	[14] *
	sediments	sonication/LLE	87	DB-1	EI	Q-MS	SIM	3.0	[35] *
	sediments	ultrasonic	101	HP-5	EI	IT-MS	SIM	3	[34]
	sediments	water extraction/ SPME (PDMS)	54	DB-5	EI	Q-MS	SIM	11	[26] *
	sediments	acetone extraction/ SPME (PDMS)	84	DB-5	EI	Q-MS	SIM	1	[26] *
Sea-nine 211	seawater	SPE (polymeric)	45	HP-5	NCI	IT-MS	MS/MS	0.05	[27] *
	seawater	SPE (polymeric)	45	HP-5	NCI	Q-MS	SIM	1.5	[28] *
	seawater	SPE (polymeric)	45	HP-5	EI	Q-MS	SIM	2.5	[28] *
	seawater, harbour	SPE (C18-disks)	66	BPX5	PCI	IT-MS	MS/MS	5.0	[36]
	seawater	SPME (PDMS-DVB)	96	DB-5		ECD		2	[21] *
	seawater	SPME (PDMS-DVB)	96	DB-1		FTD		5	[21] *

* simultaneous determination: ([22]: I, D; [21]: I, S; [24]: I, S; [20]: I, D; [14]: I, C, D; [35]: I, C, D; [23]: C, D, S; [27]: I, C, D, S, T; [28]: I, D, S, T; [26]: I, C, D, S). I: Irgarol 1051, C: Chlorothalonil, D: Dichlofluanid, T: TCMTB, S: Sea-nine 211.
[a] reported recovery of internal standard TPT (tripropyl tin chloride)

Table 1 (continued)

Antifouling booster biocides	Matrix	Extraction method (sorbent)	Recovery (%)	GC Capillary column	Ionization mode	Detection system	Mode operation	LOD (LOD, ng L^{-1} in seawater and ng g^{-1} in sediment)	Refs.
	seawater	SME	91	DB-5		ECD		0.2	[23] *
	seawater	HS-SPME	118	DB-5	EI	Q-MS	SIM	20	[24] *
	sediments	shaking/dichloro-methane	98	BPX5	PCI	IT-MS	MS/MS	20	[36]
	sediments	water extraction/ SPME (PDMS)	58	DB-5	EI	Q-MS	SIM	13	[26] *
	sediments	acetone extraction/ SPME (PDMS)	88	DB-5	EI	Q-MS	SIM	1.5	[26] *
TCMTB	seawater	SPE (polymeric)	98	HP-5	NCI	IT-MS	MS/MS	0.05	[27] *
	seawater	SPE (polymeric)	98	HP-5	NCI	Q-MS	SIM	1.5	[28] *
	seawater	SPE (polymeric)	98	HP-5	EI	Q-MS	SIM	3.0	[28] *

* simultaneous determination: ([22]: I, D; [21]: I, S; [24]: I, S; [20]: I, D; [14]: I, C, D; [35]: I, C, D; [23]: C, D, S; [27]: I, C, D, S, T; [28]: I, D, S, T; [26]: I, C, D, S). I: Irgarol 1051, C: Chlorothalonil, D: Dichlofluanid, T: TCMTB, S: Sea-nine 211.

[a] reported recovery of internal standard TPT (tripropyl tin chloride)

Table 2 Summary of developed LC-MS methods for the determination of antifouling booster biocides in marine water and sediment samples. LC conditions and MS operation modes. Extraction methods and sensitivity (limits of detection, LOD)

Antifouling booster biocides	Matrix	Extraction method	Recoveries (%)	LC separation Column	Mobile phase	Ionization mode	Detection system	Limit of detection (LOD, ng L^{-1} in seawater and ng g^{-1} in sediment)	Refs.
Irgarol 1051	seawater	SPE (C18)	88	VYDAC C18	Aq/ACN		DAD	3	[42]
	sediment	centrifugation (hexane:acetone)	103	VYDAC C18	Aq/ACN		DAD	2.1	[42]
	seawater	SPE Prospekt (C18)	90–95	Shandon C8	Aq/ACN		DAD	1	[44, 48]
	sediment	sonication (MeOH: water)	90	Shandon C8	Aq/ACN	APCI (+)	Shandon C8	no reported	[48]
	seawater	SPE (C18)	96	Zorbax C18	Aq/ACN		DAD	11	[30] *
	seawater	SPE (polymeric)	106	Shandon C8	Aq/ACN	APCI (+)		5	[19, 48] *
	seawater	SPE (graphitized carbon black)	81	LiChrosorb RP18	Aq/MeOH	APCI (+)	MS-Q (SIM)	4	[47] *
	seawater	SPE (polimeric)	88	Waters Sphrisorb ODS2	Aq.AcNH_4/ MeOH	ESI (+)	MS-TQ (MRM)	0.2	[32]
	sediment	sonication (MeOH: EtAc)	99	Bakerbon ENV	Aq/ACN	APCI (+)	MS-IT (SIM)	0.001	[50] *
	seawater	SPE (polymeric)	95	Kromasil 100 C18	Aq.AcH/ ACN	APCI (+)	MS-Q (SIM)	5	[51]
	sediment	sonication (MeHO/Acetone)	69/74	LiChrosorb RP18	Aq/MeOH	APCI (+)	MS-Q (SIM)	0.2	[52]

* simultaneous determination: ([19, 30, 31, 44, 47, 49, 50, 53])

Table 2 (continued)

Antifouling booster biocides	Matrix	Extraction method	Recoveries (%)	LC separation: Column	LC separation: Mobile phase	Ionization mode	Detection system	Limit of detection (LOD, ng L^{-1} in seawater and ng g^{-1} in sediment)	Refs.
	sediment	sonication (MeOH)	95	Zorbax SB-C18	Aq/ACN		DAD	2.8	[53] *
Diuron	seawater	SPE Prospekt (C18)	not reported	Shandon C8	Aq/ACN		DAD	2	[44] *
	seawater	SPE (C18)	87	Zorbax C18	Aq/ACN		DAD	7	[30] *
	seawater	SPE (C18)	100.3	Bakerbon ENV	Aq/ACN	APCI (+)	MS-IT (SIM)	1	[31] *
	seawater	SPE (polymeric)	99	Shandon C8	Aq/ACN	APCI (+) APCI (–)	MS-Q (SIM)	10	[19] *
	seawater	SPE (graphitized carbon black)	89	LiChrosorb RP18	Aq/MeOH	APCI (+)	MS-Q (SIM)	1	[47]
	seawater	SPE (polimeric)	93	Waters Sphrisorb ODS2	Aq.$AcNH_4$/ MeOH	ESI (+)	MS-TQ (MRM)	0.5	[32]
	sediment	sonication (MeOH: EtAc)	102	Bakerbon ENV	Aq/ACN	APCI (+)	MS-IT (SIM)	0.1	[50] *
	seawater	SPE (polimeric)	97	Kromasil 100 C18	Aq.AcH/ ACN	APCI (+)	MS-Q (SIM)	10	[51]
	sediment	sonication (MeHO/Acetone)	87/79	LiChrosorb RP18	Aq/MeOH	APCI (+)	MS-Q (SIM)	0.4	[52]
	sediment	sonication (MeOH)	92	Zorbax SB-C18	Aq/ACN		DAD	3.1	[53] *
TCMTB	seawater	SPE (C18)	91.2	Bakerbon ENV	Aq/ACN	APCI (–)	MS-IT (SIM)	1	[31] *

* simultaneous determination: ([19, 30, 31, 44, 47, 49, 50, 53])

Table 2 (continued)

Antifouling booster biocides	Matrix	Extraction method	Recoveries (%)	LC separation Column	Mobile phase	Ionization mode	Detection system	Limit of detection (LOD, ng L^{-1} in seawater and ng g^{-1} in sediment)	Refs.
	seawater	SPE (polymeric)	111	Shandon C8	Aq/ACN	APCI (–)	MS-Q (SIM)	8	[19] *
	seawater	SPE (graphitized carbon black)	87	LiChrosorb RP18	Aq/MeOH	APCI (–)	MS-Q (SIM)	20	[47] *
Chloro-	seawater	SPE (polymeric)	96	Shandon C8	Aq/ACN	APCI (–)	MS-Q (SIM)	2	[19] *
thalonil	seawater	SPE (graphitized carbon black)	76	LiChrosorb RP18	Aq/MeOH	APCI (–)	MS-Q (SIM)	1	[47]
Dichlo-	seawater	SPE (polymeric)	<10	Shandon C8	Aq/ACN	APCI (–)	MS-Q (SIM)	not reported	[19] *
fluanid	seawater	SPE (graphitized carbon black)	95	LiChrosorb RP18	Aq/MeOH	APCI (–)	MS-Q (SIM)	4	[47]
	seawater	SPE (polymeric)	89	Kromasil 100 C18	Aq.AcH/ ACN	APCI (–)	MS-Q (SIM)	5	[51]
	sediment	sonication (MeHO/Acetone)	109/108	LiChrosorb RP18	Aq/MeOH	APCI (–)	MS-Q (SIM)	1.6	[52]
Sea-nine 211	seawater	SPE (graphitized carbon black)	96.4	LiChrosorb RP18	Aq/MeOH	APCI (+)	MS-Q (SIM)	1	[47]
	seawater	SPE (C18)	100.4	Bakerbon ENV	Aq/ACN	APCI (+)	MS-IT (SIM)	1	[31] *
	sediment	sonication (MeHO:acetone)	99/104	LiChrosorb RP18	Aq/MeOH	APCI (+)	MS-Q (SIM)	0.4	[52]

* simultaneous determination: ([19, 30, 31, 44, 47, 49, 50, 53])

anid [19, 20, 27, 28, 34, 47, 51] or Sea-nine 211 [27, 28, 36] (from 45 to 53%) using polymeric material (divinylbenzene-*N*-vinylpyrrolidone copolymer, hydrophilic-lypophilic balance, HLB) as the sorbent. However, improvement in recoveries (66–72%) are obtained when C18-bonded silica or polymeric material (styrenedivinylbenzene copolymer) have been selected for both compounds.

The goal of reducing the time needed and the quantities of organic solvents has led also to the application of current extraction approaches including SPME, HS-SPME or SME. Up to now and to our knowledge, few analytical methods have been reported using these techniques [21–24]. SPME is carried out by direct dipping of the fibre into the aqueous sample [21, 22] but also by exposing the fibre in the headspace of the sample contained in a vial maintained at high temperature [23]. Compared with other sample preparation techniques, the advantages of SPME can be the disposal costs and the potential to improve detection limits. On the other hand, the application of SPME-GC is limited to analytes and volatiles that are thermally stable at the injection temperature. The coated fibres typically assayed are polyacrylate (PA), polydimethylsiloxane (PDMS), carbowax-divinylbenzene (CW-DVB) or polydimethylsiloxane-divinylbenzene (PDMS-DVB). The optimum extraction efficiency has been the aim of studies [21, 23] evaluating various parameters: sample pH, addition of salt (NaCl) to the sample, temperature in the absorption process and time in both the absorption and desorption process. Extraction is usually enhanced with increasing salt concentration and increased polarity of the compound (the salting-out effect). However, decreases in the extraction can be observed for high or even moderate salt concentrations, which can lead to a reduction of the diffusion rate of the analytes. Sample pH or time and temperature of absorption depend on the compound studied, and in general, compromise solutions are considered for simultaneous extraction of various compounds [21, 23]. The relative recoveries (instead of absolute recovery as used in exhaustive extraction procedures) of Irgarol 1051 [21, 22, 24] and Sea-nine 211 [21, 24] in seawater with PDMS-DVB, is higher than 90%, indicating the effectiveness of extraction of this technique.

SME is also an environmentally safe extraction technique (it generates little waste and requires 1–2 μL of organic solvents). Advantages offered by this technique are simplicity, speed and potential for easy automation. In addition, it does not suffer from carryout (exposure time of fibre in the GC injector to completely desorb the compounds from the stationary phase) between extractions that are observed in SPME procedures. SME has been found to be effective for biocides such as Dichlofluanid, Chlorothalonil or Sea-nine 211 [23] with high recoveries values in seawater (88–94%). In terms of recovery, limits of detection and precision, SME also exhibits a favourable performance and it is comparable to the SPME technique for the extraction of biocides.

2.2 Sediments

Until now, most analytical methods have been developed for water samples and less attention has been paid to sediments. In this sense, less survey data is available concerning the occurrence of biocides in sediments. Reported concentrations of biocides that have been found in sediments are at the low ng g^{-1} level [14, 35, 37, 38]. Extraction from sediments has been developed with conventional and current techniques. Recovery results for different extraction methods are shown in Tables 1 and 2. LLE is usually preformed by shaking or sonication [14, 34–36]. DCM, acetone or a mixture of both is commonly selected as the organic solvent. For the cleanup step, deactivated Florisil is typically used as the adsorbent. With this procedure, the recoveries are satisfactory (> 82%) for compounds such as Irgarol 1051 [14], Chlorothalonil [14, 35], and Dichlofluanid [14, 34, 35]. Similar recoveries (> 80%) have been also obtained for Diuron from sediments, for which most extraction methods have only been based on LLE and sonication [50, 52, 53]. Soxhlet extraction is also applied but only with volatile organic solvents (acetone). The extraction efficiency of this technique was lower for Irgarol 1051 [37] with recoveries of 61%.

SFE has been evaluated as an approach to minimize solvent usage and compared to conventional extraction techniques for its validation [38]. Parameters such as pressure and temperature have moderate effects on extraction efficiency, while the modifier content in CO_2 is significant. SFE of Irgarol 1051 is feasible with CO_2 modified with 20% methanol in the presence of TFA (0.65 M in MeOH). The reported recoveries of Irgarol 1051 are similar to those obtained with conventional extraction techniques (87%) [38].

In general, satisfactory recovery values have been reported when current extraction techniques have been applied. With SPME for sediments, the efficiency of extraction is comparable to the results in water samples. So, for example, Irgarol showed a 90% recovery using a coated fibre of PDMS-DVB [26]. In this study, the use of extraction solvents with different polarity (water, methanol, acetone and acetonitrile) was also evaluated. The best results were obtained with acetone while with water, the recoveries decreased to 66%. This effect was observed for the extraction of Chlorotahlonil, Dichlofluanid and Sea-nine 211 decreasing to the range 30–37%, using water [26].

As an alternative to conventional extraction methods, MAE has been assessed for the extraction of Irgarol 1051 and its main degradation product M1. This technique is based on the adsorption of microwave energy by extraction solvents, which results in an increase of temperature and pressure. In this way, diffusion of the compounds from the sediments to the solvent can be achieved. The presence of water in the sediments can be a benefit in the MAE technique, given that it helps non-polar organic solvents to absorb the microwave energy. In addition, the use of water was found to be a satisfactory solvent for the extraction of these compounds. A solvent evaporation step is not required, so the aqueous extract obtained can be pre-concentrated by a SPE procedure. Mean recoveries obtained with this procedure are higher than 80% and precision, lower than 40%, indicating the suitability of its use for the extraction of Irgarol 1051 and M1 [25].

3 Gas Chromatography-Mass Spectrometry

3.1 General Remarks

Gas chromatography is a commonly used technique in the analysis of booster biocides such as Irgarol 1051, Chlorothalonil, Dichlofluanid, Sea-nine 211 or TCMTB (Table 1). In general, the separation of most of these compounds is performed using non-polar GC-capillary stationary phases such as methyl polysiloxane or phenyl-methylpolysiloxane (DB-1, HP-1, DB-5, HP-5, ZB-5, DB-225, BPX-5, SE-54), and gradient temperatures from (60–80 °C to 280–320 °C). The splitless injection mode is generally preferred because of its robustness but on-column [20] and PTV [29] have also been used. On the other hand, the use of splitless injection has the limitation of low sample capacity (up to 2 μL). However, non-volatile co-injected compounds can be retained in the liner altering the sensitivity and usually an adsorbent is placed on the liner (i.e. carbon). To avoid this limitation of the splitless mode, on-column or PTV injectors can be an alternative because they allow a sample volume of up to 5 μL.

Typically, common detection systems such as ECD or FTD have been used for coupling to GC. In particular, ECD has been applied for the detection of halogenated compounds (i.e. Chlorothalonil, Dichlofluanid) in environmental samples offering high sensitivity and good reproducibility. However, interference can be frequently observed as well as low identification capability. MS detectors provide unambiguous component identification and also the use of library spectra. In this sense, the use and acceptance of MS systems has increased over the last years. The number of applications using GC-MS is the result of the efficiency of GC separation as well as the good qualitative information and high sensitivity of MS systems (Table 1). GC-MS methods have been developed for the simultaneous determination of all GC amenable booster biocides in seawater and sediment samples; also, some publications include the simultaneous determination by GC-MS of degradation products, M1 (2-methylthio-4-tert-butylamino-6-amino-s-triazine) which is a stable transformation product of Irgarol 1051 [15] or DMSA (*N′*-dimethyl-*N*-phenyl-sulphamide), a degradation product of Dichlofluanid [34]. Applications of GC-MS using ionization techniques based on EI, negative chemical ionization (NCI) and positive chemical ionization (PCI) are discussed in the following sections for the booster biocides, which have been the focus of attention in several publications as a consequence of their frequency of detection in the marine environment.

3.2 Irgarol 1051

Analysis of Irgarol 1051 in marine waters and sediments has normally been focused on survey studies for determining in particular the presence of this compound and its main degradation product (M1) or the simultaneous presence of other booster biocides or organotins. The presence of Irgarol 1051 has been well documented reflecting its wide distribution in coastal and estuarine waters, harbours and marinas [3] from the Côte d'Azur (France) [9], western Mediterranean [38], UK coastal [14], western Japan [15] to Biscayne Bay in Florida [16]. In addition, to measure environmental residue concentrations, analytical skills have also been used to investigate the degradation, environmental transport and fate of Irgarol 1051. Irgarol 1051 is a triazine herbicide and is amenable to analysis by GC and similarly, the degradation products of triazines, such as M1, are generally also amenable to separation by GC. However, LC separation and APCI ionization are also favoured as an analytical approach, in particular for M1, compared to its GC analysis [44].

The typical EI GC-MS spectrum of Irgarol 1051 has a base peak M^+ ion and abundant fragment ions that correspond to the $[M-NC(CH_3)_3]^+$ (m/z, 182) and the $[M-CH_3]^+$ (m/z, 238) ions (Table 3). Most of the triazines that have methyl, ethyl or larger alkylamino groups as ring substituents, have either base peaks or major ions caused by the loss of a small alkyl radical from the M^+ ion by the alpha-cleavage reaction [39]. The three GC-MS ions have a relative abundance of 98% at m/z 182, 75% at m/z 238 and a base peak at m/z 253. The spectrum operating in PCI mode, exhibits the characteristic M^+ ion as a base peak and a fragment ion that corresponds to $[MH-NC_4H_7]^+$ (m/z, 198). In addition, adduct ions that correspond to $[M+C_2H_5]^+$ (m/z, 282) and $[M+C_3H_5]^+$ (m/z, 294) were observed with low abundance (10–27%). With the NCI GC-MS spectrum, however, the $[M-H]^-$ ion is the base peak (m/z, 252) and only one low-abundance molecular ion $[M]^-$ (m/z, 253) is also observed. In this sense, considering the identification power that is offered by the EI spectrum on the basis of the number of fragment ions and relative abundance, EI has been used by most authors. The lack of mass spectrum libraries for CI conditions is also a limitation, especially for the identification of degradation products.

From the point of view of sensitivity, pre-concentration of the samples is a decisive step to achieve lower limits of detection (sub-to-low ppt). A common pre-concentration procedure included in the development of analytical methods using different detection systems (ECD, FTD, AFID, MS) that has

Table 3 Main ions for antifouling booster biocides (Irgarol 1051, Chlorothalonil, Dichlofluanid, TCMTB and Sea-nine 211) observed in GC methods and MS modes of operation using EI, NCI and PCI ionization techniques

Antifouling booster biocides	Ionization mode	Mw	Molecular ion/Precursor ion/main fragment (m/z)	MS/MS-MS operation	Fragment ions (m/z)/product ions (m/z)	Refs.
Irgarol 1051	EI	253	253 $[M]^+$	SIM	182 $[M-NC(CH_3)_3]^+$, 196 $[M-C(CH_3)_3]^+$, 238 $[M-CH_3]^+$	[20, 22, 26] [28, 29, 33]
	NCI		252 $[M-H]^-$	SIM	253 $[M]^-$	[28]
	PCI		254 $[M+H]^+$	MS/MS	198 $[MH-NC_4H_7]^+$, 282 $[M+C_2H_5]^+$, 294 $[M+C_3H_5]^+$	[27, 28]
Chlorothalonil	EI	264	266 $[M+2]^+$	SIM	229 $[M-Cl]^+$, 264 $[M]^+$, 268 $[M+4]^+$ (cluster)	[26, 28]
	NCI		266 $[M+2]^-$	MS/MS	229 $[M-Cl]^-$, 231 $[M+2-Cl]^-$	[27]
	NCI		266 $[M+2]^-$	SIM	264 $[M]^-$, 230	[28]
	PCI		267 $[M+2+H]^+$	SIM	231 $[M+2-Cl]^+$, 295 $[M+2+C_2H_5]^+$, 307 $[M+2+C_3H_5]^+$	[28]
Dichlofluanid	EI	332	123 $[PhNS]^+$	SIM	167 $[PhNSO_2]$, 224 $[M-(CH_3)_2NSO_2]^-$, 332 [M]	[20, 22] [26, 28]
	NCI		199 $[MH-SCCl_2F]^-$	MS/MS	155 $[M-SCCl_2F]^-$, 91 $[MH-SCCl_2FSO_2N(CH_3)_2]^-$	[27]
	NCI		99	SIM	155 $[M-SCCl_2F]^-$, 199 $[MH-SCCl_2F]^-$	[28]
	PCI		201	SIM	99, 224 $[M-(CH_3)_2NSO_2]^+$, 313	[28]
Sea-nine 211	EI	282	246 $[M+H-Cl]^-$, 169 $[M-C_8H_{17}]^-$	SIM	169 $[M-C_8H_{17}]^-$, 182 $[M-C_7H_{14}]^-$, 283 [M + H]	[26, 28]
	NCI		245 $[M-H-Cl]^-$	MS/MS	160 $[M-C_6H_{14}]^-$, 162 $[M+2-C_6H_{14}]^-$	[27]
	NCI		281 $[M-H]^-$	SIM		[28]

Table 3 (continued)

Antifouling booster biocides	Ionization mode	Mw	Molecular ion/Precursor ion/main fragment (m/z)	MS/MS-MS operation	Fragment ions (m/z)/product ions (m/z)	Refs.
	PCI		282 $[M]^+$	SIM	284 $[M+2]^+$, 310 $[M+C_2H_5]^+$, 322 $[M+C_3H_5]^+$, 170 $[MH-C_8H_{16}{}^+]$	[28, 36]
TCMTB	EI	238	180 $[M-SCN]$	SIM	238 $[M]$, 136 $[M-SHCH_2SHCN]^-$	[28]
	NCI		166 $[M-CH_2SCN]^-$	MS/MS	134 $[M-SCH_2SCN]^-$	[27]
	NCI		166 $[M-CH_2SCN]^-$	SIM	58 $[SCN]^-$	[28]
	PCI		182 $[M-SHCNH]^+$	SIM	136 $[M-SHCH_2SHCN]^+$	[28]

been reported is the SPE method (Table 1). In the group of detection techniques such as ECD, FTD and AFID applied to marine water samples, the detection limits are at the low ng L^{-1} level (2–20 ppt) [21, 37]. For sediments, the available documentation which is related to AFID, show detection limits at the ng g^{-1} level (50 ppb) [37].

Concerning Q-MS techniques for water analysis, the sensitivity can be considered similar to selective detectors (ECD). The detection limits obtained in EI ionization and SIM mode are at the ppt level (0.5–10 ppt) [9, 20, 28]. Sub-to-low ng L^{-1} levels (0.1 – 1 ppt) are also the reported detection limits using IT-MS in MS/MS mode [29]. To obtain optimal sensitivity in the analysis of antifouling compounds, most of the developed analytical methods have used the EI ionization technique. Alternatives such as CI in both positive and negative modes have been evaluated to give an idea of the sensitivity obtained with ionization techniques [27, 28]. NCI is not a suitable ionization technique for Irgarol 1051, the sensitivity decreases by four orders of magnitude with respect to EI. In addition, analysis in the PCI mode does not significantly increase the sensitivity, and PCI has a detection limit at the ng L^{-1} level (50 ppt). SIM procedures have also been preferred with the objective of achieving very low residue determinations because remarkable differences have been observed between sensitivities obtained by full scan and SIM modes. The SIM mode provides response factors from 10 to 150 times higher than full scan mode (see Fig. 2, where sensitivity obtained with EI and NCI techniques was evaluated).

When extraction methods such as SPME, HS-SPME or LLE have been used, comparable sensitivity has been obtained in GC-MS water analyses (30; 50; 0.5–2 ppt respectively) [2, 16, 22, 24, 33]. On the other hand, sensitivity in sediment sample analysis with extraction methods and GC-MS analysis

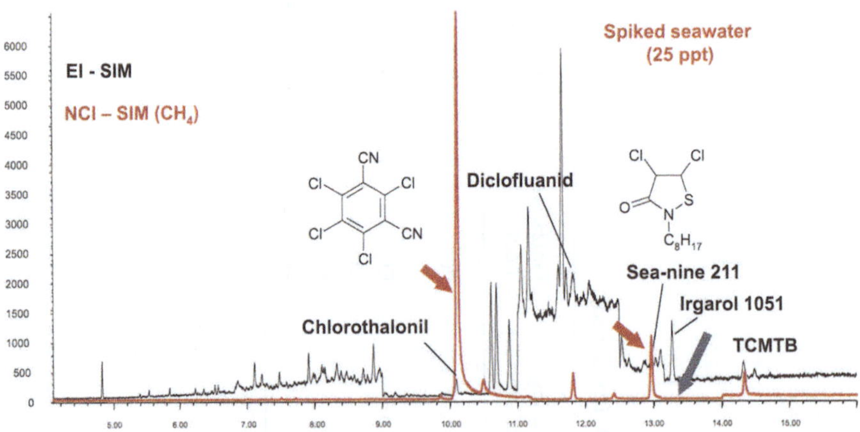

Fig. 2 GC-EI-MS and GC-NCI-MS chromatograms corresponding to a spiked seawater sample at a concentration of 25 ng L^{-1}

(SPME, MAE, SFE, soxhlet or LLE previous sonication) is typically at the $\mu g\,L^{-1}$ level (0.5–8 ppb) [25, 26, 35, 38].

3.3 Chlorothalonil

The detector most frequently used for chlorinated compound analysis has been the ECD [23], but MS has become common because of its high sensitivity, versatility and selectivity [14, 26–28, 35]. Thus, selective detectors have been progressively replaced by GC-MS mainly using EI and CI ionization techniques. EI has been widely used as the ionization method and most of the analytical methods published for the determination of Chlorothalonil as an antifouling compound in water and sediment samples, have been developed using this ionization technique [14, 26, 28, 35]. However, NCI techniques are also suitable for the analysis of organochlorine compounds because they offer higher sensitivity, with methane as the reagent gas, similar to ECD [27]. An improvement in sensitivity has also been observed when NCI and PCI have been compared. The enhancement of the signal-to-noise (S/N) ratio has been estimated at 10–20 times, with respect to PCI or EI [27]. This is well-known behaviour for chlorinated compounds.

Moreover, the use of IT-MS has allowed for improvements in the selectivity and the best sensitivity conditions in NCI have been obtained in the MS/MS mode. Limits of detection down to sub-low $ng\,L^{-1}$ (0.05 ppt) have been obtained in MS/MS mode [27] and lower sensitivity operating in the SIM mode (approx. 10 times lower than for the MS/MS mode with DL of 0.5 ppt) [28] (Table 1).

Using methane as the reagent gas, the predominant peak in the mass spectrum under NCI conditions was m/z 266 for $[M+2]^-$, one of the ions from the cluster generated by the chlorine isotopes (Table 3). This ion is selected as the precursor ion to be fragmented in product ions, which are indicative of the structure of the analyte. A product ion with m/z 229 was formed, together with a fragment ion at m/z 231, also present in the NCI-MS/MS spectrum. These ions have a relative abundance of approximately 100 and 90%, respectively, and correspond to $[M-Cl]^-$ and $[M-2-Cl]$. The parent ion is also present in the MS/MS spectrum with a relative abundance between 20–30%. MS-MS conditions are usually adjusted to provide unequivocal identification criteria based on the presence of product and precursor ions in the spectra. The EI-MS spectrum also offers suitable information for the identification of Chlorothalonil. However, in environmental samples, such as sediment samples, a high background level can make difficult the confirmation of a compound by EI-MS detection, this is usually eliminated when MS/MS conditions are applied. In the EI-MS spectrum, the base peak is also the ion m/z 266 $[M+2]^-$, and ions corresponding to m/z 264 and 268, which have a relative abundance of 76 and 49%, respectively. In addition a fragment ion corresponding to the loss of a Cl atom is present at m/z 229 with low abundance.

3.4 Dichlofluanid

The analysis of Dichlofluanid is also amenable to GC and selective detectors such as ECD or CI ionization techniques that are suitable for achieving an improvement in the sensitivity of the analytical method. In spite of this, the EI ionization technique has been applied for the determination of Dichlofluanid in water or sediments in several publications [14, 20, 22, 26, 34, 35] (Table 1). In fact, although increased sensitivity is expected for NCI techniques, the reported limit of detection obtained by EI and NCI techniques are comparable, both are at the low $ng\,L^{-1}$ level (usually between 1.5 to $5\,ng\,L^{-1}$) in water samples, taking into account the SPE pre-concentration step applied [27, 28]. For sediment samples, the sensitivity is lower, with detection limits of $\mu g\,L^{-1}$ (1–11 ppb) using other extraction methods such as SPME or LLE with sonication [26, 35]. Under EI conditions, the spectrum does not have a molecular ion at m/z 332 as a peak base but Dichlofluanid shows an abundant fragment ion (100%) at m/z 123 that is assigned to the $[PhNS]^+$ ion. Two fragment ions at m/z 224 and 167 characterize the EI spectra together with a low abundance of the molecular ion $[M]^-$ (m/z, 332). A fragment ion at m/z 224 that has an abundance of 60% originates from the loss of the $(CH_3)_2NSO_2{}^-$ radical (Table 3).

On the basis of the degradation of Dichlofluanid to the degradation product DMSA (*N′*-dimethyl-*N*-phenyl-sulphonamide), analytical methods have been developed for their simultaneous determination. The hydrolytic degradation rate of Dichlofluanid is very high, with a half-life in seawater of several hours and no accumulation of Dichlofluanid in sediments is expected. However, scarce documentation is available about the methodology for the analysis of DMSA [34]. Hamwijk et al. (2005) developed an analytical method for DMSA and Dichlofluanid for sediments using GC-MS under EI conditions. Three ions characterize the EI spectrum of DMSA. Among these, the ion at m/z 200 is used for quantification and the ions at m/z 92 and 108 are used for confirmation of DMSA.

NCI-MS/MS analysis has also been developed for the analysis of Dichlofluanid [27], where the base peak is an abundant fragment ion at m/z 199 using methane as the reagent gas. The fragmentation of this ion, which is selected as the precursor ion provides an NCI-MS/MS spectra with suitable information for confirmation purposes. The fragment ions obtained are at m/z 91 and 155, which correspond to the $[M-SCCl_2F]^-$ and $[MH-SCCl_2FSO_2N(CH_3)_2]^-$ ions, respectively. The relative abundance of the precursor and product ions in the MS/MS spectra is also a useful criteria for the confirmation.

3.5 Sea-nine 211

Sea-nine 211 is an antifouling agent whose active ingredient is a halogenated compound with two Cl atoms in its structure (4,5-dichloro-2-*n*-octyl-4-isothiazolin-3-one). Similarly to Dichlofluanid and Chlorothalonil, the analysis with ECD or NCI is expected to have an improved sensitivity with respect to the EI technique. However, analytical methods developed with NCI and EI techniques in the SIM mode do not show significant differences, with detection limits at the same order of magnitude (1.5–2.5 $ng\,L^{-1}$) and using a pre-concentration step (SPE) [28]. With ECD, the sensitivity is equivalent to NCI and offers detection limits at the low-$ng\,L^{-1}$ level (2 ppt) or sub-low-$ng\,L^{-1}$ level (0.2 ppt) depending on the extraction method (SPME and SME, respectively) [21, 23] in water samples. For sediment samples the developed analytical methods applying SPME or LLE with shaking procedures have a detection limit at ppb [26, 36], so the lower sensitivity can be related to the extraction method (Table 1).

The EI mass spectra shows three ions (between 80 to 100% of relative abundance) at m/z 169 as the base peak that corresponds to the $[M-C_8H_{17}]^-$ ion; the other two ions are at m/z 246 and 182. These fragment ions can be attributed to the loss of a Cl atom given the $[M-Cl]^-$ ion and the loss of the alkyl chain obtaining the $[M-C_7H_{14}]^-$ ion, respectively. The M^- ion with m/z 283 has a lower abundance (approx. 11%). A total of four ions characterize the EI spectrum, which offers capabilities for quantification and confirmation purposes. In addition, the PCI technique using methane, gives enough spectral information to confirm the detection of this compound, providing $[M+H]^+$ at m/z 284 and adduct ions at m/z 310 and 322 [28]. In addition, the sensitivity using PCI, in the MS/MS mode, is similar to that obtained with NCI or EI in the SIM mode [28, 36]. However, when NCI-MS/MS is applied, the sensitivity is improved with respect to NCI-MS in SIM mode obtaining detection limits of 0.05 $ng\,L^{-1}$ and 1.5 $ng\,L^{-1}$, respectively. In MS/MS experiments, the fragmentation of the precursor ion at m/z 245 gives two main fragment ions at m/z 160 and 162, which correspond to the cluster $[M-ClH]^-$ ions (Table 3).

3.6 TCMTB

The analysis of TCMTB (2-thiocyanomethylthiobenzothiazole) has been mainly focused on its determination in marine waters using MS detectors. In spite of this compound being suitable for separation by LC, analytical methods using GC separation have been developed with good sensitivity for detection in environmental samples [27, 28]. The ionization techniques, NCI and EI that have been applied have comparable sensitivity, when the SIM mode was used. Low ng L^{-1} levels are obtained as detection limits (1.5–3 ppt) considering the pre-concentration step of the marine water samples (Table 1). The EI spectrum does not provide as the base peak the M^- ion, which has an abundance of 25%. Most of the fragment ions obtained under these conditions, have a low abundance, between 20–30% for the m/z. The base peak is at m/z 180, which is the $[M - SCN]^-$ ion. The m/z 136 can be attributed to $[M - SHCH_2SHCN]^-$. When MS/MS conditions have been applied, improvement in the selectivity and sensitivity are obtained achieving detection limits of 0.05 ppt, also after SPE treatment. However, with NCI-MS/MS conditions and using methane, low fragmentation is observed from the precursor ion at m/z 166 $[M - CH_2SCN]^-$, giving only one fragment ion at m/z 134 that is the $[M - SCH_2SCN]^-$ ion [27, 28] (Table 3).

4 Liquid Chromatography-Mass Spectrometry

4.1 General Remarks

As has been previously noted, GC is common technique, but the amenability of some booster biocides to be sperated by LC, has also led to the application of LC analysis, as alternative of GC with a derivatization process. To avoid the need for derivatization, several applications utilizing LC analysis have been reported [19, 51] (Table 2). Reversed-phase LC (RPLC) with an octadecylsilane stationary phase (C18) has been used for the separation of these compounds in most of the studies in the literature [19, 30, 42, 45, 47, 48]. The mobile phase generally consists of acetonitrile or methanol together with water. However, modification of the mobile phase has been performed to improve the sensitivity of MS detection. Acid acetic or ammonium acetate is added to enhance the ionization, in particular of Irgarol 1051 and Diuron [32, 44, 51]. Gradient elution is performed starting from a low percentage of organic solvent, generally 20–30% and increasing linearly to 100%. With this gradient, a time of 10–25 min is suitable to separate mixtures of compounds and degradation products. The column and pre-column are maintained at 30–50 °C with a block heater. To improve the detectability of the target analytes, the volume of injection is typically increased from 20 to 50 μL.

The coupling of LC to DAD and mainly MS detectors is frequently used for the analysis of phenylureas or triazines [1, 19, 45]. The identification by LC-DAD is accomplished on the basis of the retention time, through comparison of the UV spectrum of the target compounds in a standard solution with the UV spectrum of the detected peak in the sample. Therefore, a limited identification capability can be achieved by UV detectors. The working wavelengths are 224–230 nm that correspond to the maximum absorbance of Irgarol 1051 and it is usually selected for its main degradation product M1 [42, 44, 45, 48, 53]. For diuron, the DAD is set at 244 nm [44, 45, 53]. In the simultaneous determination of various biocide compounds (Diuron, TCMTB, Irgarol 1051 and Chlorothalonil), quantification is carried out with UV detection at 220 nm [19].

For further identification purposes by LC-MS analysis, different ionization techniques—electrospray source ionization (ESI) and atmospheric pressure chemical ionization (APCI)—ionization modes (negative ion NI and positive ion PI) and monitoring modes, SIM and selected reaction monitoring (SRM), have been employed. LC-MS techniques have been instrumental in detecting mixtures of biocides and degradation products [46, 47]. One of the limitations of LC-MS/MS is the susceptibility of interfaces to co-extract matrix components from complex samples that can result in the suppression or less frequently in the enhancement of the analyte signal. However, it is also demonstrated that these effects can be minimized by good sample preparation, good chromatographic separation and optimized MS/MS operating conditions. Analytical methods using MS techniques are outlined for each compound in the following sections. Special attention is paid to the compounds Irgarol 1051 and Diuron, which have been of concern in several reports.

4.2 Irgarol 1051

For the determination of triazines, better sensitivity is often achieved using APCI in PI mode. A higher response is observed as a result of the easy protonation of the amine nitrogen atom. In this sense, most of the LC-MS methods have been reported using the APCI interface for the determination of Irgarol 1051 as well as its main degradation product M1 [19, 46–48, 51] (Table 2). This degradation product has been observed in different degradation experiments such as photodegradation under controlled conditions or biodegradation experiments by Phanerochaete chrysosporium [40, 41].

Although APCI techniques are soft-ionization modes, ions can be sometimes fragmented to produce structural information about a particular molecule. Fragmentation can be performed with a single quadrupole by increasing fragmentor or cone voltage. This voltage affects the transmission and fragmentation of molecules, which are related. Thus, with a high fragmentor voltage, more fragmentation occurs; in the case where a compound is not fragmented, better ion transmission can often succeed. The application of a low fragmentor value (70–75 V) commonly provides a simple mass spectrum characterized by the presence of the protonated molecule $[M + H]^+$ (Table 4). So this fragmentor value does not provide useful structural information. At a high fragmentor voltage (120–125 V), mass spectra provide good structural information and also enough sensitivity. Under these fragmentation conditions, in addition to the $[M + H]^+$ ion, the spectra shows a fragment

Table 4 Main ions for antifouling booster biocides (Irgarol 1051, Chlorothalonil, Dichlofluanid, TCMTB, Sea-nine 211 and Diuron) observed in LC methods and MS modes of operation using ESI and APCI ionization techniques

Antifouling booster biocides	Ionization mode	Molecular weight	Molecular ion/Precursor ion (m/z)/main fragment	MS/MS-MS operation	Fragment ions (m/z)/ product ions (m/z)	Refs.
Irgarol 1051	APCI (+)	253	254 $[M+H]^+$	MS (SIM)		[46, 47]
	APCI (+)		198 $[MH-C_4H_8]^+$	MS (SIM)	254 $[M+H]^+$	[19, 48, 51]
	ESI (+)		254 $[M+H]^+$	MS-MS (MRM)	198 $[MH-C_4H_8]^+$	[32]
Chlorothalonil	APCI (-)	264	245 $[M+OH-HCl]^-$	MS (SIM)		[19, 47]
Dichlofluanid	APCI (-)	332	155 $[M-SCClFN(CH_3)_2]^-$	MS (SIM)	199 $[M-SCClF]^-$	[19]
	APCI (-)		199 $[M-SCCl_2F]^-$	MS (SIM)		[47, 52]
	APCI (-)		91 $[MH-SCCl_2FSO_2N(CH_3)_2]^-$	MS (SIM)	155 $[M-SCClFN(CH_3)_2]^-$ 199 $[M-SCCl_2F]^-$	[51]
Sea-nine 211	APCI (+)	281	282 $[M+H]^+$	MS (SIM)	284 $[M+H+2]^+$	[31, 46, 47, 52]
TCMTB	APCI (-)	238	166 $[M-CH_2SCN]^-$	MS (SIM)		[19, 31, 47]
Diuron	APCI (-)	232	231 $[M-H]^-$	MS (SIM)	186 $[M-H-HN(CH_3)_2]^-$	[46]
	APCI (+)		233 $[M+H]^+$	MS (SIM)	235 $[M+H+2]^+$	[31, 47, 52]
	ESI (+)		233 $[M+H]^+$	MS-MS (MRM)	72 $[MH-C_6H_5Cl_2N]^+$	[32]
	APCI (+)		72 $[MH-C_6H_5Cl_2N]^+$	MS (SIM)	233 $[M+H]^+$	[19, 51]

ion corresponding to $[MH - C_4H_8]^+$ at m/z 198 [48, 51] with a relative abundance of approx. 70%. However, it is often the case that at the high voltages that result in more structural information, there is a simultaneous decrease in sensitivity.

For the determination of the degradation product of Irgarol, M1, a similar behaviour has also been observed in the optimization of the best conditions of fragmentation. So with a fragmentor voltage of 120 V, the mass spectra is characterized by two ions, a base peak that is the $[M + H]^+$ at m/z 214 and a less abundant ion at m/z 158 (30% relative abundance). This ion is the loss of the tert-butyl group, which is a typical fragment of triazine compounds. With this information, the presence of M1 or Irgarol 1051 in marine water and sediment samples is often confirmed by an identical match of retention time and the spectrum of a standard [48, 51].

Other current mass spectrometry approaches have been reported [32] to confirm the presence of Irgarol 1051 in complex environmental matrices. In spite of the structural information obtained with a single quadrupole by increasing fragmentor voltage, it is more selective to perform fragmentation with tandem MS to achieve the CID of a specified precursor ion. MS-MS analysis performed in analyzers such as triple quadruple (QqQ-MS) show an

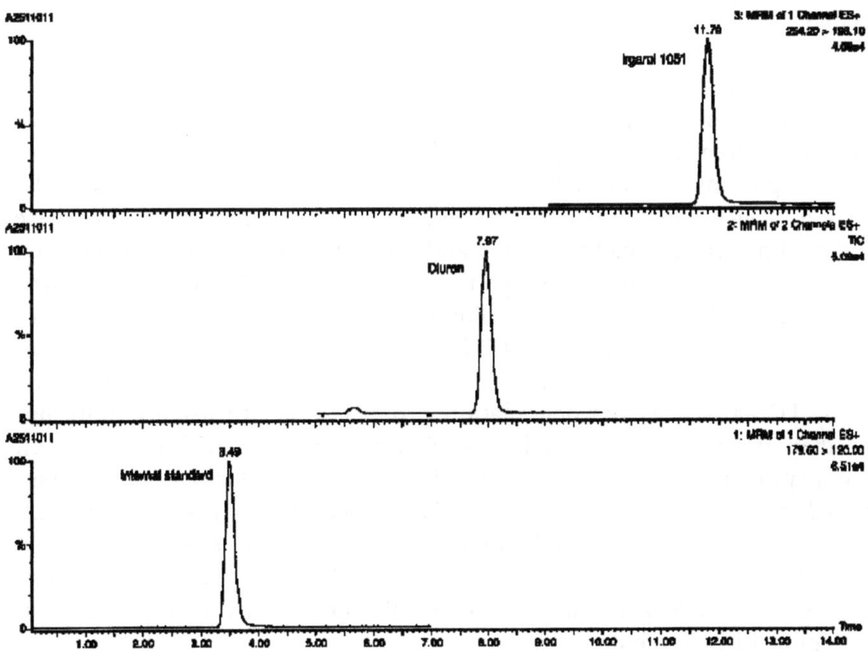

Fig. 3 LC-ESI-MS/MS analysis MRM chromatograms corresponding to the detection of Irgarol 1051 and Diuron in marine water samples. MRM transitions monitored for Irgarol 1051 (254 → 198) and Diuron (233 → 72 and 233 → 46)

increased selectivity. This is based on its two stages of mass analysis—one to pre-select an ion and the second to analyze fragments induced. The setting of the MRM transitions channel for the monitoring of target analytes is generally selected considering the signal intensities and structure-specificities of the product ions. Figure 3 shows a chromatogram corresponding to the monitored MRM transitions for Irgarol 1051. In the analysis of Irgarol 1051 by LC-QqQ-MS two main ions characterize the mass spectrum and one transition, which corresponds to the $[M+H]^+ \rightarrow [MH-C_4H_8]^+$ ions, can be monitored [32] (Table 4). Although the confirmation criteria of Irgarol 1051 can be based on the monitoring of one MRM transition and the retention time (in environmental analysis), two transitions provide greater capacity for confirmation of target analytes. Typically, the first is used for quantification and the second for confirmation.

Regarding the sensitivity achieved with enrichment through the SPE step, ESI-MRM analysis was found to afford detection limits for Irgarol 1051 one order of magnitude higher than those achieved with APCI-MS in the SIM mode. In the ESI-MRM mode the limits of detection are at the sub-low ng L^{-1} level (0.5 ppt), while in APCI and SIM mode, the limits of detection are at the low ng L^{-1} level (4–5 ppt) [32, 47, 51].

4.3 Diuron

Diuron is a phenyl-substituted urea herbicide whose analysis has been described in several publications using LC-DAD or LC-MS. However, it is known that some urea herbicides generally do not have extremely strong UV or visible light absorption that would allow selective and high-sensitivity detection. The application of GC to phenyl-ureas is difficult because these compounds are thermally unstable and rapidly degrade to isocyanates and amines. Derivatization to compounds more thermally stable will make them amenable to GC analysis. For confirmatory and quantitative trace analysis of Diuron, the use of LC-MS is appropriate and several applications have been reported. Phenylurea derivatives can be determined easily by LC as well as the degradation products that are usually more polar than its parent compound. LC-MS has been developed for the main degradation products of Diuron resulting from transformation processes in sediments and water: DCPMU [1-(3,4-dichlorophenyl)-3-dimethylurea], DCPU [1-(3,4-dichlorophenyl) urea], DCA [3,4-dichloroaniline], CPDU [1-(3-chlorophenyl)-3,1-dimethylurea] [45, 46, 53]. Other degradation products, such as demethyldiuron, 1-(3,4-dichlorophenyl)urea, have also been determined in sediments [47, 52].

Structural information is obtained by LC-MS with APCI and ESI interfaces that allow the soft ionization of the analytes in both positive (PI) or negative (NI) modes (Table 4). Using single quadrupole MS detection, APCI has become the interface more generally applied in the analysis of Diuron, providing sensitive, robust and accurate methods [31, 46, 47, 52] while few works have reported on the use of ESI [32]. The degree of confidence in identification can be enhanced by using a single quadrupole combined with collision-induced dissociation (CID). CID is obtained depending on the LC-MS system by applying an octapole or an orifice voltage that fragments the molecules. Under such conditions, the full-scan mode provides structural information which is generated in the mass spectra from protonated molecules and fragment ions. To evaluate the best mass spectral data, generally, flow injection analysis is performed at different fragmentor voltages (usually from 20 to 100 V) to find the maximum response under the optimum ESI or APCI conditions. The base peak observed for Diuron when it is analyzed by APCI in PI mode, in most of the reports, corresponds, as was previously mentioned to the protonated molecule $[M+H]^+$ at m/z 233 [31, 47, 52]. With the NI mode, the deprotonated $[M-H]^-$ molecule is the predominant peak in the spectra, at m/z 231 [46]. Under these conditions, typically low rate fragmentation is obtained. For example, the ion at m/z 235 and fragment ion at m/z 186 are obtained in both PI and NI modes, respectively [31, 46]. The fragment ion in NI mode can be assigned to $[M-H-HN(CH_3)_2]^-$. On this basis, diuron is identified using the retention time and the monitoring of ions (SIM mode). These criteria have also been applied for the identification of its degradation products (i.e.: DCPMU m/z, 221; DCPU m/z, 205 + 207; CPMU m/z, 199) [50].

For trace determinations, SPE is the sample preparation method of choice for enrichment of pollutants from water samples [19, 30–32, 47, 50–52]. Comparable sensitivity has been reported with the SPE-LC-MS (SIM mode) or SPE-LC-MS/MS (MRM mode) methods that have a limit of detection at the $ng\,L^{-1}$ level (0.1–10 ppt) [32, 52] (Table 3).

In spite of the identification power of LC-MS systems, this technique has its limitations due to the low rate of ion fragmentation, and in this sense, MS-MS detection is of great interest. This is particularly true for the identification of trace amounts of compounds in environmental samples or in the determination of the structure of degradation products. In LC-ESI-MS analysis, the predominant ion that corresponds to the protonated molecule $[M+H]^+$ is used as the precursor ion in the MRM mode. In the fragmentation of Diuron, the major product ions observed are at m/z 72 and 46. Therefore, in the analysis of Diuron, two MRM transitions $233 \rightarrow 72$ and $233 \rightarrow 46$ can be monitored as confirmatory criteria for identification purposes as well as the retention time [32]. Figure 3 shows a chromatogram corresponding to the monitored MRM transitions for Diuron. The first MRM transition was selected for quantification and the second one for confirmation.

4.4
Chlorothalonil

Few applications using LC-MS have been reported in the literature. To the best of our knowledge this technique has scarcely been applied for the determination of Chlorothalonil in marine waters and sediments. Limited information is also available for the determination of the biocide boosters, that are reviewed in this chapter, by LC-MS analysis (Table 3).

For structural information of Chlorothalonil, as is usual in optimization procedures, the operational parameters of interfaces and fragmentor voltage are optimized. APCI has been reported as the selected interface for the analysis of Chlorothalonil. Typically, the capillary voltage and the corona current are the most relevant operational parameters of APCI that are observed in both PI and NI modes. Comparing both operation modes, Chlorothalonil has been shown to exhibit a higher sensitivity under PI mode than under NI mode. Contrary to compounds such as Diuron or Irgarol 1051, that show a low fragmentation rate at low fragmentor voltage (70–75 V), the mass spectra of Chlorothalonil is characterized by the presence of two predominant ions. One of them is the molecular ion $[M]^-$; in addition the spectra show a fragment ion at m/z 246. This ion, which has a high abundance, corresponds to $[M + OH - HCl]^-$ (Table 4). An interpretation of this pattern of fragmentation is that Chlorothalonil is known to degrade under high temperature conditions leading to the substitution of a Cl atom by the hydroxyl group in the aromatic ring [19, 47]. Under SIM conditions and SPE pre-concentration, it is feasible that the analytical method can detect Chlorothalonil at the ng L^{-1} level (1–2 ppt) in seawater samples [19, 47].

4.5
Dichlofluanid

Data concerning the determination of Dichlofluanid by LC-MS using the APCI interface have been published [19, 47, 51, 52]. On-line SPE methods have been developed for trace detection in seawater samples and sediments with limits of detection of 4–5 ng L^{-1} and 1.6 ng g^{-1}, respectively [19, 51] (Table 3).

The optimum conditions of sensitivity were working with APCI in negative mode. To study the structural information for the determination of Dichlofluanid, different fragmentor voltages have been applied, in the range of 25–200 V. When fragmentor voltage values were applied in the range 70–110 V, Dichlofluanid suffers an appreciable fragmentation. The pattern of fragmentation shows a main ion at m/z 199, which has been assigned to $[M - SCCl_2F]^-$ (Table 4). This result is in agreement with most of the reports where fragmentation experiments for Dichlofluanid have been carried out [19, 47, 52]. An additional fragment ion at m/z 155 with a low relative abundance (10%) is also reported to be present in the mass spectra, with a fragmentor voltage of 75 V [51]. A higher contribution (34%) is shown by this ion, when the fragmentor voltage value is 120 V. Under these fragmentation conditions, useful structural information is obtained without loss of sensitivity. A predominant ion at m/z 91, which corresponds to the $[M - SCCl_2FSO_2N(CH_3)_2]^-$ ion together with another two ions at m/z 199 and 155, are shown in the spectra. For the m/z 155, the ion $[M - SCCl_2FNCH_3]^-$ has been assigned [51].

4.6 Sea-nine 211

For optimal conditions to determine Sea-nine 211 by LC-MS using APCI, the PI mode is often chosen. This compound easily accepts a proton in the chemical structure, so that, it exhibits a higher sensitivity under PI mode [46, 47, 52] (Table 4). However, under different fragmentation conditions (70, 110 and 120 V), no fragmentation is often observed for Sea nine 211 and only the molecular ion $[M + H]^+$ is present in the mass spectra at m/z 282. In spite of the poor structural information provided by the mass spectra, this compound has been generally identified and quantified in seawater samples by monitoring of the protonated molecular ion $[M + H]^+$ [46, 47, 52]. With this methodology, levels of $ng\,L^{-1}$ can be detected (with a detection limit of 1 ppt) in seawater. For sediments, the sensitivity is at the $ng\,g^{-1}$ level (Table 3).

4.7 TCMTB

For the selection of the mode of ionization using the APCI interface—the usual procedure—PI and NI modes have been studied for TCMTB [31] (Table 3). The NI mode is much more sensitive than the PI mode that produces a protonated molecule at m/z 239. However, the NI mode does not yield a deprotonated molecule $[M - H]^-$. Under the application of different fragmentor voltage values (70–110 V), TCMTB has a similar behaviour showing total fragmentation that can yield an intense characteristic ion at m/z 166. It is suggested that this fragment is formed due to the loss of the thiocyanomethylthio substituent [31] that corresponds to the $[M - CH_2SCN]^-$ ion (Table 4). Higher fragmentor voltages have also been applied (120 V), but no further structural information is obtained. Similar to compounds with a low fragmentation rate, such as Sea-nine 211, the SIM mode for m/z 166 and the retention time, are the identification criteria used for their determination in environmental samples. Comparable sensitivity is also obtained with the developed analytical method for the determination of TCMTB, usually at the $ng\,L^{-1}$ level (1–20 ppt) [19, 31, 47].

5 Conclusions and Future Trends

The application of mass spectrometry techniques has paid special attention to the analytical skills related to the achievement of good sensitivity for control, quantification of these contaminants as well as their identification including the main degradation products. Pre-concentration of antifouling booster biocide residues is a decisive step for the determination of trace levels. The extraction efficiency of LLE, SPE and current techniques involving SPME, SME, MAE or SFE makes these a feasible approach for booster biocide analysis. LC-MS methods have experienced progress in terms of application in the analysis of booster biocides, and are especially suitable for environmental sample analysis for routine confirmation of identity as well as structural elucidation or ultra trace analysis. However, up to now, the development of methods with current instrumentation such as MS/MS for antifouling compounds, which are amenable to separation by GC or LC, has not profusely applied. However, we can expect an increase in the use of these LC-MS tandem techniques as well as LC-TOF-MS as methods to improve identification and quantification performance.

Acknowledgements This work has been supported by Spanish Ministerio de Ciencia y Tecnología (EVITA project). M.D. Hernando acknowledges the research contract from Consejería de Educación y Ciencia, Junta de Andalucía (Spain).

The authors thanks the permission to reproduce from Elsevier the reprinted from Journal of Chromatography A, 970, M.H. Lamoree, C.P. Swart, A. van der Horst, B. van Hattum, Determination of diuron and the antifouling paint biocide Irgarol 1051 in Dutch marinas and coastal waters. Pages 183–190. Copyright 2002.

References

1. Thomas KV, Fileman TM, Readman JW, Waldock MJ (2001) Mar Pollut Bull 42:677
2. Connely DP, Readman JW, Knap AH, Davies J (2001) Mar Pollut Bull 42:409
3. Konstantinou IK, Albanis TA (2004) Environ Int 30:235
4. Díez S, Lacorte S, Viana P, Barceló D, Bayona JM (2005) Environ Pollut 136:525
5. Alzieu C, Thibaud Y, Heral M, Boutier B (1980) Revue de Travaux de l'Institut de Peche Maritime 44:301
6. Goldberg ED (1986) Environment 28:17
7. Maguire RJ (1987) Appl Organom Chem 1:475
8. Voulvoulis N, Scrimshaw MD, Lester JN (2002) Chemosphere 47:789
9. Tolosa I, Readman JW, Blaevoet A, Ghilini S, Bartocci J, Horvat M (1996) Mar Pollut Bull 32:335
10. Liu D, Pacepavicius GJ, Maguire RJ, Lau YL, Okamura H, Aoyama I (1999) Wat Res 33:2833
11. Thomas KV, McHugh M, Waldock M (2002) Sci Total Environ 293:117
12. Readman JW, Kwong LLW, Grondin D, Bartocci J, Villeneue J, Mee LD (1993) Environ Sci Technol 27:1940

13. Sargent CJ, Bowman JC, Zhou JL (2000) Chemosphere 41:1755
14. Voulvoulis N, Scrimshaw MD, Lester JN (2000) Mar Pollut Bull 40:938
15. Okamura H, Aoyama I, Ono Y, Nishida T (2003) Mar Pollut Bull 47:59
16. Gardinali PR, Plasencia M, Mack S, Poppell C (2002) Mar Pollut Bull 44:781
17. Scarlett A, Donkin ME, Fileman TW, Donkin P (1997) Mar Pollut Bull 34:645
18. Basheer C, Tan KS, Lee HK (2002) Mar Pollut Bull 44:697
19. Ferrer I, Barceló (1999) J Chromatogr A 854:197
20. Pocurull E, Brossa L, Borrull F, Marcé RM (2000) J Chromatogr A 885:361
21. Lambropolou DA, Sakkas VA, Albanis TA (2002) J Chromatogr A 952:215
22. Peñalver A, Pocurrull E, Borrull F, Marcé RM (1999) J Chromatogr A 839:253
23. Lambropolou DA, Albanis TA (2004) J Chromatogr A 1049:17
24. Lambropolou DA, Sakkas VA, Albanis TA (2002) Anal Chim Acta 468:171
25. Gatidou G, Zhou JL, Thomaidis NS (2004) J Chromatogr A 1046:41
26. Lambropolou DA, Sakkas VA, Albanis TA (2003) J Chromatogr A 1010:1
27. Hernando MD, Piedra L, Belmonte A, Agüera A, Fernández-Alba AR (2001) J Chromatogr A 938:103
28. Agüera A, Piedra L, Hernando MD, Fernández-Alba AR (2000) J Chromatogr A 889:261
29. Steen RJCA, Leonards PEG, Brinkman UATh, Cofino WP (1997) J Chromatogr A 766:153
30. Gatidou G, Kotrika A, Thomaidis NS, Lekkas TD (2005) Anal Chim Acta 528:89
31. Thomas KV (1998) J Chromatogr A 825:29
32. Lamoree MH, Swart CP, van der Horst A, van Hattum B (2002) J Chromatogr A 970:183
33. Tolosa I, Readman JW (1996) Anal Chim Acta 335:267
34. Hamwijk C, Schouten A, Foekema EM, Ravensberg JC, Collombon MT, Schmidt K, Kugler M (2005) Chemosphere 60:1316–1324
35. Albanis TA, Lambropoulou DA, Sakkas VA, Konstantinou IK (2002) Chemosphere 48:475
36. Steen RJCA, Ariese F, van Hattum B, Jacobsen J, Jacobson A (2004) Chemosphere 57:513
37. Biselli S, Bester K, Hühnerfuss H, Fent Karl (2000) Mar Pollut Bull 40:233
38. Carrasco PB, Díez S, Jiménez J, Marco MP, Bayona JM (2003) Wat Res 37:3658
39. Budde WL (2004) Mass Spectrom Rev 23:1
40. Peñuela G, Ferrer I, Barceló D (2000) J Environ Anal Chem 78:25
41. Liu D, Maguire RJ, Lau YL, Pacepavivius GJ, Okamura H, Aoyama I (1997) Wat Res 31:2363
42. Tóth S, Becker-van Slooten K, Spack L, de Alencastro LF, Tarradellas J (1996) Bull Environ Contam Toxicol 57:426
43. Penalva J, González-Martínez MA, Puchades R, Maqueira A, Marco MP, Barceló D (1999) Anal Chim Acta 387:227
44. Ferrer I, Ballesteros B, Marco MP, Barceló D (1997) Environ Sci Technol 31:3530
45. Gatidou G, Kotrikla A, Thomaidis NS, Lekkas TD (2005) Anal Chim Acta 528:89
46. Thomas VK, McHugh M, Waldock M (2002) Sci Total Environ 293:117
47. Maritnez K, Ferrer I, Barceló D (2000) J Chromatogr A 879:27
48. Ferrer I, Barceló D (2001) J Chromatogr A 926:221
49. Bou Carrasco P, Díez S, Jiménez J, Marco MP, Bayona JM (2003) Wat Res 37:3658
50. Thomas KV, Blake SJ, Waldock MJ (2000) Mar Pollut Bull 40:739
51. Gimeno RA, Aguilar C, Marcé RM, Borrull F (2001) J Chromatogr A 915:139
52. Martínez K, Barceló D (2001) Fres J Anal Chem 370:940
53. Gatidou G, Kotrikla A, Thomaidis NS, Lekkas TD (2004) Anal Chim Acta 505:153

Hdb Env Chem Vol. 5, Part O (2006): 155–170
DOI 10.1007/698_5_053

Published online: 23 November 2005

Antifouling Paint Booster Biocides: Occurrence and Partitioning in Water and Sediments

Nikolaos Voulvoulis

Centre for Environmental Policy, Imperial College London, Prince Consort Road, London SW7 2BP, UK
n.voulvoulis@imperial.ac.uk

Abstract Following the ban on the use of tributyltin (TBT) in antifouling, several organic booster biocides have been used in conjunction with copper in antifouling paints as alternative treatments. With the further restrictions on the use of TBT in commercial shipping, it is expected that these biocides will be used to a greater extent in the future. Limited data are available on the environmental occurrence, fate, toxicity and persistence of these biocides. Relatively little monitoring has been carried out, possibly owing to the comparatively recent introduction, limited usage and perceived lower toxicity of these biocides in comparison with TBT. The fate of these compounds in the aquatic environment is closely linked to their partitioning between aqueous media and particulate matter and sediment. Partitioning depends on physicochemical parameters dictated by water chemistry and geographical considerations. Sorption is responsible for reducing booster biocides' concentration and toxicity in the water column and is also the principal pathway for accumulation in sediments. Available data on the occurrence and partitioning of these biocides are reviewed and implications to policies are assessed in the light of presented findings.

Keywords Antifoulants · Fate · Organotins · Sorption · Tributyltin

Abbreviations

K_{ow} Octanol/water coefficient
ppb Parts per billion
TBT Tributyltin
TCMS(2,3,5,6-Tetrachloro-4-methylsulphonyl)
TCMTB2-(Thiocyanomethylthio)benzothiazole

1 Introduction

Antifouling products play an important role in the shipping industry and are of significant economic importance. It is estimated that, on average, fuel consumption increases 6% for every 100-μm increase in the average hull roughness caused by fouling organisms [1,2]. After 6 months of fouling, a ship expends 40% extra fuel to maintain a normal speed; other costs include cleaning, repainting and the loss of revenue due to the time taken for these actions [3]. Many antifouling systems have been used over the centuries, from copper and lead sheathing, to the development of paints in the twentieth century, containing compounds of copper, lead, arsenic and mercury. Organotins were first used in the mid-1960s, owing to their acute toxicity to target organisms, with tributyltin (TBT) the most popular compound. Concern over the effects of organotins used in antifouling paints first arose in France, where severe problems were encountered in commercial oyster fisheries in areas where there was intense boating activity and poor tidal exchange [4–6]. Since then, the distribution, fate and effects of organotins on the marine [7–11] and freshwater environments [12–16] have been studied extensively. Research evidence of the damaging effect of triorganotins on the reproduction and growth of various marine life forms has prompted action by many countries to regulate or ban their use in antifouling products [17–19]. Following these legislative controls, TBT contamination of boat harbour water and freshwaters has decreased [20,21], however a number of studies indicate that even though regulations were effective in reducing TBT levels, contamination of sediments by organotin compounds is still widespread and has ecotoxicological consequences [22,23].

TBT products have been superseded by products based on copper, containing organic booster biocides to improve the efficacy of the formulation. In the absence of the antifouling potency of TBT, a copper compound such as cuprous oxide (Cu_2O), copper thiocyanate (CuSCN) or metallic copper is utilised as the principal biocide. Copper exhibits antifouling activity against organisms such as barnacles, tube worms and the majority of algal fouling species. However, several algal species (e.g. *Enteromorpha* spp., *Ectocarpus* spp., *Achnanthes* spp.) show marked physiological tolerance to copper. In order to achieve protection against these tolerant species, a number of booster biocides are used in conjunction with copper to control copper-resistant fouling organisms [24]. Organic booster biocides that have recently been used in the UK include chlorothalonil, dichlofluanid, diuron and Irgarol 1051 (Irgarol), and, less frequently, Kathon 5287, maneb, TCMS pyridine, 2-(thiocyanomethylthio)benzothiazole (TCMTB), thiram, zineb, ziram and zinc pyrithione.

Irgarol is highly effective against freshwater and marine algae. It belongs to the *s*-triazine group of compounds which act as photosystem-II inhibitors, with the inhibition of photosynthetic electron capture and transport in chloroplasts as their biochemical mode of action [25, 26]. Diuron, one of the major urea herbicides in use since the 1950s, also inhibits photosynthesis. It is predominantly used on land for general weed control on noncrop areas. Chlorothalonil and dichlofluanid are protective fungicides widely used in agriculture, with a wide range of action against a number of organisms. Chlorothalonil is highly toxic to fish and aquatic invertebrates, but is not phytotoxic [27, 28]. Dichlofluanid is one of the most widely used antifouling biocides in the country [29]. However, it has been detected in saline coastal waters at concentrations higher than in freshwaters, suggesting that its use in antifouling products may be of significance [30]. Kathon 5287 is a highly effective, broad-spectrum biocide. It is an isothiazolone compound which is licensed for use as an active component in antifouling products in the UK and elsewhere. It was the first organic booster biocide to be registered for use by the USA EPA. A number of other booster biocides are also used in antifouling

Table 1 Physiochemical data for selected antifouling biocides

	Mol Wt	Vapour pressure (Pa)	Aqueous solubility ($g\,m^{-3}$)	$\log K_{ow}$	Melting point (°C)	Refs.
Irgarol	253.37	0.00001488	7	2.38	130	[26]
Diuron	233.10	0.00041	36.4	2.85	159	[36]
Chlorothalonil	265.92	0.0013	0.6	2.64	250.5	[27]
Dichlofluanid	333.2	0.000133	0.006	3.7	106	[37]
Sea-Nine 211/ Kathon	213.3	0.0004	0.0065	2.85	0.01	[38]
TCMTB (Busan)	238.4	1	0.00033	3.3	<– 10	[39]
Zinc pyrithione	317.68	0.000133	77.96	2.13	262	[38]
Zineb	275.8	< 0.00001	10	≤ 1.3	157	[40]

Table 2 Other applications of booster biocides. (From Ref. [24])

Compound	Other applications
Chlorothalonil	Agriculture (fungicide), paints, adhesives
Dichlofluanid	Agriculture (fungicide)
Diuron	Agriculture (herbicide)
Irgarol 1051	Agriculture (herbicide)
Kathon 5287 (Sea-Nine 211)	No other applications
Maneb	Agriculture (fungicide in fruits and vegetable crops)
TCMS pyridine	No other applications
TCMTB (Busan)	Agriculture (fungicide), wood preservative, leather industry
Thiram	Agriculture (fungicide, leaf and seed treatment)
Zinc pyrithione	Agriculture (bactericide, fungicide), shampoos (at concentrations of around 1%)
Zineb	Agriculture (fungicide, fruits and vegetable crops)
Ziram	Agriculture (fungicide, fruits and vegetable crops)

applications: TCMTB, TCMS pyridine, zinc pyrithione and the dithiocarbamates thiram, ziram, zineb and maneb. These are protective fungicides with a wide range of action against a number of organisms.

For booster biocides available data on their occurrence, fate and toxicity, with regards to antifouling applications, have been reviewed [24, 31–33]. Table 1 shows some chemical properties of these biocides. As a result, recently Irgarol and diuron were banned in the UK [34] and chlorothalonil has been banned from amateur use [35].

In addition, agricultural and industrial uses of these biocides represent significant inputs to the marine environment. Chlorothalonil is a fungicide used widely and in large quantities in agriculture, often to control the fungal disease Septaria, which is encouraged by the misty, moist conditions found around estuaries such as the Blackwater region of Essex [41]. As it is applied in spring, runoff is likely to occur up to a month later (depending on rainfall) when there is a pulse of input into the aquatic environment [42]. Monitoring data support this, indicating that agricultural runoff is an important source of chlorothalonil to the marine environment [41]. Table 2 summarises other applications of booster biocides.

2
Occurrence and Fate in Waters

Booster biocides were introduced into antifouling paint formulations only after the restrictions imposed on the use of organotins. Limited monitoring data are available for Irgarol [43–46], while of the other booster biocides, diuron, chlorothalonil, dichlofluanid and Kathon/Sea-Nine have received the most attention. Since almost all of these compounds have agricultural uses their presence in the aquatic and estuarine environments cannot be attributed solely to the use of antifoulants [47].

There are many factors that influence the degradation and persistence of these biocides in the marine environment. These include their chemical and physical properties as well as ecosystem-specific parameters such as the nature and concentration of microbial populations, dissolved and suspended material and temperature. For example, in a study investigating biodegradability in seawater, the breakdown of four biocides (diuron, Irgarol, Kathon and chlorothalonil) was monitored over 8 weeks by bioassay using *Amphora coffeaeformis* (a ship-fouling diatom). The results demonstrated that abiotic pathways were unlikely to be a major route for degradation of these substances in the sea, although photolysis may be significant in the upper layers of water [48].

2.1 Irgarol

A number of studies on the occurrence of Irgarol in coastal waters [41, 43, 45, 46, 49–59] and freshwaters [44, 51, 60] indicate its widespread occurrence in aquatic environments. Irgarol is the most frequently detected antifouling biocide worldwide [31].

Water samples collected from marinas, estuaries and coastal waters along the UK south coast and sediment samples from the Hamble estuary in the UK showed that Irgarol residues were present in most marine and estuarine samples, although not in freshwaters [45]. The highest concentrations were found in areas of high boating activity, particularly marinas and the Hamble estuary, indicating a correlation with its use in antifouling paints. Monitoring of water samples taken from Plymouth Sound (UK) showed Irgarol at all sampling sites. The highest levels were found in close proximity to areas of high boat density, especially where water flow was restricted within marinas; the highest detected value was 127 $ng\,L^{-1}$ [46]. Monitoring of subsurface waters from the French Mediterranean (Côte d'Azur) coastline showed substantial levels of Irgarol in all marinas, with concentrations reaching 640 $ng\,L^{-1}$ [61]. Assessment of the contamination of different compartments of Lake Geneva (water, sediments, zebra mussels, macrophytes and algae) over a period of 9 months found concentrations comparable to those observed in marine studies [44].

The highest concentrations of Irgarol have been associated with marinas, while ports tend to have lower concentrations. Irgarol was also detected at high concentrations in fishery harbours in one study in Japan [51]. The restrictions on the use of Irgarol on small boats may change this distribution.

Concentrations of Irgarol vary with season; higher concentrations recorded in marinas and boatyards outside the boating season are attributed to maintenance of vessels—scrubbing and repainting activities—and at the beginning of the season elevated concentrations coincide with the introduction of newly antifouled boats to the water [52].

No correlations have been found between dissolved Irgarol 1051 concentrations and pH, temperature or salinity [52].

2.2 Diuron

Diuron has also been detected in saline coastal waters at concentrations higher than in freshwaters, suggesting that its use in antifouling products may be of significance [30, 62–64]. It has been detected in the UK, Sweden, Spain, the Netherlands, Portugal and Japan [31]. As with Irgarol, concentrations of diuron in water vary depending on the boating season [62].

The impact of the agricultural uses of diuron has been demonstrated in the detection of herbicide residues in water samples from draining streams and pumping stations of the agricultural area of Thessaloniki, Greece. Considerable amounts of diuron were found to be released from the agricultural fields and transferred through rivers, draining streams and pumping stations to the coast [65]. In the UK, Environment Agency data demonstrate that one of the pesticides which most frequently exceeds the environmental quality standard of 1 mg L^{-1} in controlled waters is diuron, predominantly owing to its use as a herbicide in agriculture [30].

2.3 Chlorothalonil

Occurrence of chlorothalonil does not follow the same pattern as that of many other biocides [31]. It can be present in high concentrations in early spring—before the boating season—and this may be attributed to agricultural use of the compound on adjacent land [41]. Other studies have shown the absence of chlorothalonil in some marinas [49, 63], which may be due to its rapid degradation in water: biodegradation is apparent after 4 weeks and proceeds faster in seawater supplemented by cultured marine bacteria [48].

2.4 Dichlofluanid

Dichlofluanid has been detected in Greece [66] and Spain [67]. In the UK it has not been detected in water [41], and it has been reported as being absent from Mediterranean marinas [49].

2.5
Sea-Nine 211/Kathon

Sea-Nine 211 has been detected in Japan [68]. It has been detected in marinas in Spain, Denmark and Greece in high concentrations but for only short periods of time [67, 69, 70]. This may be due to its rapid degradation [31]; its half-life in sea water has been reported as between 24 h [70] and 8.5 days [48]. In another study on the fate of Kathon in the aquatic environment, it was found that its half-life was less than 1 h in both an aerobic and an anaerobic microcosm consisting of marine sediment and seawater [71].

Kathon concentrations in five UK marinas were recorded as being below detection limits, and this was attributed to the fact that sampling took place outside the main boating season, although the extent to which Kathon is used in the UK was also questioned [69]. Indeed it is no longer permitted for amateur use in the UK [35].

2.6
TCMTB (Busan) and TCMS Pyridine (Densil 100)

The presence of both TCMTB and TCMS pyridine is also reported as being below detection limits [69]; as with Kathon, TCMTB is not permitted for amateur use in the UK [35].

2.7
Zinc Pyrithione

Zinc pyrithione concentrations have been below detection limits in the UK [69, 72], possibly because it biodegrades rapidly or accumulates in sediments [31]. However, in conditions where light is limited zinc pyrithione does not biodegrade rapidly, and may therefore persist [73].

2.8
Zineb

Data on the occurrence of zineb in the marine environment are limited, perhaps because the methodology for its monitoring is less clearly defined than for other booster biocides [74]. Despite the paucity of information it is one of the remaining organic booster biocides that is permitted for both professional and amateur use [35].

3
Partitioning and Fate in Sediments

The fate of alternative biocides in the aquatic environment is closely linked to their partitioning between aqueous media and particulate matter and sediment. The partitioning depends upon physicochemical parameters dictated by water chemistry and geographical considerations. Factors include sediment characteristics, presence of suspended sediment and the chemical characteristics of the water column [75] and the specific characteristics of the area sampled, for example the volume of the harbour [50].

Organic pollutants can either be sorbed onto particulate matter or exist in solution. Adsorption onto particulate matter and flocculation—in which the species are trapped by newly formed particulate matter—are often termed as "sorption" [21]. Sorption is responsible for reducing both the concentration and the toxicity of biocides in the water column, and is also the principal pathway for accumulation in sediments.

Influences on the transfer of Irgarol, chlorothalonil, dichlofluanid and diuron from the soluble to the particulate phase in estuarine environments have been investigated [76]. For these four biocides the removal from water increases with the concentration of particulate matter. The sorption of chlorothalonil and dichlofluanid is enhanced at high pH (8.0) and increased suspended matter, whilst for Irgarol and diuron the influence of pH is not obvious. Salinity does not seem to play a significant role in the partitioning behaviour of the biocides. Comparisons of different sediment types show that silty-clay sediments exhibit a larger capacity for biocide adsorption than silty-sandy sediments. There are clearly different patterns of sorption for each biocide as controlled by the physicochemical environment.

Sediment disturbances may release biocides back into the water column. Where the initial sediment concentrations of the biocides are low, the concentrations in water, resulting from desorption, are below detection limits. Dichlofluanid is among the most strongly bound to sediments, with a release of less than 1% (0.6–0.9%) from contaminated sediments. Irgarol and chlorothalonil demonstrate a higher degree of desorption (1.9–2.4 and 1.2–1.7%, respectively) [76].

3.1
Irgarol

Irgarol has been shown to have strong adsorption characteristics [77]. It has been detected in marine sediments in Greece [78] Germany [50], the UK [52], Spain [79] and Sweden [80] and in freshwater sediments in Lake Geneva [44].

Irgarol levels in sediment after the boating season are significantly higher than those before and after the boating season and levels of Irgarol in marinas and mooring areas are significantly higher than those in coastal sites, indicating that the major source of Irgarol is antifouling paints on boats [76]. Concentrations in estuarine sediments vary little vertically, indicating a homogenous, reworked sediment probably due to bioturbation or disturbance by activities such as dredging, tide energy or waves. Certain sites, for example boatyards, are dredged at least once a year and the sediment is likely to have been disturbed [41]. Sediment contamination with Irgarol was found to be related to high concentrations in the water column [45].

Concentrations of Irgarol reported in the Blackwater estuary (ranging from 0.15–0.68 ppb in water and 3.3–222 ppb in sediment) are similar to values previously reported for water but are significantly higher for sediments [41]. Because water is the primary medium for the exposure of most aquatic organisms to these types of biocides, almost all studies have been carried out to determine concentrations of Irgarol in waters and not in sediments. In addition, because few data exist on the partitioning of Irgarol in the aquatic environment, it is difficult to determine if the observed lower concentrations in waters in comparison with sediments are associated with Irgarol's preference towards sorptive partitioning. For example, this could be attributed to the tidal exchange with uncontaminated waters. It has been suggested, however, that partitioning onto settled or suspended particulate matter may prove to be a critical process in determining the compound's fate [45].

Information regarding the persistence of Irgarol in the aquatic environment is limited. Generally, degradation of Irgarol seems to be slow but is more likely to occur in water than in sediment as a consequence of photodegradation. Irgarol is not biodegradable in seawater; it has been recorded that no degradation occurs in 8 weeks [48]. Solar irradiation degrades more than 80% of Irgarol in water after 15 weeks [81]. In sediments, it has been shown that degradation is slow even under aerobic conditions, with half-lives of 100 and 200 days for marine and freshwater sediment, respectively, and is considerably slower for anaerobic conditions [44]. Given the toxicity of Irgarol there may be cause for concern that leaching of Irgarol from boats in marinas may have effects on receiving water communities. Concentrations recorded after the boating season in waters (0.15–0.68 ppb) may inhibit the growth rate and photosynthetic activity of several species. For example the growth of *Enteromorpha intestinalis*, a species known to carpet the surface of the mudflats in the Blackwater estuary [82], would probably be inhibited under the concentrations of Irgarol observed in the Blackwater study [41]. The no-effect concentration for this species has been reported at 0.022 ppb, with a 72-h EC_{50} of 2.5 ppb [46]. The concentrations of Irgarol found are not high enough to have acute toxic effects directly on higher species, but its chronic effects at lower concentrations are unknown and difficult to determine.

3.2 Diuron

Diuron has been detected in low concentrations in sediments in the UK [29, 63, 83], Spain [79] and the Netherlands [83] both during and after the boating season. Diuron shows the least preference to sorption of Irgarol, chlorothalonil and dichlofluanid at just 20–32% [76], and this is supported by the data on relative concentrations in water and sediments.

3.3 Chlorothalonil

Chlorothalonil has been detected in marine sediments in Greece, including in areas where agricultural inputs are minimal, implying that antifouling paints represent a significant input of the compound [78].

In the UK, chlorothalonil has been detected in an area where agricultural inputs are significant, with high concentrations observed in June, seemingly related to agricultural runoff. All sediment samples collected during that period showed contamination with the biocide. Contamination of both water and sediment was considerably less in October than in June, probably owing to the decline in the amount of runoff of the pesticide into the aquatic environment after the initial pulse subsequent to agricultural application [41].

The low concentrations in water may also be due to the lack of persistence of the compound in the water column [48]. Chlorothalonil has been found to degrade after 4 weeks in natural seawater and even faster in water supplemented by cultured marine bacteria, indicating that biodegradation of the compound can be considerable. Degradation still occurs when the biocide is present in low concentrations [84]. Even faster degradation has been witnessed in freshwater where chlorothalonil exhibits a half-life of only a few hours. A number of studies have shown that the biocide is known to exhibit greater persistence in soils, with a half-life of 1–2 months [27, 84, 85]. Maximum concentrations of chlorothalonil reported in waters and sediments in the Blackwater estuary (1.38 and 34.3 ppb, respectively) were below values noted as acutely toxic to a number of species [41]. For example, tests on the in vivo toxicity of chlorothalonil to *Ictalurus punctatus* (channel catfish) have shown the chemical to be toxic with a 96-h LC_{50} value of 52 ppb [86]. An interim water quality guideline for the protection of marine life has been set at 0.36 ppb of chlorothalonil [27]. For all of the samples where chlorothalonil was detected in the Blackwater study, the concentrations exhibited were above this value.

3.4 Dichlofluanid

Dichlofluanid has been detected in marine sediments in Greece [66, 78], the UK [41] and Spain [79]. Levels in sediment have been reported as significantly higher after the boating season than before, and after the boating season dichlofluanid levels are significantly higher in marinas and mooring areas than in coastal sites, indicating that antifouling paints on boats could be a major source of this compound [76].

Dichlofluanid tends not to be found in waters; however, considerable concentrations of the biocide have been observed in sediments [41]. It is unstable in water [87], which may explain the absence of contamination of water samples. However, it has been reported to have relatively high concentrations in sediments after the boating season, which may be due to its very low solubility in water and the high octanol/water partition coefficient. Dichlofluanid has a strong affinity for particulate material [88] and exhibits the greatest tendency towards solid-phase partitioning when compared with Irgarol, chlorothalonil and diuron, accounting for between 90 and 99% of dichlofluanid available [76]. However, it is susceptible to anaerobic degradation and this may play an important role in its removal from the environment [88].

3.5 Sea-Nine 211/Kathon

Sea-Nine binds strongly to sediment, and once bound is essentially immobile [89]. It has been detected in sediments in southeastern Spain [79], but its detection has not been reported in other studies [66, 78]. This may be because it degrades rapidly; it has a half-life in seawater of less than 1 hour [89].

3.6 TCMTB (Busan) and TCMS Pyridine (Densil 100)

Limited data are available on the sorption characteristics of TCMTB and it has proved impractical to theorise on its behaviour [87]. Data on its presence in marine sediments are also limited.

3.7 Zinc Pyrithione and Zineb

Along with dichlofluanid, zinc pyrithione and zineb are among the few remaining antifouling products that are permitted for amateur use in the UK [35]. Despite this, few data exist on their occurrence in sediments, and their partitioning behaviour has not been studied.

4 Discussion

It has been demonstrated that the use of organic booster biocides in antifouling paints on boats can lead to a significant presence in the coastal environment, in both waters and sediments. Sources of booster biocides have been identified as antifouling paints and agricultural and industrial inputs, and their occurrence has been recorded at levels that may have significant impacts on coastal and estuarial environments. Concentrations appear higher in sediments, where anoxic conditions can lead to their persistence.

The complete ban on the use of TBT in antifouling paints on all ships could result in an increase in the occurrence of booster biocides in the marine environment, despite the subsequent restrictions on the use of a number of them. A greater understanding of the long-term fate of these compounds in the environment is necessary to facilitate a more informed decision-making process when regulating antifoulant use.

While the use of organotin-based antifouling formulations is to end, care must be taken to avoid the substitution by other biocides which may be just as damaging to the marine environment. This could be the case with a number of the replacement biocides, and, as has been demonstrated, levels of these biocides in both waters and sediments could be high enough to warrant concern.

Although new antifoulants must be as effective against the fouling organisms as the organotins, they need to perform better in other areas, the foremost of which is persistence. While toxicity to target organisms is required, persistence and sorptive behaviour must be minimised. The severity of the problem of TBT lay in the combination of high toxicity and high persistence in sediments, and some of the alternative biocides could demonstrate environmental characteristics similar to those of TBT [90]. Booster biocides must be licensed not to combine high toxicity with high persistence and investigating partitioning behaviour is instrumental in making this choice.

One of the lessons learned from the TBT experience is that policies should be developed to regulate biocides on the basis of testing prior to licensing rather than after the biocide has been used, and monitoring studies have suggested that regulation is necessary. The results presented here show that several booster biocides partition strongly to sediments. While this reduces the concentrations in the water and therefore the risk to aquatic organisms, it leads to greater persistence of the compounds in the environment; hence, the risk to sediment-dwelling organisms is increased. Monitoring should take place in order to confirm the results of testing that takes place prior to licensing, rather than as a substitute for this testing. It should be possible to predict in advance whether a particular biocide will accumulate to a hazardous level in sediments and to use this information when regulating its use. Although there is now evidence on the levels of these biocides in waters, the lack of information on their fate and behaviour makes accurate risk assessment difficult.

With policies worldwide almost eliminating the use of TBT, the use of booster biocides in antifouling products has increased; hence, environmental occurrence in the future may be more significant. Restrictions on the use of diuron, Irgarol, chlorothalonil, TCMTB and Kathon in the UK may lead to an associated fall in concentrations in water, but the partitioning behaviour of some of these compounds is such that their occurrence in sediments could remain high. The strong binding to sediments and the low desorption rates of some of these substances could lead to increased persistence and subsequent ecotoxicological consequences.

Policy decisions in the future will need to be based on detailed testing of antifouling agents, including measures not only of toxicity but also of persistence and sorptive behaviour. They must also take into account the uncertainty that is inherent in applying information from laboratory-scale tests to whole ecosystems.

5
Conclusions

The use of antifouling paints has been identified as an important source of booster biocides in the marine environment, along with agricultural and industrial inputs. For most of the biocides, evidence of their levels in waters now exists; however, the lack of information on their fate, behaviour and toxicity makes accurate risk assessment difficult. Concentrations appear significantly higher in sediments, where anoxic conditions can lead to their persistence.

With the elimination of the use of TBT, it is expected that booster biocides will be used in increasing amounts in antifouling products and thus environmental occurrence in the future may be even more significant. Policies must be based on scientific evidence on the overall performance of these compounds, especially persistence and partitioning, if antifoulants are to be used sustainably and responsibly in the future.

References

1. Liu D, Maguire RJ, Lau YL, Pacepavicius GJ, Okamura H, Aoyama I (1997) Water Res 31:2363
2. Townsin R (1987) In: Celena 25th anniversary symposium, Genoa
3. Terlizzi A, Fraschetti S, Gianguzza P, Faimali M, Boero F (2001) Aquat Conserv-Mar Freshwater Ecosyst 11:311
4. Ruiz JM, Bachelet G, Caumette P, Donard OFX (1996) Environ Pollut 93:195
5. Quevauviller P, Donard OFX, Etcheber H (1994) Environ Pollut 84:89
6. Alzieu C, Sanjuan J, Deltreil J, Borel M (1986) Mar Pollut Bull 17:494
7. Clark EA, Sterritt RM, Lester JN (1988) Environ Sci Technol 22:600

8. Maguire RJ (1992) Water Sci Technol 25:125
9. Sudaryanto A, Takahashi S, Iwata H, Tanabe S, Muchtar M, Razak H (2005) Mar Pollut Bull 50:227
10. Devier MH, Augagneur S, Budzinski H, Le Menach K, Mora P, Narbonne JF, Garrigues P (2005) J Environ Monit 7:224
11. Fernandez MA, Wagener ADR, Limaverde AM, Scofield AL, Pinheiro FM, Rodrigues E (2005) Mar Environ Res 59:435
12. Dowson PH, Bubb JM, Lester JN (1992) Mar Pollut Bull 24:492
13. Dowson PH, Pershke D, Bubb JM, Lester JN (1992) Environ Pollut 76:259
14. Yang F, Maguire RJ, Chau YK (2001) Water Qual Res J Can 36:805
15. Regoli L, Chan HM, de Lafontaine Y, Mikaelian I (2001) Aquat Toxicol 53:115
16. Looser PW, Bertschi S, Fent K (1998) Appl Organomet Chem 12:601
17. Lau MMM (1991) Arch Environ Contam Toxicol 20:299
18. de Mora S (1996) Tributyltin: case study of an environmental contaminant. Cambridge University Press, Cambridge
19. Champ MA (2000) Sci Total Environ 258:21
20. Dowson PH, Bubb JM, Lester JN (1994) Chemosphere 28:905
21. Dowson PH, Bubb JM, Lester JN (1993) Mar Pollut Bull 26:487
22. Dowson PH, Bubb JM, Williams T, Lester JN (1993) Water Sci Technol 28:133
23. Chau Y, Maguire R, Brown M, Yang F, Batchelor S (1997) Water Qual Res J Can 32:453
24. Voulvoulis N, Scrimshaw MD, Lester JN (1999) Appl Organomet Chem 13:135
25. Dahl B, Blanck H (1996) Mar Pollut Bull 32:342
26. Hall LW, Giddings JM, Solomon KR, Balcomb R (1999) Crit Rev Toxicol 29:367
27. Caux PY, Kent RA, Fan GT, Stephenson GL (1996) Crit Rev Environ Sci Technol 26:45
28. Ernst W, Doe K, Jonah P, Young J, Julien G, Hennigar P (1991) Arch Environ Contam Toxicol 21:1
29. Boxall A, Conrad A, Reed S (1998) Environmental problems from antifouling agents: survey of manufacturers, chandlers (suppliers) and treatment sites. Report P215. Environment Agency
30. NRA (1995) Pesticides in the aquatic environment. HMSO, London
31. Konstantinou IK, Albanis TA (2004) Environ Int 30:235
32. Yebra DM, Kiil S, Dam-Johansen K (2004) Prog Org Coat 50:75
33. Omae I (2003) Appl Organomet Chem 17:81
34. Health and Safety Executive (2002) Summary report of a survey to investigate the use of antifouling products (AFPs) in UK freshwaters
35. Health and Safety Executive (2000) Pestic Newslett 49
36. Howard PH (1991) Handbook of environmental fate and exposure data for organic chemicals. Lewis, Chelsea, MI
37. Bayer AG (1996) Preventol A 4-S: Product information. BG Organic Chemicals
38. The Department of Chemistry at the University of Akron (2005)
39. National Institute for Occupational Safety and Health (1997) International Chemical Safety Cards
40. BCPC (2004) The pesticide e-manual
41. Voulvoulis N, Scrimshaw MD, Lester JN (2000) Mar Pollut Bull 40:938
42. Fletcher CA, Bubb JM, Lester JN (1994) Sci Total Environ 155:47
43. Tolosa I, Readman JW, Blaevoet A, Ghilini S, Bartocci J, Horvat M (1996) Mar Pollut Bull 32:335
44. Toth S, Becker van Slooten K, Spack L, de Alencastro LF, Tarradellas J (1996) Bull Environ Contam Toxicol 57:426
45. Gough MA, Fothergill J, Hendrie JD (1994) Mar Pollut Bull 28:613

46. Scarlett A, Donkin ME, Fileman TW, Donkin P (1997) Mar Pollut Bull 34:645
47. HMSO (1995) Pesticides. HMSO, London
48. Callow ME, Willingham GL (1996) Biofouling 10:239
49. Ferrer I, Barcelo D (1999) J Chromatogr A 854:197
50. Biselli S, Bester K, Huhnerfuss H, Fent K (2000) Mar Pollut Bull 40:233
51. Liu D, Pacepavicius GJ, Maguire RJ, Lau YL, Okamura H, Aoyama I (1999) Water Res 33:2833
52. Bowman JC, Readman JW, Zhou JL (2003) Mar Pollut Bull 46:444
53. Gardinali PR, Plasencia MD, Maxey C (2004) Mar Pollut Bull 49:1072
54. Hall LW, Killen WD, Gardinali PR (2004) Mar Pollut Bull 48:554
55. Harino H, Kitano M, Mori Y, Mochida K, Kakuno A, Arima S (2005) J Mar Biol Assoc UK 85:33
56. Okamura H, Aoyama I, Ono Y, Nishida T (2003) Mar Pollut Bull 47:59
57. Okamura H, Sugiyama Y (2004) Chemosphere 57:739
58. Ferrer I, Ballesteros B, Marco MP, Barcelo D (1997) Environ Sci Technol 31:3530
59. Steen R, Leonards PEG, Brinkman UAT, Cofino WP (1997) J Chromatogr A 766:153
60. Berard A, Dorigo U, Mercier I, Becker van Slooten K, Grandjean D, Leboulanger C (2003) Chemosphere 53:935
61. Readman JW, Kwong LLW, Grondin D, Bartocci J, Villeneuve JP, Mee LD (1993) Environ Sci Technol 27:1940
62. Lamoree MH, Swart CP, van der Horst A, van Hattum B (2002) J Chromatogr A 970:183
63. Thomas KV, McHugh M, Waldock M (2002) Sci Total Environ 293:117
64. Comber SDW, Gardner MJ, Boxall ABA (2002) J Environ Monit 4:417
65. Albanis TA (1991) Chemosphere 22:645
66. Sakkas VA, Konstantinou IK, Lambropoulou DA, Albanis TA (2002) Environ Sci Pollut Res 9:327
67. Martinez K, Ferrer I, Barcelo D (2000) J Chromatogr A 879:27
68. Harino H, Mori Y, Yamaguchi Y, Shibata K, Senda T (2005) Arch Environ Contam Toxicol 48:303
69. Thomas KV (1998) J Chromatogr A 825:29
70. Thomas KV (2001) Biofouling 17:73
71. Jacobson A, Mazza L, Lawrence L, Lawrence B, Jackson S, Kesterson A (1993) Pesticides in urban environments: fate and significance, ACS symposium series 522. American Chemical Society, Washington, DC, p 127
72. Thomas KV (1999) J Chromatogr A 833:105
73. Maraldo K, Dahllof I (2004) Mar Pollut Bull 48:894
74. Voulvoulis N, Scrimshaw MD, Lester JN (1999) Chemosphere 38:3503
75. Comber SDW, Franklin G, Gardner MJ, Watts CD, Boxall ABA, Howcroft J (2002) Sci Total Environ 286:61
76. Voulvoulis N, Scrimshaw MD, Lester JN (2002) Mar Environ Res 53:1
77. Lambropoulou DA, Sakkas VA, Albanis TA (2004) Int J Environ Anal Chem 84:47
78. Albanis TA, Lambropoulou DA, Sakkas VA, Konstantinou IK (2002) Chemosphere 48:475
79. Martinez K, Barcelo D (2001) Fresenius J Anal Chem 370:940
80. Haglund K, Pettersson A, Peterson M, Kylin H, Lord SC, Dollenmeier P (2001) Bull Environ Contam Toxicol 66:50
81. Okamura H (2002) Chemosphere 48:43
82. English Nature (1998) Essex estuaries candidate marine special conservation, littoral and sublittoral biotope mapping and data capture exercise. English Nature research report, project no E3034/01/001. English Nature

83. Thomas KV, Blake SJ, Waldock MJ (2000) Mar Pollut Bull 40:739
84. Davies PE (1988) Bull Environ Contam Toxicol 40:405
85. Tu CM (1992) J Environ Sci Health Part B Pestic Contam Agric Wastes 27:695
86. Gallagher EP, Cattley RC, Digiulio RT (1992) Chemosphere 24:3
87. van Wezel AP, van Vlaardingen P (2004) Aquat Toxicol 66:427
88. Thomas KV, McHugh M, Hilton M, Waldock M (2003) Environ Pollut 123:153
89. Jacobson A, Willingham GL (2000) Sci Total Environ 258:103
90. Voulvoulis N, Scrimshaw M, Lester JN (2002) Chemosphere 47:789

Hdb Env Chem Vol. 5, Part O (2006): 171–200
DOI 10.1007/698_5_054

Published online: 22 December 2005

Photochemical Fate of Organic Booster Biocides in the Aquatic Environment

Vasilios A. Sakkas · Ioannis K. Konstantinou · Triantafyllos A. Albanis (✉)

Chemistry Department, Laboratory of Industrial Chemistry, Ioannina University, 45110 Ioannina, Greece
talbanis@cc.uoi.gr

Abstract Considering the relevance and importance of photochemical processes in the environmental fate and behavior of organic micropollutants, the present review describes the state-of-the-art knowledge regarding the photodegradation of antifouling biocides in the aquatic environment. It includes data on photodegradation rates, primary and end photoproducts, and the pathways and mechanisms for most of the organic booster biocides (i.e., irgarol 1051, Sea-Nine 211, dichlofluanid, diuron, chlorothalonil, TCMTB, zinc and copper pyrithione, maneb, zineb, ziram, thiram, and triphenylboron-pyridine) used in antifouling paints.

Light-induced degradation took place both with direct or indirect (photosensitized) mechanisms via first-order kinetics. Direct photolysis in most cases seemed to be a minor event compared to photosensitized processes that usually enhanced the degradation. The composition of the water matrix is a key factor for the photofate of biocides in various natural waters. Half-lives ranged from a few minutes (zinc/copper pyrithiones) to several days (diuron and irgarol 1051) depending on the irradiation conditions as well as the constitution of the irradiated media. In most cases identified photoproducts were less toxic and innocuous than the parent compounds, however, the formation of more toxic compounds at the first degradation steps or synergistic effects among the transformation products should be taken into consideration.

Keywords Biocides · Direct/photosensitized degradation · Pathways · Photoproducts

Abbreviations

3,4-DCA	3,4-Dichloroaniline
APCI	Atmospheric pressure chemical ionization
BT	Benzothiazole
CDOM	Colored dissolved organic matter
CPDU	1-(3,4-Dichlorophenyl)-3,1-dimethylurea
CuPT	Copper pyrithione
DCPMU	1-(3,4-Dichlorophenyl)-3-methylurea
DCPU	1-(3,4-Dichlorophenyl)urea
DMDC	Dimethyldithiocarbamates
DMSA	*N,N*-Dimethyl-*N′*-phenyl-sulfamide
DOM	Dissolved organic matter
DTCs	Dithiocarbamates
EBDC	Ethylene(bis)dithiocarbamates
EBIS	Ethylene(bis)isothiocyanate sulfide
EI-MS	Electron impact–mass spectra
ETU	Ethylenethiourea
EU	Ethyleneurea
FA	Fulvic acids
GC	Gas chromatography
GS26575	2-Methylthio-4-*tert*-butylamino-6-amino-*s*-triazine
HA	Humic acids
IMO	International Maritime Organization
LC	Liquid chromatography
MBT	2-Mercaptobenzothiazole
MBTS	(bis)benzothiazolyl disulfide
MEPC	Marine Environment Protection Committee
MS	Mass spectrometry
MTBT	2-(methylthio)benzothiazole
NHS	Natural humic substances
NOS	Natural occurring substances
OBT	2(3H)-Benzothiazolone
OHBT	2-hydroxybenzothiazole
PSA	Pyridine-2-sulfonic-acid
SIM	Selected ion monitoring
SPE	Solid phase extraction
TBT	Tri-butyl tin
TCMS pyridine	2,3,5,6-Tetrachloro-4-(methyl sulphonyl) pyridine
TCMTB	2-(thiocyanomethylthio)benzothiazole
TPBP	Triphenylboron-pyridine
USEPA	United States Environmental Protection Agency
UV	Ultraviolet
ZnPT	Zinc pyrithione

1 Introduction

As a result of the restriction imposed by the European Union, regulations on antifouling paints containing organotin biocides have been proposed due to their negative impact on marine environment [1,2]. New alternative formulations have been developed based mainly on the metal copper or its oxide. However, these formulations are less efficient at inhibiting fouling by algae. To remedy this deficiency, booster biocides [3] (chlorothalonil, dichlofluanid, diuron, irgarol 1051, Sea-Nine 211, TCMS pyridine, TCMTB, zinc pyrithione, and zineb) have been added to some products in order to prevent the growth of bacteria, macroalgae, mussels, and other invertebrates. Due to their widespread use considerable coastal concentrations of these biocides have been found in areas of high yachting activity, particularly in marinas and sportive harbors [4–7].

In order to understand and predict their fate in the natural environment and to assess their risk, it is necessary to improve our knowledge of their chemical reactions and degradation rates under environmental conditions. Photochemical transformation is one of the main abiotic degradation pathways occurring in natural waters and has received increasing interest in recent years [8]. The photochemical fate of the above-mentioned booster biocides is reviewed herein. Phototransformation of antifouling biocides under natural conditions may be a complex process. In order to understand the mechanisms involved it is necessary to investigate both direct photolysis and indirect (photosensitized) transformations under relevant experimental conditions. Direct photolysis occurs when a given pollutant absorbs UV-visible light energy and undergoes transformation. Indirect or sensitized photolysis occurs either by direct energy transfer from the excited species [9] to the pollutant or by leading to the formation of reactive species such as singlet oxygen or hydroxy radical, which enter into a series of reactions [10–12]. Since the steady-state concentrations of the above-mentioned reactants are predominantly a reflection of the concentrations of nitrate, cDOM, and bicarbonate levels, the rate of photodecomposition and fate of an organic pollutant could vary as a function of the composition of water.

Photodegradation is also affected by factors controlling the spectral distribution, intensity, and duration of sunlight. Such factors include latitude, cloud cover, date etc. and UV-B radiation absorption by atmospheric ozone. Moreover the penetration of near UV radiation in natural waters is influenced by factors such as depth of mixing, turbidity, presence of dissolved material absorbing in near UV, etc. In moderately turbid coastal waters, incident light with wavelengths of 380 nm or less is almost completely attenuated at depths of 1–2 m but, in very clear parts of the ocean, 20 m may be required to remove 90% of the radiation entering the surface.

2 Photochemical Transformation Processes of Pollutants in Aqueous Environment

The appearance of trace amounts of micropollutants that occur in surface water and groundwater has caused an increasing public and scientific concern. Their fate in the aqueous environment is often unknown, however, direct and indirect photochemical processes may contribute to the phototransformation/photodecomposition of these compounds in natural waters. Mostly these reactions can occur simultaneously in natural waters, therefore it is essential to consider both processes when examining the photochemical behavior of micropollutants. Besides the degradation kinetics, literature reports on the photodegradation products of micropollutants is relatively abundant [12]. However, little information is available on the reaction mechanisms involved in the photolysis under typical environmental conditions. For environmental considerations it is important that they can eventually be converted to innocuous, and preferably mineral, photoproducts.

2.1 Direct Photolysis

Most biocides absorb light at relatively short UV wavelengths. Since sunlight reaching the Earth's surface (mainly UV-A, with varying amounts of UV-B) contains only a very small amount of short wavelength UV radiation [13, 14], the direct photodegradation of biocides by sunlight is expected to be, in most

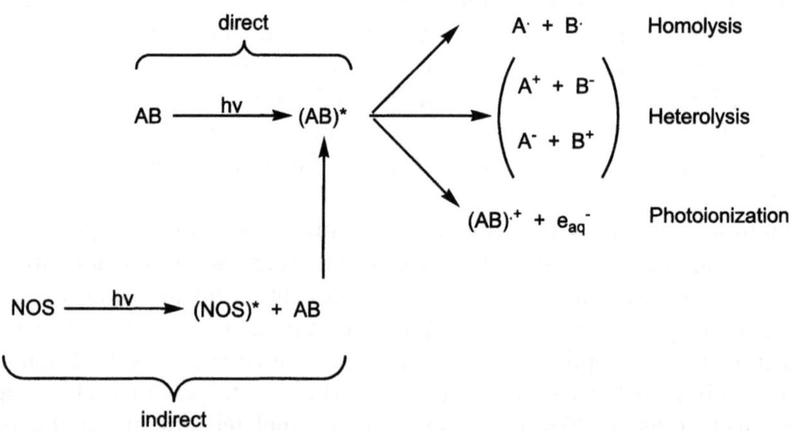

Fig. 1 Direct and photosensitized transformation processes of micropollutants in the aqueous environment

cases, of only limited extent. Direct photolysis can occur if the considered pollutant absorbs light. However, the ability to undergo chemical changes after the absorption of photons is an intrinsic property of the molecule and can vary drastically among compounds. Direct irradiation will lead to the promotion of the molecules to their excited singlet states, which may then through intersystem crossing produce triplet states. Such excited states can then undergo, among other processes: (i) homolysis, (ii) heterolysis, and (iii) photoionization (Fig. 1). The reactions occurring as a consequence of direct light absorption by the pollutant could be fragmentation, isomerization, hydrogen atom abstraction, intramolecular rearrangement, dimerization-addition reactions, cyclization, and electron transfer reactions.

2.2
Sensitized/Indirect Photolysis

Photosensitized photodegradation is based on the absorption of light by a chemical substance other than the pollutant, usually naturally occurring substances (NOS) of the aquatic environment. This may then transfer energy from its excited state to the micropollutants, which can undergo different degradation processes as shown in Fig. 1. Photosensitized or indirect photolytic reactions are thought to proceed due to the presence of chemical transients such as hydroxyl (OH), alkyl peroxy (ROO), hydroperoxyl (HO_2) and carbonate (CO_3^-) radicals, superoxide ion (O_2^-), singlet oxygen ($^1O_2$), hydrogen peroxide (H_2O_2), aquated electrons (e^-_{aq}), and colored dissolved organic matter (cDOM) in its excited triplet state. Waters containing sufficiently high metal ion concentration (through ligand-to-metal charge transfer reaction and photo-Fenton chemistry) and semiconductor particles (through electron–hole pair formation) have been also shown to sensitize the phototransformation of organic contaminants in aqueous solutions [15–17].

Previous research has shown that the hydroxyl radical (produced through the photolysis of nitrate, nitrite, and cDOM) plays a significant role in the transformation of organic contaminants in natural waters due to its reactivity and non-selectiveness [18–20]. In addition carbonate radicals, generated from the reaction of OH with either carbonate or bicarbonate ions, react rapidly with electron-rich compounds and have been shown to play a significant role in limiting the persistence of pesticides [21, 22]. Moreover, alkyl peroxy radicals, produced through the reaction of groundstate oxygen with excited cDOM chromophores, and singlet oxygen formed upon the absorption of sunlight by cDOM and subsequent energy transfer to ground-state oxygen ($^3O_2$) [23, 24] may also promote the photodegradation of the micropollutant. Short-lived triplet states of cDOM ($^3cDOM^*$) may contribute significantly to the decomposition of micropollutants through electron abstraction, hydrogen transfer, or both. Also, aquated electron e^-_{aq}, a highly reactive and strongly reducing species, has previously been reported to be produced upon the photolysis of cDOM in sunlit natural waters and scavenged by nitrate [25]. All these reactive transients are considered to play a significant role in limiting the persistence of many chemical pollutants. The ability of each reactive agent to contribute toward the phototransformation of the contaminant under natural solar light is likely to be affected by the composition of natural water, particularly on the concentrations of nitrate, cDOM, and bicarbonate/carbonate ions. Humic and fulvic substances in natural waters were found to be present at concentrations ranging from 0.3 to 30 mg L^{-1}, while nitrate and bicarbonate (also considered principal components of surface water) are present with concentrations ranging from 3 to 323 μM and 0.4 to 4.4 mM, respectively. An important advantage of photosensitized photodegradation is the possibility of using light of wavelengths longer than those corresponding to the absorption characteristics of the pollutants.

On the other hand, a decrease in a contaminant photodegradation rate in natural waters could be observed. DOM present in natural waters absorbs most of the available photons emitted (since it is one of the most important sunlight-absorbing components of the aquatic environment [26]) thereby slowing down direct photochemical reaction (optical filter effect). Another reason for the observed filter effect may be the binding of micropollutants to DOM by hydrophobic partitioning or weak van der Waal forces, thus affecting photodegradation. The retarded photodegradation rate, especially in seawater, is also consistent with $^{\bullet}OH$ scavenging by chloride ions [27].

As a result, phototransformation of pollutants under natural conditions may be a complex process. In order to evaluate the persistence of pollutants such as antifouling biocides, both direct and photosensitized transformation experiments under relevant laboratory and field conditions should be performed using natural waters of different composition.

More detailed information of the photoreactivity of each biocide under natural and/or simulated solar irradiation is given in the following sections.

3 Kinetics, Photoproducts, and Reaction Pathways of Antifouling Booster Biocides

3.1 Chlorothalonil

Chlorothalonil (2,4,5,6-tetrachloroisophtalonitrile) has been found to degrade after 4 weeks in natural sea water and even faster in water supplemented by cultured marine bacteria. Research by Davies [28] found degradation still occurred when the biocide was present in low concentrations. Even faster degradation has been witnessed in freshwater where chlorothalonil exhibited a half-life of only a few hours ($t_{1/2} = 4-150$ h). Walker et al. [29] reported degradation half-lives of between 1.8 and 8 days in natural estuarine water and sediment–water test systems. In general biodegradation is lower than photolytic processes [29].

Chlorothalonil can undergo photodegradation in the water column demonstrating a half-life of few hours [30, 31], while faster degradation has been witnessed in ground water under simulated solar irradiation with a half-life less than an hour [32]. Millet et al. [33] calculated the half-lives in the aquatic environment of chlorothalonil photodegradation using the program GC-Solar as 22–206 days (from summer to winter). Chlorothalonil photodecomposition was carried out in different natural waters [31, 34] and it was observed that the degradation rate was significantly enhanced compared to distilled water, with the exception of seawater in both natural and simulated solar irradiation. More than 99% of chlorothalonil degrades within 60 h in lake and river water while 33% and 28% still remain in distilled and seawater, respectively, under natural solar irradiation. Simulated solar irradiation for 10 h resulted in 63% and 59% decline of chlorothalonil concentration in distilled water and seawater, respectively, while more than 99% was consumed in both river and lake waters. The degradation rates have shown that the presence of DOM in natural waters enhances the photodegradation rate of chlorothalonil. The enhanced photodegradation in both outdoor and laboratory experiments is attributed to the presence of naturally occurring photosensitizers in natural waters such as dissolved organic matter, nitrates, and carbonate-bicarbonate ions.

As far as the formation of photoproducts is concerned, six compounds could be detected as possible degradation intermediates, however, only five have been identified including one pair of isomers. The main degradation products were benzamide (1), chloro-1,3-dicyanobenzene (2), dichloro-1,3-dicyanobenzene (3), trichloro-1,3-dicyanobenzene (4) [32]. A tentative degradation pathway is proposed for chlorothalonil photodegradation in natural waters (Fig. 2).

Chromatographic data indicate the presence of one additional product, compound 5. This compound involves a benzyl and methyl group in the molecule, while the absence of isotopic pattern of chlorine atoms supports the homolytic cleavage of the C – Cl bond. No nitrogen atoms seem to be present in the molecule, demonstrating the lack of nitrile groups that could be photohydrolyzed to the corresponding benzoic acids through benzamide intermediates. However, the data were insufficient to propose a structure.

Compounds 1 (benzamide) and 5 (unidentified) have been detected only during the photodegradation of chlorothalonil in natural and humic waters [31, 34]. It was the first time that benzamide had been observed during chlorothalonil photodegradation in natural waters indicating that dissolved organic matter afforded both an increase in photosensitization and in $^{\bullet}$OH processes.

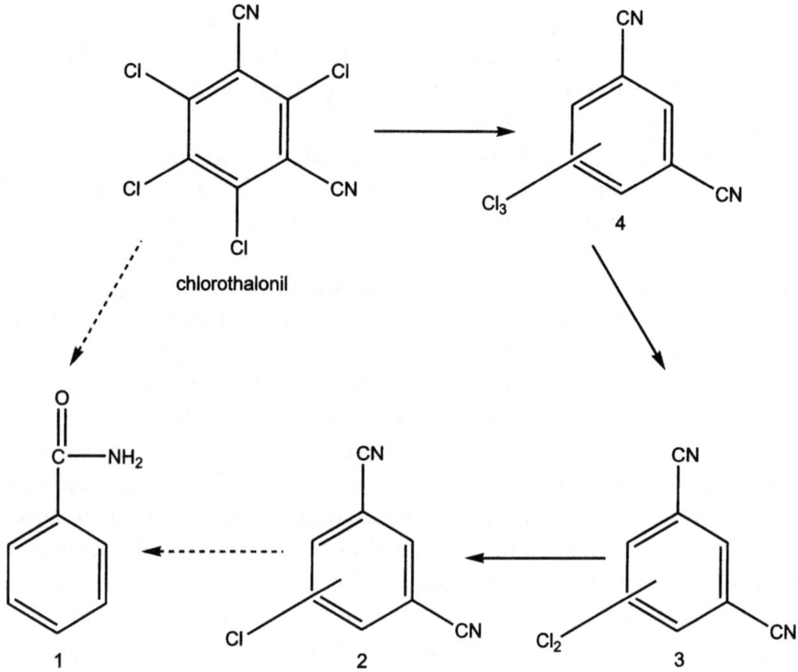

Fig. 2 Photolytic degradation pathways and main phototransformation products of chlorothalonil in aqueous environment [31, 34]

The distribution pattern and the total number of degradation products differed between the various water samples and were dependent on the constitution of the irradiated media. In general the number of degradation products was greater in natural and humic water than in distilled water. However, the presence of common transformation products (compounds 2, 3, and 4) in distilled and in natural waters indicate that chlorothalonil is susceptible to both direct and indirect photolytic reactions in natural waters.

3.2 Dichlofluanid

The photodegradation half-life in natural seawater was 53 h [35] and a half-life of 18 h was reported using a bioassay method [36].

The photodegradation rate of dichlofluanid (*N*-dichlorofluoromethylthio-*N*′,*N*′-dimethyl-*N*-phenylsulfamide) was lower in natural waters than in distilled water following the order: lake water < river water < sea water < distilled water, showing a strong dependence on the constitution of the irradiated media and especially on the concentration of dissolved organic matter [35]. As the TOC concentration increases in natural waters the rate of photolysis decreases.

Photolysis was faster under simulated irradiation (Xe lamp, 1.1 KW) than under natural sunlight, producing first-order rate constants of 0.0817 and 0.0163 h^{-1} for distilled water, 0.0518 and 0.0132 h^{-1} for sea water, 0.0394 and 0.0117 h^{-1} for river water, and 0.0335 and 0.0084 h^{-1} for lake water. This observation seems normal when considering that sunlight intensity varies depending on the time of day and on cloud cover, while the suntest apparatus keeps the intensity constant. Experiments conducted by Sakkas et al. [35] under the presence of various concentrations of DOM (HA and FA), also produced first-order degradation curves, allowing for the calculation of the dichlofluanid photolysis rate constants. In all cases the presence of DOM

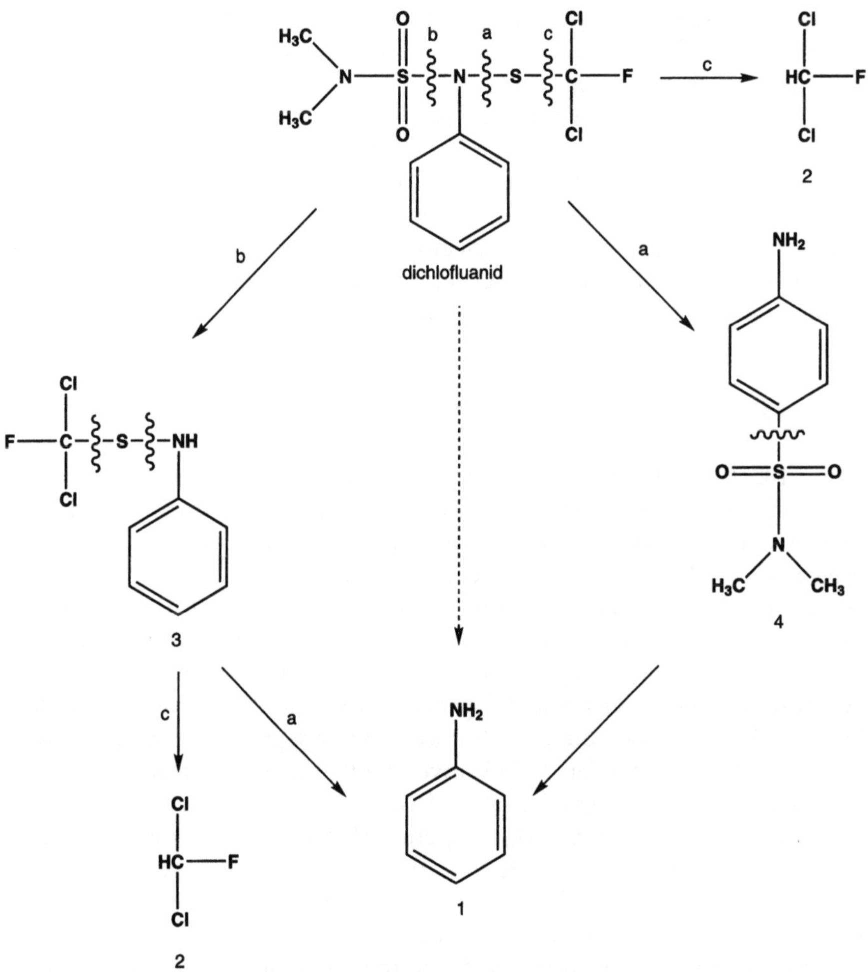

Fig. 3 Photolytic degradation pathways and main phototransformation products of dichlofluanid in aqueous environment [34, 35]

slowed the rate of photolysis. For example, experiments performed with concentrations 4, 8, 16, and 24 mg L^{-1} of HA produced rate constants of 0.0392, 0.0360, 0.0324, and 0.0281 h^{-1}, respectively. The same tendency has been observed for FA. The rate constants decreased as the concentration of FA increased: 0.0277, 0.0233, 0.0195, and 0.0171 h^{-1} at concentration of 4, 8, 16, and 24 mg L^{-1}, respectively. The decrease in the rate of photodegradation could be due to either DOM competing with dichlofluanid for the available photons or to binding between DOM and dichlofluanid. The very low solubility of the compound in water (2 mg L^{-1}) and its high octanol/water partition coefficient ($K_{OW} = 3.7$), indicate that this biocide has a tendency to associate with particulate matter.

The main degradation products occurred from photodegradation, hydrolysis, and anaerobic degradation were *N,N*-dimethyl-*N′*-phenyl-sulfamide (DMSA), *N*-dichlorofluoromethylthio-aniline, and aniline [34, 35, 37]. Their structures are shown in Fig. 3.

DMSA is the major degradation product of dichlofluanid in biotic and abiotic processes, arising from the N – S bond cleavage and the rearrangement of the *N,N* dimethyl sulfonyl group in the *para* position. In addition, another minor peak was identified and associated to dichlorofluoromethane, which arises from the N – S bond cleavage of dichlofluanid. Loss of the *N,N* dimethylsulfonamyl $SO_2N(CH_3)_2$ group from dichlofluanid results in the formation of *N*-dichlorofluoromethylthio-aniline.

3.3
Irgarol 1051

Irgarol 1051(2-methylthio-4-*tert*-butylamino-6-cyclopropylamino-*s*-triazine) is considered to be non-biodegradable [38] and its degradation in seawater and freshwater is slow, with a half-life of about 100 and 200 days, respectively [39]. However, the detection of irgarol degradation product (2-methylthio-4-*tert*-butylamino-6-amino-*s*-triazine, known also as GS26575) demonstrates that it could undergo environmental transformation in aquatic ecosystems. Solar photodegradation of irgarol 1051 has been proposed as the most possible environmental process while studies have shown that this biocide undergoes direct or indirect photodegradation with the formation of the *N*-dealkylated derivative (GS26575) as the major degradation product [40–43]. Irgarol 1051 and GS26575 are shown to demonstrate a similar seawater half-life [44], suggesting that GS26575 has a greater environmental persistence than irgarol 1051 [45]. However, these data conflict with other studies which have shown that GS26575 was at generally lower concentrations than the parent compound [46], which suggests that the environmental transformation rate of irgarol 1051 is relatively slow and that the rate of GS26575 degradation is greater than that of its formation.

Irgarol photodegradation has been studied in a variety of natural waters under both natural and simulated solar irradiation [34, 47]. The data demonstrate that photolysis of irgarol 1051 was lower in environmental waters compared to distilled water, showing a strong dependence on the constitution of the irradiated media. After 66 calendar days of daylight exposure 66%, 60%, 57%, and 65% of irgarol 1051 was degraded in distilled, sea, river, and lake water, respectively, while more than 66%, 57%, 55%, and 53% degraded within 48 h under simulated solar irradiation. The presence of DOM in natural waters inhibits the degradation rate of irgarol 1051. The retarded photodegradation indicate that organic matter absorbed most of the photons emitted thereby slowing down direct photochemical reaction of irgarol 1051 (optical filter effect). Photoreaction of the colored dissolved organic matter (cDOM) in sunlight caused a decrease mostly in the UV-B (280–315 nm) region. This is very important in the case of irgarol 1051 because it does not absorb strongly above 320 nm. Another reason may be that irgarol 1051 was partially bound to dissolved organic matter ($\log K_{OC} = 3.0$, $\log K_{OW} = 3.9$) by hydrophobic partitioning or weak van der Waal forces, and so this fraction was never available for photolysis action.

A recent study of Okamura and Sugiyama [48] investigated irgarol 1051 photolysis in the presence of some photosensitizers. Aqueous solutions containing irgarol or its metabolite GS26575, with or without photosensitizers, were irradiated using a light source from a UV-A fluorescent lamp for 48 h. GS26575 was more stable than irgarol when irradiated in the presence of photosensitizers such as acetone, benzophenone, tryptophan, and rose bengal. Four types of fulvic acids (FA) and two types of humic acids (HA) purified from natural soils and river waters were used as photosensitizers at a concentration of $10\ \mathrm{mg\,L^{-1}}$. Degradation of irgarol 1051 and simultaneous production of GS26575 were observed in the presence of each of the natural humic substances tested. The half-lives of irgarol ranged from 6.8 to 39 h while the maximum concentration of GS26575 ($0.35\ \mathrm{mg\,L^{-1}}$) was observed in the presence of HA-I. This shows that the photochemical reaction assisted by natural humic substances (NHS) may lead to accumulation of GS26575 in natural aquatic environments in agreement with Sakkas et al. [47], who reported that GS26575 was a major photoproduct among five intermediates identified in HA/FA solutions under simulated solar light. NHS also accelerated the photodegradation of irgarol. However, GS26575 was more persistent than the parent compound, which was completely degraded. In the same study [48] after 48 h of irradiation, irgarol 1051 in river water was degraded with simultaneous production of GS26575. On the other hand, when irgarol 1051 was irradiated in pure water no significant degradation was observed and GS26575 was not formed, which is in agreement with the observations of previous studies [42, 47]. Therefore, indirect photosensitized reactions in natural waters may result in the accumulation of GS26575 in aquatic environments.

Two routes of degradation pathways are observed during the photodegradation of irgarol 1051 according to the proposed reaction scheme (Fig. 4). The first route involves the oxidation of irgarol's sulfur atom leading to the formation of sulfone (compound 5). The cleavage of the sulfur group of the triazine ring results in the formation of 2-hydroxy-4-*tert*-butylamino-6-cyclopropylamino-*s*-triazine (compound 2). This observation is in agreement with other studies describing the direct photolysis of *s*-triazines where photoreaction must not involve the alkyl group but the methylthio or the chlorine group, supporting the idea that hydroxy derivative formation is a major pathway in direct photolysis [49, 50]. The same observation has been also reported during the photocatalytic degradation of irgarol 1051 [51]. Thus, the formation of the mono-dealkylated derivative, i.e., 2-methylthio-4-*tert*-butylamino-6-amino-*s*-triazine (GS26575), was greatly favored over the hydroxylated and the sulfonyl derivative since it occurred for more than 90% of photoreactions (according to relative abundance of the compounds) and was attributed to indirect processes. Okamura et al. [42] have also reported the presence of this byproduct as the main one during irgarol 1051 photolysis in distilled water, while Torrents et al. [52] have observed the formation of chlorodealkylated derivatives in the case of atrazine during direct photolysis. Irradiation of aqueous irgarol 1051 solutions containing dissolved organic matter during simulated solar irradiation, resulted in the formation of diaminohydroxy-*s*-triazine (compound 1),

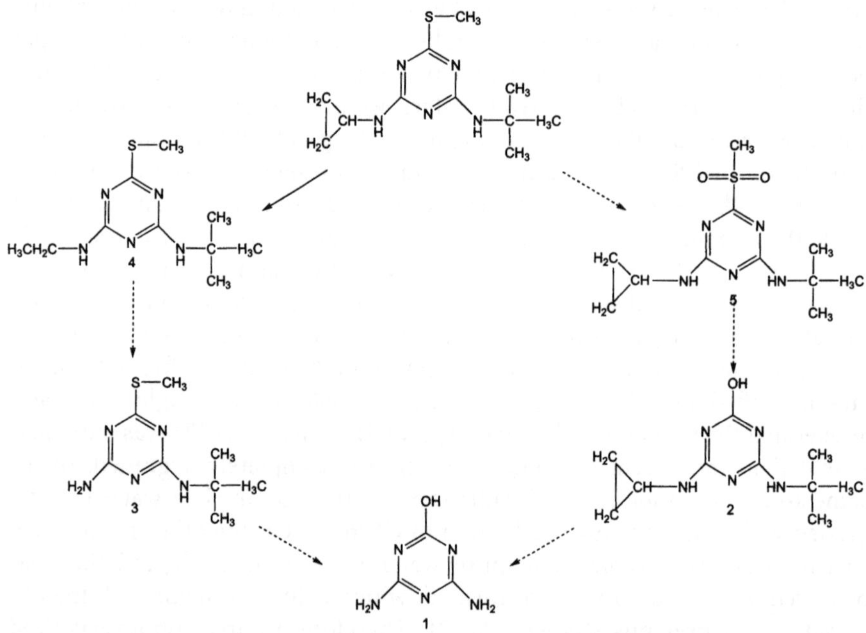

Fig. 4 Photolytic degradation pathways and main phototransformation products of irgarol 1051 in aqueous environment [34, 47]

indicating both an increase in photosensitization and in •OH processes. Singlet oxygen formed during the photolysis reacts with dissolved organic matter to form peroxide. It is possible that the peroxide formed may generate the hydroxyl radicals from the humic solution, which in turn causes dealkylation [52]. 2-Methylthio-4-*tert*-butylamino-6-ethylamino-*s*-triazine (compound 4) was also formed during these series of experiments, apart from the derivatives mentioned in distilled water (compounds 2, 3 and 5), showing that in the presence of DOM both direct and indirect reactions can occur.

The product profile of irgarol photodegradation in natural waters was very similar to that of the HA/FA system, which strongly suggests that DOM present in natural waters sensitize the photolysis of irgarol 1051, as indicated by the formation of the dealkylated derivative, diaminohydroxy-*s*-triazine. However, in this case too, 2-methylthio-4-*tert*-butylamino-6-amino-*s*-triazine was the main byproduct formed, proving its occurrence in surface waters.

Hall et al. [44] showed the presence of the irgarol 1051 metabolite (GS26575), increasing from 0.4% at the start of the experiment to 3.6% of the total concentration after 42 days. This good mass balance supports previous reports that GS26575 is the major degradation product of irgarol 1051 [45]. An experiment to determine the rate of GS26575 removal showed only 74% of the compound to be present after 42 days. Once again, if a first-order degradation reaction is applied then a seawater half-life of 82 days can be calculated. This suggests that the rate of GS26575 degradation is much greater than that of irgarol 1051 and that under the experimental conditions employed, GS26575 is not as environmentally persistent as irgarol 1051. On the other hand, Okamura [43] demonstrated that irgarol was rapidly photodegraded, in contrast to its major metabolite under the same irradiation conditions. The results indicated that 78–93% of GS26575 was degraded after exposure to solar irradiation for 10 months, and thus that it is more persistent than the parent compound.

In the study of Lam et al. [53] during the $HgCl_2$-catalyzed hydrolysis of irgarol-1051, a new degradation product was identified. The EI-MS of this unknown product, hereby designated as M2, is similar to that obtained by Ogawa et al. [54]. With careful chromatographic separation and purification, M2 was successfully isolated and identified as 3-[4-*tert*-butylamino-6-methylthiol-*s*-triazin-2-ylamino]-propionaldehyde – the *N*-propionaldehyde derivative of M1. The presence of this compound was also detected in the coastal waters of Hong Kong, indicating that the environmental degradation of irgarol 1051 is more complicated than expected.

3.4
Sea-Nine 211

Sea-Nine 211 (4,5-dichloro-2-*n*-octyl-4-isothiazolin-3-one), degrades rapidly both biologically and chemically in natural seawater [55]. Isothiazolone has a half-life of 8.4 days according to Callow and Willingham [38] and has a residence time of 5 days according to Ranke [56]. The hydrolytic half-life was > 720 h for a pH 7 buffered solution while degradation in natural seawater containing microorganisms showed a half-life of less than 24 h. Anaerobic half life of < 0.5 days was calculated by Thomas et al. [37], increasing when the compound was introduced to sediments associated with paint particles. A summary of Sea-Nine 211 half-lives in various environmental matrices was reported by Willingham and Jacobson [57]. Biological degradation is considered to be over 200 times faster than hydrolysis and photolysis [57], which contributes to a lesser extent to its degradation in the environment. Photolysis half-life experiments, conducted at pH 7, gave values of $t_{1/2} = 322$ h and $t_{1/2} = 315$ h, respectively [34, 58]. The rapid degradation reduces the concentration to significantly below toxic levels. Its metabolites are ring-opened structures and their toxicity is reduced by 4–5 orders of magnitude [59, 60].

Phototransformation of Sea-Nine 211 in natural waters under natural sunlight [34, 60] was significantly enhanced following the order: lake water > river water > sea water > distilled water showing a strong dependence on the composition of the irradiated media. Simulated solar irradiation for 30 h resulted in a 77%, 87%, 92%, and 97% decline of Sea-Nine 211 concentration in distilled, sea, river, and lake water, respectively. Experiments with concentrations of 4, 8, 16, and 24 mg L^{-1} of HA produced rate constants of 0.0690, 0.0776, 0.0835, and 0.1008 h^{-1}, respectively, resulting in 89%, 91%, 93%, and 96% decline in Sea-Nine 211 concentration. The same tendency has been observed for FA. The rate constants increased as the concentration of FA increased: 0.0624, 0.0664, 0.0726, and 0.0771 h^{-1} at concentrations of 4, 8, 16, and 24 mg L^{-1}, respectively. As the DOM and nitrate concentration in natural waters increases, a faster degradation rate is observed. This rapid phototransformation in all experiments is attributed to the presence of naturally occurring photosensitizers in natural waters. Energy, electron, and hydrogen atom transfer reactions as well as reactions with photochemically generated free radicals may be very significant in the environmental phototransformation of Sea-Nine 211, which does not absorb strongly above 290 nm.

Two main transformation pathways are observed during the phototransformation of Sea-Nine 211 according to the proposed reaction scheme (Fig. 5). The first pathway (a) involves cleavage of the isothiazolone ring and subsequent oxidation of the resulting alkyl metabolites [61] thus, *N*-*n*-octyl acetamide is formed. The cleavage of the weakest bond of the molecule (N – S) of Sea-Nine 211 and subsequent dechlorination and hydroxylation results in *N*-*n*-octyl hydroxypropionamide. Further oxidation yields the formation of *N*-*n*-octyl malonamic acid, which is then decarboxylated to give the corresponding *N*-*n*-octyl acetamide. The presence *N*-*n*-octyl hydroxypropionamide and *N*-*n*-octyl malonamic acid was not identified in either sample [34, 60]. The above transformation pathway was also observed during analysis of the dark control samples in an outdoor experiment, indicating that

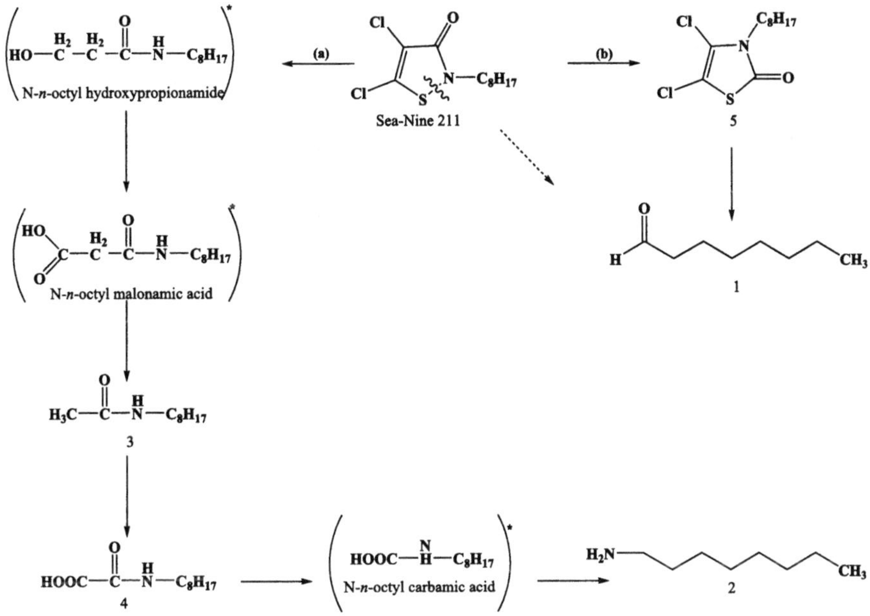

Fig. 5 Photolytic degradation pathways and main phototransformation products of Sea-Nine 211 in aqueous environment [34, 60]

this route also accounts for other processes such as hydrolysis and biodegradation [39, 61]. In the case of distilled water, degradation owing to hydrolysis (dark experiment) accounted for 30% of the decline in Sea-Nine 211 concentration. In lake water, dark controls analyses (hydrolysis and biodegradation) have shown a 65% contribution to the transformation of the biocide, indicating that biodegradation is considerable [55]. Further oxidation of *N*-*n*-octyl acetamide leads to the formation of *N*-*n*-octyl oxamic acid, which could be then phototransformed to the corresponding *N*-*n*-octyl carbamic acid as indicated by Thomas [39] in the metabolic pathway of Sea-Nine 211 under aerobic conditions. The presence of *n*-octyl isocyanate identified by Sakkas et al. [60] may support the formation of *N*-*n*-octyl carbamic acid since it reacts rapidly with water in a nucleophilic addition step generating the corresponding *N*-*n*-octyl carbamic acid. Although unstable in water its presence may be attributed to the loss of water of *N*-*n*-octyl carbamic acid, probably due to a thermal process prior to ionization in the GC-MS chamber. *N*-*n*-Octyl carbamic acid decarboxylates yielding *n*-octyl amine.

The second pathway (b) consists on the phototransposition of Sea-Nine 211. Photochemical interconversion of 1,2 into 1,3 isomers has been reported for dihetero compounds, e.g., isoxazole to oxazole, imidazole to pyrazole, as well as isothiazolone to thiazolones [62–64]. Photoexcitation of isothiazolones is suggested to result in cleavage of the N – S bond, resulting in the formation of a species that can be viewed as diradicals, which can then lead to the formation of thiazolones. In the case of Sea-Nine 211 the photoisomeration of the parent molecule yields 4,5-dichloro-3-*n*-octyl-thiazolin-2-one. The mass spectra of 4,5-dichloro-3-*n*-octyl-thiazolin-2-one was similar to Sea-Nine 211 and the fact that it appears at lower retention time than Sea-Nine 211 supports the proposed structure, being less polar than the starting molecule. N – C alkyl bond cleavage of 4,5-dichloro-3-*n*-octyl-thiazolin-2-one and the oxidation of the alkyl group results in *n*-octanal.

Chromatographic data [34, 60] indicated the presence of three more compounds that could not be identified but which were regarded as possible transformation products because they were detected in all experiments and their concentration increased and decreased as a function of the reaction time. The evolution of chloride ions that reach only 63% of the stoichiometric value after 30 h of simulated solar irradiation in distilled water clearly demonstrates that chlorinated transient organics are present in the solution after that time.

3.5 Diuron

Diuron (3-(3,4-dichlorophenyl)-1,1-dimethylurea) is reported to be relatively persistent in seawater [38], and considerably stable to hydrolysis. As far as the degradation of diuron is concerned under UV irradiation conditions in natural seawater, photolysis appeared to be a very minor process [43]. A quite long half-life of diuron has been reported, ranging from one month to one year [43, 65]. Mazellier et al. [66] studied the photochemical behavior of diuron in the presence of different iron(III) species. The half-lives of diuron when submitted to such process in the environment was estimated to be 1–2 h or a few days, depending on the concentration of iron(III). Faure and Boule [67] calculated the quantum yields at a concentration of 10^{-4} mol L^{-1}, and reported $\Phi_{(254\ nm)} = 0.02$ and $\Phi_{(256\ nm)} = 0.01$.

Diuron photohydrolysis [68] occurs in the *meta* or *para* position with respect to urea group, their proportion depending on the irradiation wavelength [69]. According to Faure [70] and Jirkovsky et al. [69], who irradiated solutions with the same lamps, the *meta*-hydroxylated product (1) is approximately nine times more concentrated than the *para* isomer (2). On the contrary, when diuron was irradiated at the visible region using a light source at 365 nm [67] the *para*-hydroxylated derivative was found to be the major degradation product.

Diuron could be hydrolyzed in water leading to the formation of 3,4-dichloroaniline. Although the latter has not been reported to be formed during diuron photolysis, it is thought to be one of the most important biotic and abiotic transformation products of the parent compound. For this reason the photochemical fate of 3,4-dichloroaniline is also discussed in this context. Aqueous photolysis of 3,4-dichloroaniline occurs through sunlight irradiation [71]. In general, for halogenated anilines this transformation rate is highly dependent on the position of the halogen on the ring. For 3,4-DCA, the faster rate of photodegradation is consistent with its high absorption band (> 300 nm) and the relatively high quantum yield ($\Phi_{(313\,\mathrm{nm})} = 0.052$) [72]. In winter the transformation rate was shown to be completed in 20 days, ten times slower than that in summer. The half-life for direct phototransformation in water was determined at 0.4 h, while natural sunlight resulted in half-lives of 6 ± 3.6 h at the water surface [73]. Dichloroanilines are reported to be transformed first into phenolic compounds and then into aminophenoxazones but no toxicological data exist for these substances.

Loss and oxidation of the alkyl chains are the dominant processes during the phototransformation of ureas [74], as depicted at Fig. 6. On the other hand, substitution of a halogen atom by a hydroxyl-group, and hydroxylation of the aromatic ring are considered to be the secondary processes. It was observed that diuron photodegradation consisted of two phases of different but almost constant rates. The degradation started at a high rate and when approximately 50% of the original substance degraded the rate of degradation fell to a twentieth of the first phase. The complete process of the degradation could be best described by a cubic function ($y = -10^{-4}x^3 + 0.030x^2 - 3.0255x + 100$; $R^2 = 0.952$) [74].

Absorption of light may indeed induce a phototransformation of diuron with a possible formation of more toxic intermediate photoproducts. In en-

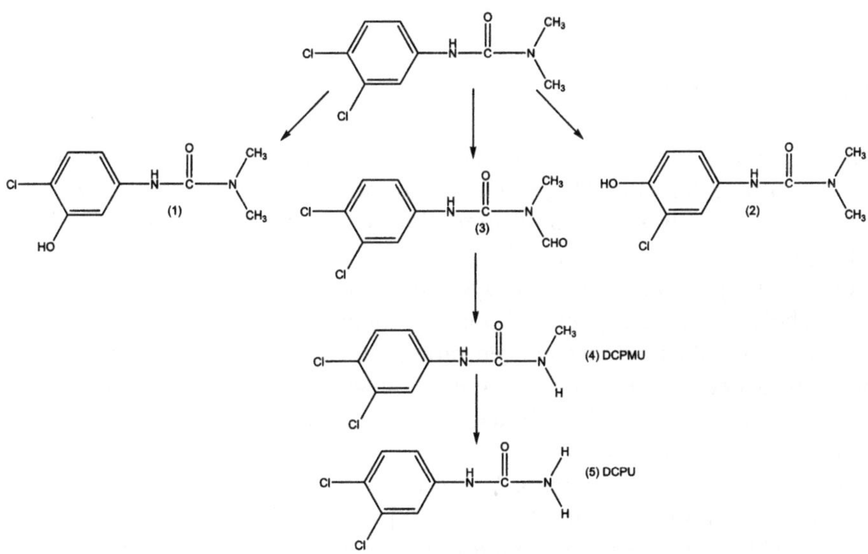

Fig. 6 Photolytic degradation pathways of diuron in aqueous environment based on reported identified phototransformation products [68–70]

vironmental conditions photolysis is usually slow and photoproducts do not accumulate appreciably. Microtox was used for the evaluation of the toxicity of the irradiated solutions of diuron [68]. The observed toxicity is due to minor photoproducts, however, it does not disappear so rapidly. Hence the decrease of herbicide concentration is not necessarily associated with a lower toxicity of the solution. The toxicity also increased with irradiation time, but it remained almost stable between 50% and 98% conversion. It was not possible to evaluate the EC_{50} of pure diuron since it is higher than its solubility in water [68]. Bonnemoy et al. [68] deduced that at least one of the photochemical intermediates is more toxic than diuron, while synergistic effects among degradation products cannot be excluded. It has been reported [68] that when diuron is irradiated in aqueous solution photohydrolysis products in *meta* and in *para* positions are the main photoproducts (compounds 1 and 2). However, the small amounts of 3-(3,4-dichlorophenyl)-1-formyl-1-methylurea (3) and 3-(3,4-dichlorophenyl)-1-methylurea (4), resulting from the oxidation or elimination of a methyl group, may be present since such products are formed by irradiation in the absence of water [68]. It was reported by Tixier et al. [75] that products (1) and (2) have almost the same toxicity to *Vibrio fischeri* as diuron, but that compounds (3) and (4) are significantly more toxic. Consequently, the observed increase in toxicity may be mainly due to minor photoproducts. It is most likely that during diuron photolysis the toxic products formed will be photoconverted in a further reaction step, but this conversion is less rapid and the complete disappearance of the herbicide is not associated with a decrease of toxicity.

3.6 Zinc and Copper Pyrithiones

Zinc pyrithione (bis(2-pyridylthio)zinc 1,1′-dioxide) (ZnPT) has a short half-life in natural seawater ($t_{1/2} \leq 4$ h; [76]) and low aqueous solubility. Zinc pyrithione photolysis and biodegradation are both very rapid procedures with photodegradation being the most effective process whilst hydrolysis half-lives range from 96 to 120 days. Zinc pyrithione is known to undergo *trans*-chelation, with copper and possibly manganese replacing the zinc ion, depending on their relative concentrations in the water or sediment. It has been suggested that in waters where UV degradation of zinc pyrithione is poor, it may accumulate in sediments as these complexes, prior to biodegradation and/or hydrolysis [39, 77].

Photolytic half-lives ranging between 2–18 min have been reported in a variety of irradiation conditions [76, 78] (filtered Xe arc lamp and natural sunlight), while less toxic photoproducts are formed [79]. Degradation of pyrithione is mainly caused by the UV component of the sunlight spectrum, and Neihof et al. [78] reports that wavelengths of 320–355 nm are the most effective in producing photodegradation. An aqueous photolysis study [80] based on US EPA guidelines under simulated natural sunlight showed that photolysis of zinc pyrithione in pH 9 buffer or artificial seawater at pH 8.2 occurred very rapidly (first-order half-life from the first hour of incubation was ~ 13–18 min). In contrast, similarly slow degradation rates to those in the hydrolysis study were found in samples incubated in the dark (half-lives extrapolated from 30 days of incubation were 83–91 days).

ZnPT and CuPT hydrolysis and photolysis [81] were also studied in sterile synthetic seawater and fresh water. Whereas the half life for hydrolysis of pyrithiones in dark sterile waters ranged from 7 to more than 90 days, the photolysis half life ranged from 15 to 30 min. Photodegradation of ZnPT is much faster than CuPT [82]. Under natural solar irradiation ($E = 880\ \mathrm{W\,m^2}$), the half-lives of ZnPT and CuPT were estimated as approximately 2 and 20 min, respectively [82]. The degradation half-life of ZnPT and CuPT was estimated to be between 7 and 9 min [79]. The degradation times for ZnPT and CuPT ranged from 7 to 9 min and did not differ significantly from one another (ANOVA, $p = 0.897$). Environmental photolysis rates for compounds with high quantum yields (e.g., the pyrithiones) depend upon various factors including angle and intensity of incident surface solar radiation, water depth and clarity, presence of photosensitizers and particulate matter, and wave action. Photolysis rates are strongly influenced by the photon flux density and spectral distribution of the light sources [83]. This may explain the differences in half-lives in the different experiments mentioned above. The amount of ZnPT and CuPT that will photodegrade is, therefore, dependent on the water depth to which the 320–355 nm wavelengths can penetrate in the water column, together with the photon flux density. It is clear, however, that radiation in the 300–355 nm range, which causes photodegradation of pyrithione, can penetrate to significant depths under favorable conditions. Therefore, it is likely that photochemical processes play an important role in the environmental degradation of pyrithiones. They can therefore be marketed as being environmentally neutral, non-persistent antifouling compounds based on the fact that they easily photolyze [76], and thus rapidly degrade into less toxic compounds [79, 80, 84, 85].

Several degradation products were formed during ZnPT photodegradation (Fig. 7). Turley et al. [76] have reported that pyridine-2-sulfonic-acid (1), is the major photolysis product of the parent compound. Other degradation products formed included pyridine sulfinic acid (2), pyrithione disulfide (3), pyridine disulfide (4), and the pyridine/pyrithione mixed disulfide (5) (Fig. 7). Some byproducts formed with light exposure, but not in the dark. An earlier work [80] also reported rapid degradation of related pyrithione biocides (pyrithione, sodium pyrithione, pyrithione disulfide, and a tertiary-butylamine pyrithione derivative, generally tested at 100 ppm) in seawater exposed to sunlight, at a range of pHs.

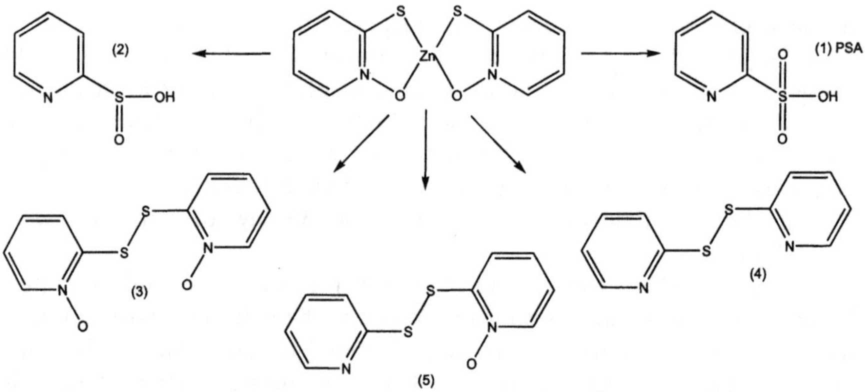

Fig. 7 Principal photodegradation products of zinc pyrithione in aqueous environment [76, 82, 84]

Maraldo and Dahllof [79] reported that ZnPT and CuPT and their eventual breakdown products lost their toxicity rapidly when exposed to light.

3.7 Dithiocarbamates

Dithiocarbamates (DTCs) can be divided into two groups: the dimethyldithiocarbamates (DMDCs), including ferbam, ziram, and thiram and the ethylene (bis)dithiocarbamates (EBDCs), such as maneb, zineb, and mancozeb. Ethylenethiourea (ETU) is one of the principal metabolites of EBDCs and is thought to be the source of most of the toxicity associated with EBDCs. ETU is also the major identifiable product of UV irradiation, according to Gruiskshan and Jarrow [86], who studied its photolysis and hydrolysis. There are only limited photodegradation studies of DTCs in aqueous media perhaps because of the lack of selective analytical methodology.

3.7.1 Thiram and Ziram

Reported photodegradation half-life of thiram (bis(dimethylthiocarbamoyl) disulfide) in water at pH 5 and 25 °C is reported to be 8.8 h while hydrolysis half-lives range between < 1 (pH 9) and 77 days (pH 5) depending on the pH [87]. As far as ziram abiotic fate is concerned, the hydrolysis and photolysis half-lives are 0.74 and 0.36 days, respectively. Upon hydrolysis and soil photolysis, ziram (zinc bis(dimethyldithiocarbamate)) quickly degrades to thiram. In the environment, the major volatile transformation byproducts of thiram and ziram are CO_2 and CS_2 [88].

3.7.2
Maneb and Zineb

Ethyleneurea (EU), ethylenethiourea (ETU), ethylene (bis)isothiocyanatesulfide (EBIS), and glycine were formed in irradiated solutions of EBDCs as well as in the dark. Their structures are shown in Fig. 8. EU is the main decomposition product of photolysis. The time when approximately half of the starting material is degraded is 10.5 h for the irradiated sample and 23.5 h for the non-irradiated sample [89].

Since EBDCs are transformed rapidly to ETU in the aqueous environment there is a need for information regarding the photolytic fate of ETU. In water, ETU is relatively stable to hydrolysis and photolysis in aqueous media. Gruiskshan and Jarrow [86] showed that aqueous solutions of ETU exposed to UV light (above 285 nm) undergo very slow photolysis. Irradiation at an intensity of 1900 $\mu W\,cm^{-2}$ for 24 h have an insignificant loss (< 5%). Irradiation at 3300 $\mu W\,cm^{-2}$ for 15 days resulted in 33% loss of ETU [90]. However, photosensitized oxidation is considered a major degradation pathway for ETU [90] in the presence of photosensitizers occurring in natural waters (i.e., riboflavin, chlorophyll). For example, the photosensitized oxidation of ETU using riboflavin leads to concentration less than 5% of that in the dark control within 4 days [91]. Generally, the photolysis half-life of ETU in natural water is reported as 1–4 days [90]. ETU is probably oxidized to EU by photochemically produced hydroxyl radicals. The identified degradates (Fig. 9) were glycine, 3-(2-imidazolin-2-yl)-2-imidazolidinethione (Jaffe's base), EU, and hydantoin (2,4-imidazolidinedione) [90].

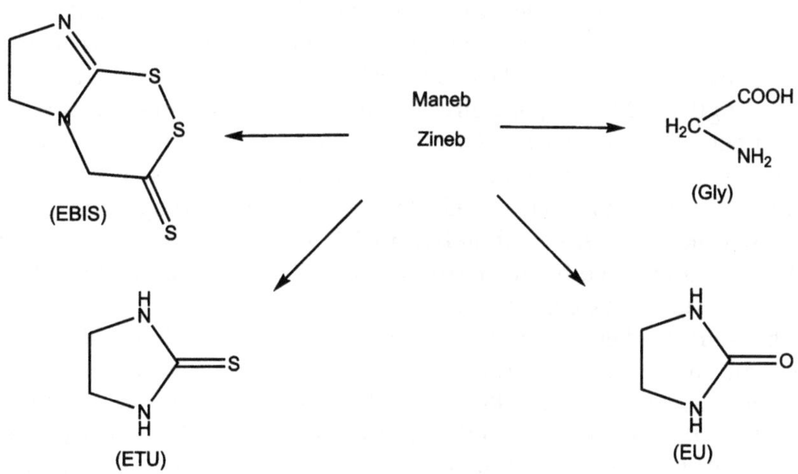

Fig. 8 Principal photodegradation products of maneb and zineb in aqueous environment [89]

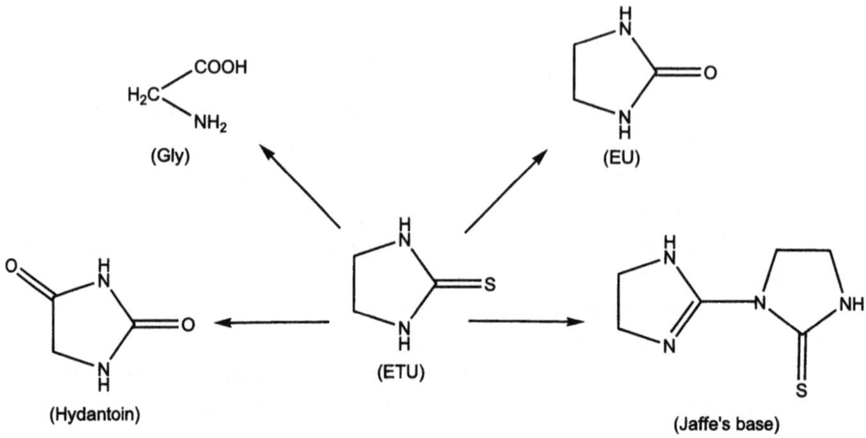

Fig. 9 Sensitized photoxidation products of ethylenethiourea (ETU) in solar light [90, 91]

3.8 TCMTB

TCMTB [(2-(thiocyanomethylthio)benzothiazole] is rapidly degraded in water by direct photolysis in sunlight with reported half-lives ranging between 1.5 and 3.9 h [92], producing 2-mercaptobenzothiazole (MBT, $pKa = 6.94 \pm 0.05$) as the dominant product (about 50% yield) and traces of benzothiazole (BT) [93]. However, photolysis rate in a laboratory or a field situation may differ significantly. It is very plausible that photolysis plays an especially important role in the toxicity studies, and that MBT is at least partly responsible for the observed toxic effects [92]. The sunlight photolysis of MBT in water systems was also investigated by Brownlee and coworkers [93], who found BT (30–45%) and 2-hydroxybenzothiazole (OHBT, 4–5%) as degradation byproducts. Since both BT and OHBT absorb sunlight only very weakly at wavelengths > 290 nm [94], direct photolysis of such compounds will be a minor fate [95]. On the basis of previous studies [93, 96], a partial photolytic pathway was formulated for TCMTB and MBT in aquatic systems and is depicted in Fig. 10. Photolysis of TCMTB leads to MBT through a hydrolytic pathway, which may either photolyze to BT and OHBT. In natural waters, MBT can also undergo biomethylation to MTBT. Thus its photodegradation is also important to be considered.

Laboratory studies [97, 98] have been reported on the photolysis of MBT in organic solvents (acetonitrile, ethanol, or benzene) using UV lamps and Pyrex vessels. Abdou et al. [98] found that the irradiation ($\lambda_{exc} = 313$ nm) of MBT in the previously mentioned solvents produced 5.6% benzothiazole (BT), 9.7% elemental sulfur, 20.8% 2(3)benzothiazolethione (the thione form of MBT), 32% bis(benzothiazolyl)-disulfide (MBTS), and 20.5% 2(3H)-benzothiazolone (OBT), the keto form of OHBT. A pathway that describes the

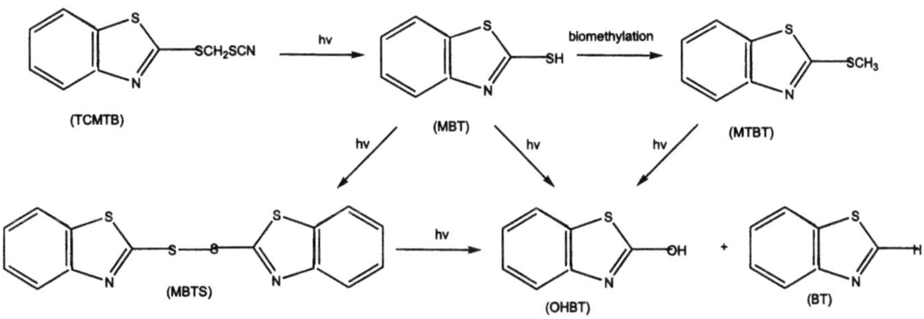

Fig. 10 Photolytic degradation pathways of TCMTB and MTB in aqueous environment, based on reported identified phototransformation products [93, 96]

formation of the obtained photoproducts according to Brownlee et al. [93] is depicted in Fig. 11.

Thus, upon sunlight irradiation, MBT cleaves at the sulfur–hydrogen bond. Subsequent recombination of the thiyl radicals formed leads to the disulfide MBTS (a) [99]. These intermediates have been proposed not only by Abdou et al. [98] but also by Párkányi and Abdelhamid [97]. Irradiation of the MBTS, besides the retro S–S cleavage into to two thiyl radicals (b) [99], another homolysis can be discussed leading to disulfan radical (c) as well as to 2-benzothizolyl radical (d). This type of competitive carbon–sulfur cleavage has been investigated earlier [100] for the photoreactions of disulfide. Radical (d) leads to benzothiazole BT (e) through a hydrogen abstraction mechanism, while 2(3H)-benzothiazolone (OBT) (h) is formed by oxygen

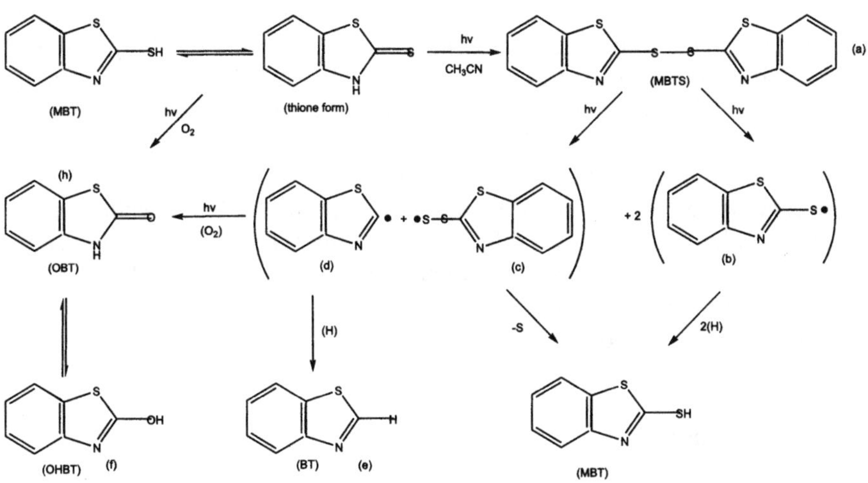

Fig. 11 Photolytic degradation pathways and main transformation products of MTB in aqueous environment (after Brownlee et al. [93])

attack [101]. On the other hand, the disulfane radical (c) can break down to elemental sulfur and MBT after H-transfer. In experiments performed by Kirouani-Harani [96], the irradiation of MBT in water at pH 7 also led to BT and OHBT. This result correlates with those obtained by Brownlee et al. [93], and is also consistent with the previous mechanism (Fig. 11). MBTS also might act as quencher for singlet oxygen as reported by Foote [102] and Foote and Peters [103]. By these means, reaction of MBT with singlet oxygen cannot be excluded since there are many $C = S$, $C = O$ exchange reactions reported for thiones after treatment with singlet oxygen [96, 104, 105]. In such cases the transformation of MBT to OHBT can also occur as a direct photochemically induced reaction with singlet oxygen. Another similar photoconversion mechanism of MBT in concentrated organic solutions of benzene, toluene, alcohols, or acetonitrile is reported elsewhere [106]. MBT was found to be phototransformed not only to MBTS dimer but also to $BTSO_2 - O_2STB$ dimer, as well as to BT and OHBT. The direct photolysis of MBT in water is also reported to take place through either the singlet excited state and the resulting primary intermediates of solvated electrons and benzothiazolyl radicals, as well the triplet excited state. However, the desulfurization steps need further clarification. Episulfide or episulfoxide might be involved as intermediates.

Solar photodegradation of MBT was performed in Milli-Q purified water at pH 8.0 and in a natural water sampled from a lake [106]. Desulfurization into benzothiazole was clearly favored in the latter case and the disappearance of MBT was reported to be fourfold faster in the natural than in the Milli-Q water. The results of this study showed that sensitized reactions mediated by the chromophoric components of natural waters significantly contribute to the transformation of MBT in the aquatic environment.

Finally, the estimated quantum yield (Φ) for MBT using 320 nm radiation was evaluated to be in the range 0.023–0.112 in pure water [96]. Malouki et al. [106] also investigated the direct photolysis of MBT using 313 nm radiation. The anionic form was found to be photoconverted into BT and OHBT when irradiated in aerated medium with similar quantum efficiency ($\Phi = 0.02$).

Solutions of BT and OHBT were further irradiated, but no degradation of these compounds was observed in aqueous solution. All bands exhibited no absorption changes over several hours, indicating the stability of these compounds [96]. The environmentally important conclusion to be drawn from these findings is that BT and OHBT are the anticipated stable products of MBT and TCMTB photolysis in aquatic environments.

3.9 Triphenylboron-pyridine

Hydrolysis, thermolysis, and photolysis are competing abiotic decomposition pathways of triphenylboron-pyridine (TPBP). TPBP is unstable in artificial

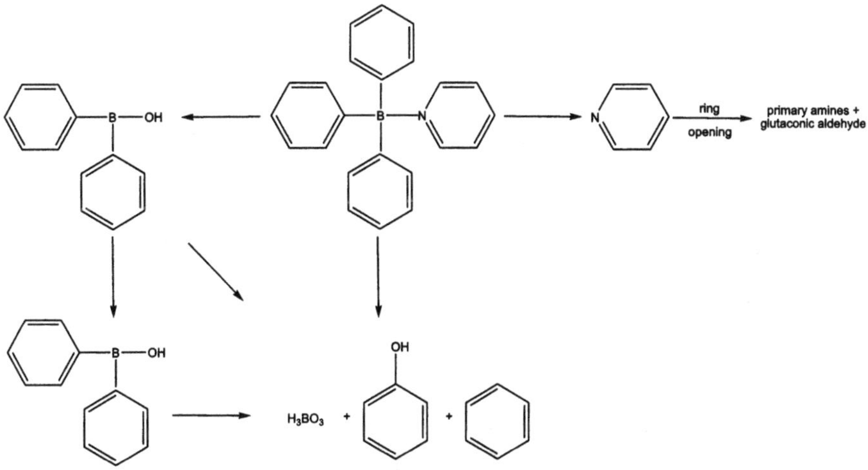

Fig. 12 Photolytic degradation pathways of TPBP in aqueous environment, based on reported identified phototransformation products [108]

seawater and dissipates quickly. Half-lives of TPBP at concentrations of 5, 50, and 500 $\mu g\,L^{-1}$, were 1, 6, and 34 days, respectively. [107].

The photodegradation of TPBP was studied by Amey and Waldron [108] in aerated artificial seawater using UV irradiation (low pressure Hg lamp, 5.5 W, $\lambda = 254$ nm). The photodecomposition of TPBP as well as primary intermediate photoproducts of mono- and diphenylboron acids was quite rapid with a photolytic lifetime of less than 1 h. Finally, pyridine, phenol, and benzene were observed as the secondary products (Fig. 12). Pyridines also decay rapidly due to their photolytic decomposition and result in ring-opening products (including primary amines and glutaconic aldehyde) or their derivatives [108]. Finally, natural sunlight photolysis experiments using sterile artificial seawater resulted in 75–80% of degradation after 3–6 h of light exposure [108]. The photolytic transformation of TPBP into non-persistent photoproducts is of particular interest for the aquatic environment.

All the data regarding the persistence of antifouling paint booster biocides in seawater as well as their photochemical half-lives mentioned in the text are summarized in Table 1.

4
Concluding Remarks

For micropollutant photodegradation studies concerned with environmental considerations it is important to focus on two main objectives: (i) the degradation kinetics (i.e., the rate at which the parent compound degrades in the

Table 1 Summary of data for persistence, main transformation modes, and photochemical half-lives of antifouling paint booster biocides

	Persistence data ($t_{1/2}$)			References	Principal degradation mode
	Seawater	Photolysis			
		Field	Laboratory		
Chlorothalonil	0.2–8.0 days	1.9–4.2 days	0.3–1.5 days	[28–33, 39]	Biotic–photolysis
Dichlofluanid	18 h	53 h	13 h	[35, 36, 39]	Hydrolysis–biotic
Irgarol 1051	100–350 days	56 days	1.6 days	[39, 43, 44, 46–48]	Photolysis
GS26575	82–200 days	56 days		[43, 44, 46]	Photolysis
Sea-Nine 211	< 1–8.4 days	13.1 days	0.42 days	[37, 38, 46, 55–57, 60]	Biotic
Diuron	30–360 days		4–91 days	[38, 43, 65, 66]	Biotic
Zinc pyrithione	< 240 min	2–18 min	17–30 min	[39, 76, 78–80, 82]	Photolysis–biotic
Copper pyrithione	< 24 h	7.1–7.4 min	29.1 min	[79, 81, 82]	Photolysis
Thiram			8.8 h	[87]	Hydrolysis–biotic
Ziram			8.6 h	[88]	Hydrolysis–biotic
Maneb			10.5 h	[89]	Hydrolysis–photolysis
Zineb	96 h			[39]	Hydrolysis
TCMTB	740 h	1.5–3.9 h		[39, 92]	Hydrolysis–photolysis
TPBP	1–34 days	< 2 h	< 1 h	[107, 108]	Photolysis–hydrolysis

aqueous environment and the pathways of transformation) and (ii) toxicity assessment.

Analytical results confirmed the occurrence of both direct and photosensitized mechanisms for the photodegradation of organic booster biocides in the aquatic environment. Kinetics for both mechanisms followed a pseudo-first-order law. All the compounds considered in the present review underwent direct photolysis in solar light. However, in most cases the reactions were slow because either the pollutants absorbed light poorly or they were poorly photoreactive. The only exceptions were zinc and copper pyrithiones, which were susceptible to direct photolysis and degraded in a few minutes. Regarding indirect processes, the rates of phototransformation for most of the biocides were higher in natural water samples than in pure water, indicating that photosensitized processes took place in natural water samples, thereby limiting the long-persistence of these chemicals in the aqueous environment. In contrast (e.g., for dichlofluanid), the photodegradation in natural waters was partially hampered by the presence of DOM. Thus, DOM could have both a photosensitizing or an optical filter effect. The identified photoproducts were of diverse structural nature, showing that phototransformation of antifouling biocides under natural conditions may be a complex process.

Finally, in most cases reduced toxicity was found after the photoinduced transformations, a fact very important for the risk assessment of biocides. The only exception was diuron, which was quite persistent under the irradiated conditions studied (both under natural and simulated solar light) demonstrating quite long half-lives; on the other hand more toxic byproducts appeared, possibly showing synergistic effects.

References

1. Hugget RJ, Unger MA, Seligman PF, Valkiris AO (1993) Environ Sci Technol 26:232
2. Alzieu C (2000) Sci Total Environ 258:99
3. Thomas KV (1998) J Chromatogr A 825:29
4. Thomas KV, Blake SJ, Waldock MJ (2000) Mar Pollut Bull 40:739
5. Boxall ABA, Comber SD, Conrad AU, Howcroft J, Zaman N (2000) Mar Pollut Bull 40:898
6. Martinez K, Ferrer I, Barcelo D (2000) J Chromatogr A 879:27
7. Sakkas VA, Kontsantinou IK, Lambropoulou DA, Albanis TA (2002) Environ Sci Pollut Res 9:327
8. Fielding M, Barcelo D, Helweg A, Galassi S, Torstensson L, Van Zoonen P, Wolter R, Angeletti A (1992) Pesticides in ground and drinking water. Water pollution research report 27. European Commission, Brussels
9. Torrents A, Anderson BG, Bilbounian S, Jonhson WE, Hapeman CJ (1997) Environ Sci Technol 31:1476
10. Zepp RG, Schlotzauer PF, Sink RM (1985) Environ Sci Technol 19:74
11. Mill T, Hendry DG, Richardson H (1980) Science 207:886
12. Barrows HD, Canle L, Santaballa JA, Steenken S (2002) J Photochem Photobiol B: Biol 67:71

13. Murov SL, Carmichael I, Hug GL (1993) Handbook of photochemistry. Marcel Dekker, New York
14. Krupa SV, Kickert RN, Jager HJ (1998) Elevated ultraviolet (UV)-B radiation and agriculture. Springer, Berlin Heidelberg New York
15. Vaughan PP, Blough NV (1998) Environ Sci Technol 32:2947
16. Durand G, Barcelo D, Albaiges J, Mansour M (1990) Chromatographia 29:120
17. Hirsch M, Hutzinger O (1989) Environ Sci Technol 23:1306
18. Barrows HD, Ernestova LS, Kemp TJ, Skurlatov YI, Purmal AP, Yermakov AN (1998) Prog React Kin 23:145
19. Zepp RG, Hoigne J, Bader H (1987) Environ Sci Technol 21:443
20. Lam MW, Tantuco K, Mabury SA (2003) Environ Sci Technol 37:899
21. Larson RA, Zepp RG (1988) Environ Toxicol Chem 7:265
22. Huang J, Mabury SA (2000) Chemosphere 41:1775
23. Mill T, Richardson H, Hendry DG (1978) Oxidation of organic compounds in aquatic systems: the free radical oxidation of cumene vol 1. Pergamon, New York
24. Zepp RG, Wolfe GL, Baughman GL, Hollis RC (1977) Nature 267:421
25. Fischer AM, Winterle JS, Mill T (1987) In: Zika RG, Cooper WJ (eds) Photochemistry of environmental aquatic systems. ACS symposium series 327. American Chemical Society, Washington, DC, p 141
26. Larson RA, Weber EJ (1994) Reaction mechanisms in environmental organic chemistry. CRC, Boca Raton, Florida
27. Liao CH, Kang SF, Wu FA (2001) Chemosphere 44:1193
28. Davies PE (1987) Bull Environ Contam Toxicol 4:405
29. Walker WW, Cripe CR, Prichard PH (1988) Chemosphere 17:2255
30. Caux PY, Kent RA, Fan GT, Stephenson GL (1996) Crit Rev Environ Sci Technol 26:45
31. Sakkas VA, Lambropoulou DA, Albanis TA (2002) Chemosphere 48:939
32. Penuela GA, Barceló D (1998) J Chromatogr A 823:81
33. Millet M, Palm WU, Zetzsch C (1996) Exotoxicol Environ Saf 41:44
34. Sakkas VA (2002) PhD Thesis, University of Ioannina
35. Sakkas VA, Konstantinou IK, Albanis TA (2001) J Chromatogr A 930:135
36. Callow ME, Finlay JA (1995) Biofouling 9:239
37. Thomas KV, McHugh M, Hilton M, Waldock M (2003) Environ Pollut 123:153
38. Callow ME, Willingham GL (1996) Biofouling 10:239
39. Thomas KV (2001) Biofouling 17:73
40. Liu D, Maguire RJ, Lau YL, Pacepavicious GJ, Okamura H, Aoyama I (1997) Water Res 31:2363
41. Liu D, Pacepavicious GJ, Maguire RJ, Lau YL, Okamura H, Aoyama I (1999) Water Res 33:155
42. Okamura H, Aoyama I, Liu D, Maguire RJ, Pacepavicius GJ, Lau YL (1999) J Environ Sci Health B 34:225
43. Okamura H (2002) Chemosphere 48:43
44. Hall LW, Giddings JM, Solomon KR, Balcomb R (1999) Crit Rev Toxicol 29:367
45. Okamura H, Aoyama I, Takami T, Maruyama T, Suzuki Y, Matsumoto M, Katsuyama I, Hamada J, Beppu T, Tanaka O, Maguire RJ, Liu D, Lau YL, Pacepavicious GJ (2000) Mar Pollut Bull 40:754
46. Thomas KV, McHugh M, Hilton M, Waldock M (2002) Sci Total Environ 293:117
47. Sakkas VA, Lambropoulou DA, Albanis TA (2002) J Photochem Photobiol A:Chem 147:135
48. Okamura H, Sugiyama Y (2004) Chemosphere 57:739

49. Frimmel FH, Hessler DP (1994) In: Helz GR, Zepp RG, Crosby DG (eds) Aquatic and surface photochemistry. CRC, Boca Raton, Florida, p 137
50. Khan SU, Gamble DS (1983) J Agric Food Chem 31:1099
51. Konstantinou IK, Sakellarides TM, Sakkas VA, Albanis TA (2001) Environ Sci Technol 35:398
52. Torrents A, Anderson BG, Bilboulian S, Johnson WE, Hapeman C (1997) Environ Sci Technol 18:1476
53. Lam KH, Lam MHW, Lam PKS, Qian T, Cai Z, Yu H, Cheung YH (2004) Mar Pollut Bull 49:336
54. Ogawa N, Okamura H, Hirai H, Nishida T (2004) Chemosphere 55:487
55. Jacobson AH, Willingham GL (2000) Sci Total Environ 258:103
56. Ranke J (2002) Environ Sci Technol 36:1539
57. Willingham GL, Jacobson AH (1996) In: Devito SC, Garett RL (eds) Designing safer chemicals: Green chemistry for pollution prevention. ACS symposium series 640. American Chemical Society, Washington, p 224
58. Shade WD, Hurt SS, Jacobson AH, Reinert KH (1994) In: Gorsuch JW, Dwer FW, Ingersoll CM, LaPoint TW (eds) Ecological risk assessment of a novel marine antifoulant. Environmental toxicology and risk assessment, vol 2. American Society for Testing and Materials, Philadelphia, p 381
59. Willingham GL, Jacobson AH (1999) In: Proceedings of the third Asia-Pacific conference on paint research association, no 14. International Center for Coatings Technology, pp 1–13
60. Sakkas VA, Konstantinou IK, Albanis TA (2002) J Chromatogr A 959:215
61. Jacobson A, Mazza LS, Thompson W (1999) Annual meeting of the Society of Environmental Toxicology and Chemistry (SETAC), Philadelphia, Pennsylvania
62. Kopecky J (1992) Organic photochemistry: A visual approach. Wiley, New York
63. Ramamurthy V, Schanze KS (1997) Molecular and supramolecular photochemistry. Marcel Dekker, New York
64. Rokach J, Hamel P (1979) JCS Chem Comm, p 786
65. Wauchope RD, Buttler TM, Hornsby AG, Augustijn-Beckers PWM, Burt JP (1992) Rev Environ Toxicol 123:1
66. Mazellier P, Jirkovsky J, Bolte M (1997) Pest Sci 49:259
67. Faure V, Boule P (1996) XXVI Congres, Groupe Français des Pesticides, Nancy, France
68. Bonnemoy F, Lavedrine B, Boulamh A (2004) Chemosphere 54:1183
69. Jirkovsky J, Faure V, Boule P (1997) Pest Sci 50:42
70. Faure V (1996) PhD Thesis, University of Clermont-Ferrand
71. Miller GC, Zisook R, Zepp R (1980) J Agric Food Chem 28:1053
72. Othmen K, Boule P (1999) J Photochem Photobiol A:Chem 121:161
73. European Commission, Scientific Committee on Toxicity, Ecotoxicity and the Environment (2002) Risk assessment of 3,4-dichloroaniline. CSTEE Report, Brussels
74. Lanyi K, Dinya B (2005) Microchem J 80:79
75. Tixier C, Bogaerts P, Sancelme M, Bonnemoy F, Twagilimana L, Cuer A, Bohatier J, Veschambre H (2000) Pest Manag Sci 56:455
76. Turley PA, Fenn RJ, Ritter JC (2000) Biofouling 15:175
77. Galvin RM, Mellado JMR, Neihof RA (1998) Eur Water Manag 4:61
78. Neihof RA, Bailey CA, Patouillet C, Hannan PJ (1979) Environ Contam Toxicol 8:355
79. Maraldo K, Dahllof I (2004) Mar Pollut Bull 48:894
80. http://www.nra.gov.au/registration/prs.shtml

81. www.pcimag.com/CDA/ArticleInformation/features/BNP__Features__Item/0,1846,12 321,00.html
82. Yamaguchi Y, Kumakura A, Ishigami M, Shibata K, Senda T, Yamada Y (2004) In: Shibata K, Senda T (eds) Proceedings of international symposium on antifouling paint and marine environment (InSAfE), Tokyo, Japan, p 228
83. Zepp RG, Cline DM (1977) Environ Sci Technol 11:359
84. Polson G, Skoulis N, Ritter J (2005) In: Proceedings of 15th annual meeting of the Society of Environmental Toxicology and Chemistry (SETAC), Lille, France, p 145
85. Evans SM, Birchenough AC, Brancato MS (2000) Mar Pollut Bull 40:204
86. Gruiskshan PA, Jarrow HC (1973) J Agric Food Chem 21:333
87. www.fao.org/WAICENT/FAOINFO/AGRICULT/AGP/AGPP/Pesticid/JMPR/Download /96_eva/thiram.pdf
88. US EPA (2002) Ziram: a preliminary HED risk assessment for the reregistration eligibility decision (RED) document. US EPA, Washington DC, 20460
89. Downing E (2000) Environmental fate of maneb. Environmental Monitoring and Pest Management, Department of Pesticide Regulation, Sacramento, CA 95814-3510
90. Xu S (2000) Environmental fate of ethylenethiourea. Environmental Monitoring and Pest Management, Department of Pesticide Regulation, Sacramento CA 95814-3510
91. Ros RD, Crosby DG (1973) J Agric Food Chem 21:335
92. van Wezel AP, van Vlaardingen P (2001) Maximum permissible concentrations and negligible concentrations for antifouling substances. National Institute of Public Health and the Environment RIVM report 601501008. The Netherlands
93. Brownlee BG, Carey JH, MacInnis GA, Pellizzari IT (1992) Environ Toxicol Chem 11:1153
94. Fife TH, Hutchins JEC, Wang MS (1975) J Am Chem Soc 97:5878
95. Konstantinou IK, Zarkadis AK, Albanis TA (2001) J Environ Qual 30:121
96. Kirouani-Harani H (2003) PhD thesis, Technical University of Berlin
97. Párkányi C, Abdelhamid AO (1985) Heterocycles 23:2917
98. Abdou WM, Sidky MM, Wamhoff H (1987) Z Naturforsch 42:1153
99. Knight AR (1974) In: Patai S (ed) Chemistry of the thiol group, vol 1. Wiley, New York, p 445
100. Coyle JD (1986) Introduction of organic photochemistry. Wiley, p 159
101. Crank G, Mursyidi A (1982) Aust J Chem 35:775
102. Foote CS (1979) In: Wasserman HH, Murray RW (eds) Singlet oxygen. Academic, New York, p 139
103. Foote CS, Peters GW (1971) J Am Chem Soc 93:3795
104. Wamhoff H, Ertas M (1982) Angrew Chem 94:800
105. Suzuki N, Sano K, Tani N, Izawa Y (1981) Heterocycles 16:1133
106. Malouki MA, Richard C, Zertal A (2004) J Photochem Photobiol A:Chem 167:121
107. Takahashi K, Yoshikawa E, Akiyama M, Kitaori K, Masuoka S (2004) In: Shibata K, Senda T (eds) Proceedings of international symposium on antifouling paint and marine environment (InSAfE), Tokyo, Japan, p 196
108. Amey RL, Waldron C (2004) In: Shibata K, Senda T (eds) Proceedings of international symposium on antifouling paint and marine environment (InSAfE), Tokyo, Japan, p 239

Hdb Env Chem Vol. 5, Part O (2006): 201–212
DOI 10.1007/698_5_055

Published online: 23 November 2005

Present Status of Antifouling Systems in Japan: Tributyltin Substitutes in Japan

Hideo Okamura (✉) · Hirohisa Mieno

Faculty of Maritime Sciences, Kobe University, Fukaeminami, 5-1-1 Higashinada, 658-0022 Kobe, Japan
okamurah@maritime.kobe-u.ac.jp

Abstract An international conference held by the International Maritime Organization in October 2001 adopted an "International convention on the control of harmful anti-fouling systems on ships." This treaty includes bans on the use of harmful organotin compounds, which act as biocides in paint products. The Japanese government ratified the treaty in July 2003. The Japan Paint Manufacturers' Association (JPMA) has commenced a program of self-control of antifouling systems in Japan and has issued on their website a list of registered paint products with and without certain biocides. The purpose of this review is to assess the status of the antifouling systems that are being used in Japan, based on the JPMA list. A total of 380 paint products from ten paint companies were registered by June 2005. Sixteen biocides are registered in 359 products, and the remaining 21 products contain no biocides. Single biocides are used in 60 products, as two-mixtures in 200 products, as three-mixtures in 80 products, and as four-mixtures in 19 products. Five biocides, such as cuprous oxide, triphenylborane-pyridine, Sea-Nine 211, copper thiocyanate, and chlorothalonil, are used as a single product. Combinations such as cuprous oxide and copper pyrithione, zinc pyrithione and triphenylborane-pyridine, cuprous oxide and Diuron, and cuprous oxide and zinc pyrithione are mainly used in two-biocide mixtures to produce synergistic antifouling effects.

Keywords Biocide · Combination · Mixture · Organotin-free · Paint

Abbreviations

CuPT	Copper pyrithione
DCF	Dichlofluanid
Densil	Densil S-100
IMO	International Maritime Organization
Irgarol	Irgarol 1051
JPMA	Japan Paint Manufacturers' Association

NACS	Naphthenic acids, copper salts
SN211	Sea-Nine 211
TBT	Tributyltin
TCPM	*N*-(2,4,6-Trichlorophenyl)maleimide
Thiram	Tetramethylthiuram disulfide
TPBP	Triphenylborane-pyridine
TPN	Chlorothalonil
ZnPT	Zinc pyrithione

1 Introduction

Not enough information on the new antifouling agents as tributyltin (TBT) substitutes has been released to public in Japan since a strict regulation on organotin compounds began in the late 1980s. The manufacture of TBT oxide has been prohibited since it was defined in 1989 as a "class 1 specified chemical substance" by the law concerning Examination and Regulation of Manufactures of Chemical Substances. Thirteen TBT compounds and seven triphenyltin compounds were defined in 1990 as "class 2 specified chemical substances," requiring a report on the manufacturing/importing of these substances, and they have not been legally used in Japan since 1996. Therefore, ship antifouling paints that have been manufactured in Japan since 1996 are considered not to contain organotins.

Limited information on the efficiency of TBT substitutes has been available worldwide since the late 1990s [1–3]. Until early 2004, there was only limited information on the TBT substitutes regularly used in Japan [4, 5]. Therefore, environmental scientists have analyzed residue levels of suspected antifouling agents to investigate their potential use in Japan. Detection of Irgarol 1051 (Irgarol), its metabolite, and Diuron in water samples collected from coastal marinas, fishery harbors, and ports has provided information on their possible use in antifouling paints on ships and boats in Japan [6–9]. Meanwhile, the "International convention on the control of harmful anti-fouling systems on ships, 2001 (IMO Convention)" was adopted at the International Maritime Organization (IMO) diplomatic conference held in October 2001. The Japanese government ratified the IMO Convention in July 2003. The convention requires publicly accessible data to be provided on TBT substitutes that have been or will be introduced into the Japanese market as antifouling paint products. The antifouling systems used in Japan are regulated by the Japan Paint Manufacturers' Association (JPMA), and "the JPMA list of registered organotin-free anti-fouling paints (the List)" was first publicly available in early 2004 [10]. It is clearly defined on their website that the aim of the JPMA self-regulatory management program is to provide IMO Convention compliance information and related information to ship owners, ship operators, government regulatory authorities and other related bodies, by examining the paints manufactured/distributed by JPMA regular members, supporting members or nonmember companies. This is the first source of official and historical information on TBT substitutes commercially used in Japan. Consumers are able to learn what biocides are in different types of commercial paints. The List has been revised several times since it was first compiled in 2004. In this work, the present status of the antifouling systems used in Japan, in terms of the biocides they contain and their combinations in paint products, was investigated using the latest version of the List.

2
TBT Substitutes in Japan

2.1
Biocides in Antifouling Paint Products

The List comprises information on the name, color and type of the antifouling system, its CAS number, its chemical name, and whether it is manufactured in Japan or by other manufacturers, but it excludes data on the amount produced, utilization rates and disposal methods, etc. According to the List, a total of 380 paint products made by ten companies are registered. Figure 1 shows the percentage of each company based on the number of products. Four companies (A, B, C, D) produce about 75% of the total in terms of the number of products.

Every commercial product is classified into one of the types of antifouling systems on the List. The antifouling systems have been classified into 12 types (Table 1), according to whether or not they contain a biocide and which

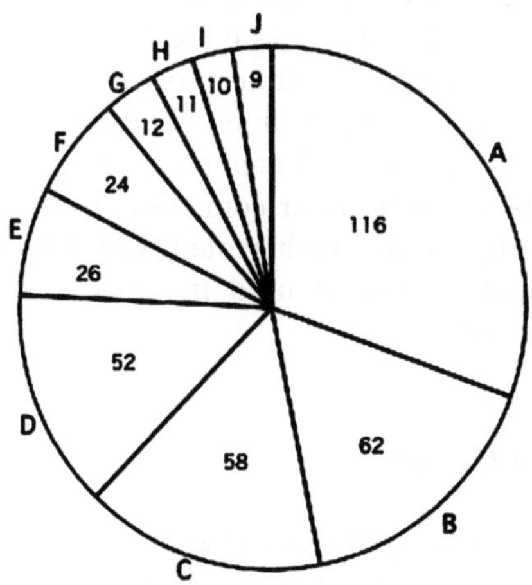

Fig. 1 Number of commercial products per company

Table 1 Number of paint products per antifouling system

Antifouling system	Number of products
Organotin-free self-polishing type	155
Organotin-free self-polishing hydrolysis type	85
Organotin-free ablative type	60
Organotin-free self-polishing hydration type	30
Organotin-free conventional type	24
Biocide-free silicon type	16
TBT-free self-polishing copolymer	4
Foul release coating	2
Conventional antifouling	1
Hybrid TBT-free self-polishing	1
TBT-free controlled depletion polymer	1
TBT-free soluble matrix	1
Total	380

TBT tributyltin

type. Sixteen products (4.2%) are of a "biocide-free silicone" type, and the other 364 products (95.8%) contain biocides free of organotin compounds. The most popular type is the "organotin-free self-polishing type" with 155 products, followed by the "organotin-free self-polishing hydrolysis type" with 85 products, and the "organotin-free ablative type" with 60 products. These three types comprise 82% of the biocidal paint products. The sum of the self-polishing-type products was 275 (76%).

The 96% of paint products in the List that contain biocides are listed in Table 2 and are shown in Fig. 2. The 16 biocides are numbered according to their frequencies in the products on the List. Cu_2O is used most frequently, and tetramethylthiuram disulfide (thiram) is used in only two products. Most of the biocides are organic or organometallic biocides. Among the 16 biocides, triphenylborane-pyridine (TPBP) (**4**), CuSCN (**10**), *N*-(2,4,6-trichlorophenyl)maleimide (**13**), and naphthenic acids, copper salts (15) are not listed as TBT substitutes in other countries [1–3]. The antifouling biocides 2-(thiocyanomethylthio)benzothiazole (TCMTB) and manganese ethylene bis(dithiocarbamate) (maneb) used in other countries [1,2] are not included in the JPMA List.

2.2 Combinations of Biocide Mixtures

Marine antifouling paints are well blended as a combination of biocide, resin, and solvent to achieve antifouling performance, color development, and repellent or biocidal effect. The biocide combinations in the 380 paint products

Table 2 TBT substitutes in the registered list approved by the Japan Paint Manufacturers' Association

No.	Abbreviation	Common name	Chemical name	CAS no.
1	Cu_2O	Dicopper oxide	Cuprous oxide	1317-39-1
2	CuPT	Copper pyrithione	Copper, bis(1,hydroxy-2(1*H*)-pyridinethionato *O*,*S*)	14915-37-8
3	ZnPT	Zinc pyrithione	Zinc-2-pyridinethiol-1-oxide	13463-41-7
4	TPBP	PK, TPBP	Triphenylborane-pyridine	971-66-4
5	Diuron	Diuron, DCMU	3-(3,4-Dichlorophenyl)-1,1-dimethyl urea	330-54-1
6	SN211	Sea-Nine 211	4,5-Dichloro-2-n-octyl-4-isothiazolin-3-one	64359-81-5
7	Irgarol	Irgarol 1051	2-Methylthio-4-*tert*-butylamino-6-cyclopropylamino-*s*-triazine	28159-98-0
8	TPN	Chlorothalonil	2,4,5,6-Tetrachloroisophthalonitrile	1897-45-6
9	DCF	Dichlofluanid	*N*,*N*-Dimethyl-*N*′-phenyl-*N*′-(dichlorofluoromethylthio)ulfamide	1085-98-9
10	CuSCN	Cuprous Thiocyanate	Cuprous thiocyanate	1111-67-7
11	Ziram	Ziram, PZ	Zinc dimethyl dithiocarbamate	137-30-4
12	Zineb	Zineb	Zinc ethylenebis(dithiocarbamate)	12122-67-7
13	TCPM	IT-354	*N*-(2,4,6-Trichlorophenyl)maleimide	13167-25-4
14	Densil	Densil S-100	2,3,5,6-Tetrachloro-4-(methylsulfonyl)pyridine	13108-52-6
15	NACS	Naphthenic acids, copper salts	Naphthenic acids, copper salts	1338-02-9
16	Thiram	Thiram, TT	Tetramethylthiuram disulfide	137-26-8

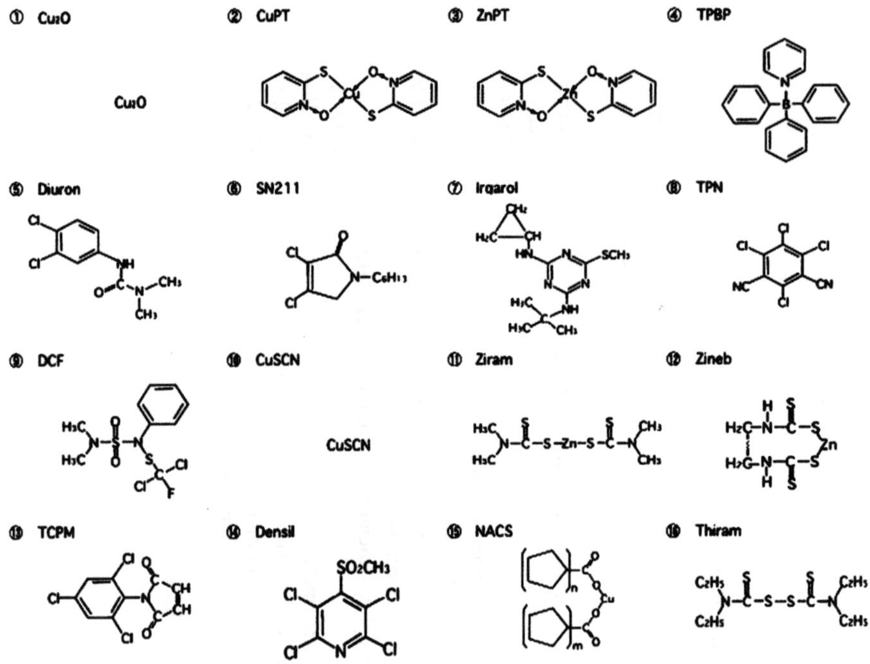

Fig. 2 Chemical structure of tributyltin substitutes in the registration list

are shown for each biocide in Table 3, and the ratio of the combinations is illustrated in Fig. 3.

Cu_2O is combined in 279 products, followed by copper pyrithione (CuPT) in 127 products, zinc pyrithione (ZnPT) in 87 products, TPBP in 63 products, Diuron in 63 products, Sea-Nine 211 (SN211) in 39 products, and Irgarol in 32 products. Two-biocide mixtures with 200 products comprise the majority

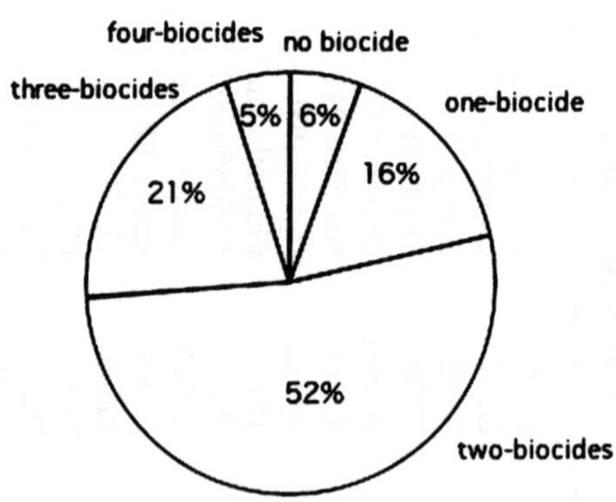

Fig. 3 Combinations of biocide mixtures

Table 3 Combinations of biocides in antifouling paint products

Biocide	Numbers of products No biocide	One biocide	Two-biocide combination	Three-biocide combination	Four-biocide combination	Total
1. Cu_2O	0	52	153	66	8	279
2. CuPT	0	0	100	24	3	127
3. ZnPT	0	0	60	18	9	87
4. TPBP	0	5	40	8	10	63
5. Diuron	0	0	23	33	7	63
6. SN211	0	1	1	30	7	39
7. Irgarol	0	0	11	10	11	32
8. TPN	0	1	1	19	2	23
9. DCF	0	0	4	13	3	20
10. CuSCN	0	1	3	5	4	13
11. Ziram	0	0	0	3	7	10
12. Zineb	0	0	4	1	2	7
13. TCPM	0	0	0	2	3	5
14. Densil	0	0	0	4	0	4
15. NACS	0	0	0	2	0	2
16. Thiram	0	0	0	2	0	2
Number of products	21	60	400/2 = 200	240/3 = 80	76/4 = 19	380

of combinations, and three-biocide mixtures occur in 80 products. The trends are clearly shown in Fig. 4.

In the 60 products containing one biocide, Cu_2O is used in 52 products, followed by TPBP in five products, and SN211, chlorothalonil (TPN) and CuSCN in one product each (Table 3).

In two-biocide mixtures, Cu_2O is used in 153 products, followed by CuPT in 100 products, ZnPT in 60 products, TPBP in 40 products, and Diuron in 23 products. Two-biocide mixtures consisted of 13 combinations with 200 products (Table 4). The major combinations are Cu_2O and CuPT with 99 products, ZnPT and TPBP with 36 products, Cu_2O and ZnPT with 23 products, and Cu_2O and Diuron with 22 products.

In three-biocide mixtures, Cu_2O is used in 66 products, followed by Diuron in 33 products, and SN211 in 30 products. Three-biocide mixtures consisted of 24 combinations with 80 products (Table 5). The major combinations are Cu_2O, Diuron, and SN211 with 13 products, Cu_2O, CuPT, and dichlofluanid (DCF) with 13 products, and Cu_2O, Diuron, and TPN with 11 products.

Four-biocide mixtures consisted of 19 products, with Irgarol used in 11 products, followed by TPBP in ten products, ZnPT in nine products, and Cu_2O in eight products. Four-biocide mixtures consisted of seven combinations, and the major one is ZnPT, TPBP, Irgarol and ziram (Table 6).

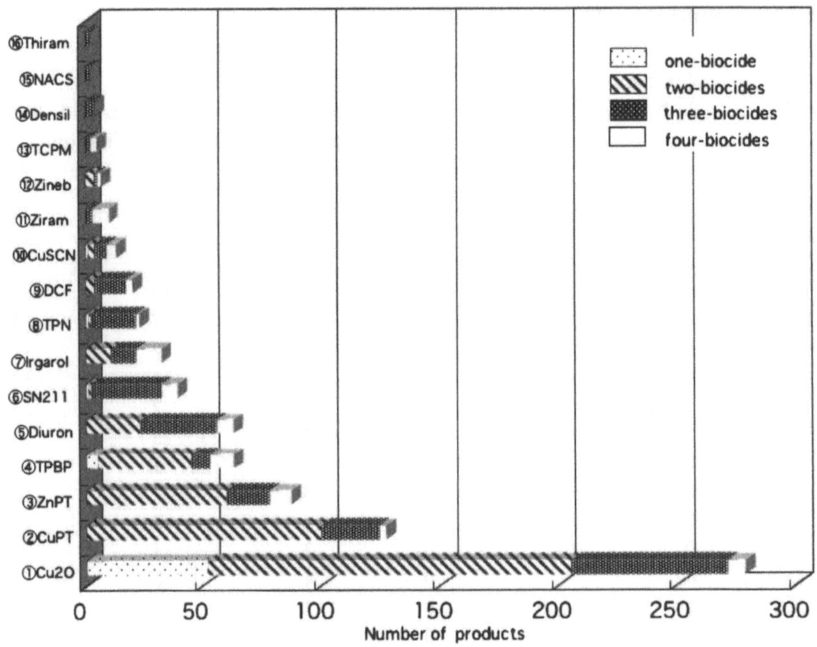

Fig. 4 Combinations of biocides in antifouling paint products

Table 4 Combinations of two-biocides with the number of products for each type

Cu_2O + CuPT: 99	ZnPT + TPBP: 36	Cu_2O + ZnPT: 23	Cu_2O + Diuron: 22
TPBP + DCF: 4	TPBP + Irgarol: 4	Cu_2O + zineb: 4	Cu_2O + Irgarol: 3
TPN + Diuron: 1	Cu_2O + CuSCN: 1	Cu_2O + SN211: 1	CuPT + CuSCN: 1
ZnPT + CuSCN: 1			

The biocides numbered **1–8** are frequently combined in antifouling paint products (Table 3). The four most frequently combined biocides, Cu_2O, CuPT, ZnPT, and TPBP, are used alone or in two-biocide mixtures, with a frequency greater than 50%. Conversely, the other 12 biocides listed in Table 3 are mainly combined in three- or four-biocide mixtures. From these combinations, it can be assumed that the role of the remaining 12 is as a cobiocide to enhance the antifouling effects of the first four biocides. Cu_2O is used with 13 cobiocides in 73% of the total products, but it is never used with TPBP, ziram, or thiram. CuPT is always used with Cu_2O, and is not used alone. CuPT or ZnPT is used in two-biocide mixtures in most cases. ZnPT and CuPT have a similar chemical structure with a different central metal in the mother moiety. It has been reported that ZnPT is converted to CuPT when it is combined with Cu_2O [5]. There is only one product containing a combination of Cu_2O, CuPT, and ZnPT in the List. TPBP is most frequently used in two-biocide mixtures with 40 products. It is reported that TPBP cannot be used with Cu_2O,

Table 5 Combinations of three-biocides with the number of products for each type

Cu_2O + Diuron + SN211: 13	Cu_2O + CuPT + DCF: 13	Cu_2O + Diuron + TPN: 11	Cu_2O + SN211 + TPN: 6
Cu_2O + CuPT + SN211: 5	Cu_2O + ZnPT + Irgarol: 4	ZnPT + TPBP + ziram: 3	Cu_2O + CuPT + CuSCN: 2
Cu_2O + CuPT + DS100: 2	Cu_2O + Diuron + Irgarol: 2	Cu_2O + SN211 + TCPM: 2	Cu_2O + ZnPT + NACS: 2
Diuron + thiram + DS100: 2	ZnPT + Diuron + Irgarol: 2	ZnPT + TPBP + TPN: 2	Cu_2O + CuPT + Diuron: 1
Cu_2O + CuPT + ZnPT: 1	Cu_2O + SN211 + Irgarol: 1	Cu_2O + ZnPT + Diuron: 1	Diuron + Irgarol + CuSCN: 1
TPBP + SN211 + CuSCN: 1	ZnPT + Irgarol + zineb: 1	ZnPT + TPBP + CuSCN: 1	ZnPT + TPBP + SN211: 1

Table 6 Combinations of four-biocides with the number of products for each type

ZnPT + TPBP + Irgarol + ziram: 6	Diuron + TPBP + SN211 + CuSCN: 3	Cu_2O + Diuron + Irgarol + DCF: 3
Cu_2O + CuPT + SN211 + TCPM: 3	Cu_2O + ZnPT + Irgarol + zineb: 2	ZnPT + TPBP + ziram + TPN: 1
Diuron + SN211 + CuSCN + TPN: 1		

because boron in the mother moiety of TPBP reacts with copper [5]. As proof of this, there is no combination of TPBP with Cu_2O in the List. Diuron is most frequently used in three-biocide mixtures with 33 products. Diuron is used with Cu_2O in most products at a frequency of 84% (53/63 products). SN211 is used in three-biocide mixtures at a frequency of 75%, and in four-biocide mixtures at a frequency of 17%. Irgarol is not used alone, but is used in two- to four-biocide mixtures. TPN is used in three-biocide mixtures at a frequency of 83%. The biocides numbered **9–16** in Table 3 are rarely used alone, but are mainly used in combination with the biocides numbered **1–8**.

The List contains the type of antifouling system that was applied by a company. All products with biocides are classified as an organotin-free self-polishing hydrolysis type or hydration type, an ablative type, or a conventional type, according to the leaching mechanisms of the biocides from the paint surface. Table 7 shows the top eight antifouling systems with biocides. The most popular is the self-polishing hydrolysis type containing Cu_2O and CuPT with 65 products. Cu_2O alone is classified as a conventional type (21 products) and an ablative type (20 products). The mixtures with Cu_2O are classified into five types, depending on the cobiocide. For example, the combination Cu_2O and CuPT is classified as a self-polishing hydrolysis type (65 products), a self-polishing type (21 products), and a self-polishing hydration type (12 products).

Table 7 Top eight organotin-free antifouling systems

Biocide mixtures	Type of antifouling system	Products
Self-polishing hydrolysis type	Cu_2O + CuPT	65
Self-polishing type	ZnPT + TPBP	25
Conventional type	Cu_2O	21
Self-polishing type	Cu_2O + CuPT	21
Ablative type	Cu_2O	20
Self-polishing type	Cu_2O + ZnPT	16
Ablative type	Cu_2O + Diuron	15
Self-polishing hydration type	Cu_2O + CuPT	12

There are 70 products for copper-free paints, and they comprise 26% of the 380 products. In copper-free products, three biocides, TPBP, SN211, and TPN, are a major component of one-biocide combinations, and the combination of ZnPT and TPBP occurs in 50 products (71%) as a two-or-more-biocide mixture.

The List includes color information on the products. Most of the red products always contained Cu_2O, and the products with a nonred color contained TPBP, ZnPT, and SN211 as major biocides.

2.3 Effectiveness of Antifouling Paints

Interaction effects between biocides are not well known to date. Synergistic and antagonistic effects of some biocides have been evaluated using nontarget test organisms [11]. One combination, consisting of the three biocides Irgarol, Diuron, and DCF, comprising 30% of the total products containing Irgarol, was found to enhance toxicity. This combination had synergistic effects on test species. On the other hand, one product containing Irgarol and SN211 was registered on the List, although it is reported that this combination had antagonistic or additive effects [11]. Antifouling paint products are carefully designed to provide the synergistic effects of biocides at lower concentrations. The combination TPBP and Cu_2O is not included in products, probably because of its antagonistic effect [5]. It may be assumed that the biocide combinations in the List have been selected because of their synergistic effects between biocides.

The data analyzed in this paper are based on the information (380 products from ten companies) released by the JPMA in June 2005. The List has been periodically revised since it was first compiled in 2004; therefore, future changes in antifouling systems in Japan can be continually assessed as the List is revised. Information on the level of production and use of the biocides, along with data on the species and biocide combinations, is expected to be publicly available in the near future.

3 Conclusions

The data registered by the JPMA that were released in June 2005 were analyzed to assess the present status of antifouling systems used in Japan. The following conclusions are drawn from this assessment.

1. The antifouling paint products registered in the Japanese market comprise 380 types manufactured by ten companies.
2. There were 21 nonbiocidal products (5%), and 16 biocides were included in 95% of the total items.

3. Among the 380 paint products, two biocides were included in 200 types, three biocides in 80 types, only one biocide in 60 types, and four biocides in 19 types.
4. The biocides used alone were limited to Cu_2O, TPBP, SN211, CuSCN, and TPN.
5. The combinations containing two biocides comprised 89% of the total in the four combinations Cu_2O and CuPT, Cu_2O and ZnPT, Cu_2O and Diuron, and ZnPT and TPBP.
6. The combinations including three and four biocides are based on the four combinations mentioned in conclusion 5, with the addition of one to two other biocides, which corresponded to 80 and 79% of the total items.
7. Synergistic effects of certain biocide combinations should result in effective antifouling action at lower concentrations.

Acknowledgements This investigation was conducted as a part of a research project funded by a Grant-in-Aid for Scientific Research (B16310026), the Iwate Yume Kendo Project, and The Shipbuilder's Association of Japan.

References

1. Voulvoulis N, Scrimshaw MD, Lester JN (1999) Appl Organomet Chem 13:135
2. Thomas KV (2001) Biofoul 17:73
3. Konstantinou IK, Albanis TA (2004) Environ Int 30:235
4. Yonehara Y (2000) Bull Soc Sea Water Sci Jpn 54:7
5. Yamada H, Kakuno A (2003) Suiken Center Rep 6:56
6. Liu D, Pacepavicius GJ, Maguire RJ, Lau YL, Okamura H, Aoyama I (1999) Water Res 33:2833
7. Okamura H, Aoyama I, Liu D, Maguire RJ, Pacepavicius GJ, Lau YL (2000) Water Res 34:3523
8. Okamura H, Aoyama I, Ono Y, Nishida T (2003) Mar Pollut Bull 47:59
9. Harino H, Mori I, Yamaguchi Y, Shibata K, Senda T (2005) Arch Environ Contam Toxicol 48:303
10. Japan Paint Manufacturers' Association (2005) http://www.toryo.or.jp/jp/anzen/index.html
11. Fernández-Alba AR, Hernando MD, Piedra L, Chisti Y (2002) Anal Chim Acta 456:303

Hdb Env Chem Vol. 5, Part O (2006): 213–226
DOI 10.1007/698_5_056

Published online: 23 November 2005

Toxicity and Preliminary Risk Assessment of Alternative Antifouling Biocides to Aquatic Organisms

Hisashi Yamada

National Research Institute of Fisheries Science,
Fisheries Research Agency, 2-12-4, Fukuura, Kanagawa, 236-8648 Yokohama, Japan
yamaq@fra.affrc.go.jp

Abstract Published literature has been reviewed regarding the toxicity of representative alternative antifouling biocides (Chlorothalonil, Dichlofluanid, Diuron, Irgarol 1051, Sea Nine 211, TCMTB, Zineb, ZnPT, CuPT and PK) in aquatic environments, and their hazardous impact on aquatic environments has been evaluated using the information obtained during the process of review. The following statements represent the conclusions. Acute toxicities of the alternative antifouling biocides are in a range of 10^{-3} to 10^{1} mg L^{-1}. Irgarol 1051 exhibits especially high toxicity to phytoplankton and seaweed. On the basis of these toxicity values the biocides are classified as "toxic to very toxic" by the OECD classification guidelines. The acute toxicity value of M1, which is a degradation product of Irgarol 1051, ranges between 10^{-2} and 10^{1} mg L^{-1}. M1 is also classified as "toxic" to "very toxic" by the OECD classification guidelines. The toxicity levels of these alternative antifouling biocides are almost the same as TBT. The reported concentrations of Irgarol 1051, M1, and Sea Nine 211 exceed the respective PNEC values especially in marinas and fishery harbors implying that Irgarol 1051, M1, and Sea Nine 211 are already causing a hazardous impact on the aquatic ecosystem in some marinas and fishery harbors. In order to assess the ecotoxicological risk of these biocides in detail, further research to clarify the toxicity and to develop the methods of estimating concentrations in natural waters are needed.

Keywords Alternative biocide · Aquatic organisms · Risk assessment · TBT free antifouling paints · Toxicity

Abbreviations

EC_{50}	median effect concentration
CuPT	bis-(1-hydroxy-2(1H)-pyridinethionate-O,S) copper

DDT	dichloro diphenyl trichloroethane
LC_{50}	median lethal concentration
LOEC	lowest observed effect concentration
M1	2-methylthio-4-*t*-butylamino-6-amino-*s*-triazine
IMO-MEPC	the Marine Environment Protection Committee of the International Maritime Organization
NOEC	no observed effect concentration
OECD	Organization for Economic Cooperation and Development
PCP	pentachloro phenol
PK	pyridine triphenyl boron
PNEC	predicted no effect concentration
TBT	tributyltin compounds
TCMTB	2-(thiocyanomethylthio)benzothiazole
TPT	triphenyltin compounds
ZnPT	bis-(1-hydroxy-2(1H)-pyridinethionate-O,S) zinc

1 Introduction

Preventing organisms from attaching to the hulls of ships is a crucial aspect of ship maintenance and the effective operation of marine vessels. There has been many developments in and wide usage of numerous antifouling paints. The discovery that organotin compounds [especially tributyltin compounds (TBT) and triphenyltin compounds (TPT)] effectively control the biofouling of a wide variety of organisms led to the replacement of the cuprous-oxide-based paints that had been used until around 1965 with these antifouling compounds. However, it has subsequently become apparent that organotin compounds are poorly biodegradable, and remain in the aquatic environment for a long time. They are eventually bioaccumulated in aquatic organisms. Moreover, they are also found to be highly toxic to aquatic organisms. These findings clearly indicate that organotin compounds are hazardous to the aquatic environment.

In Japan: (1) the law concerning the regulation of the examination and manufacture, etc., of chemical substances prohibits the manufacture, import, and sale of organotin compounds; and (2) notifications and other forms of administrative guidance together with the controls on use that are voluntarily imposed by the various organizations have served to regulate the use of these compounds in antifouling paints. Meanwhile, (3) the Marine Environment Protection Committee of the International Maritime Organization (IMO-MEPC) has been exploring a world-wide prohibition of the application of organotins as antifouling biocides. In October 2001 IMO-MEPC adopted the "International Convention on the Control of Harmful Antifouling Systems on Ships, 2001".

This convention resulted in a ban on the application of organotin-based antifouling paints from 2003. Since then a number of antifouling paints, that are free of organotin compounds, have been developed. In addition to cuprous oxide, several organic chemical substances, such as 2-mercaptopyridine *N*-oxide zinc salt, are used as biocides in organotin-free antifouling paints. However, few studies have been undertaken on either the fate in the aquatic environment or the harmful effects on aquatic organisms of these alternative antifouling biocides (referred to as "new biocides" hereafter).

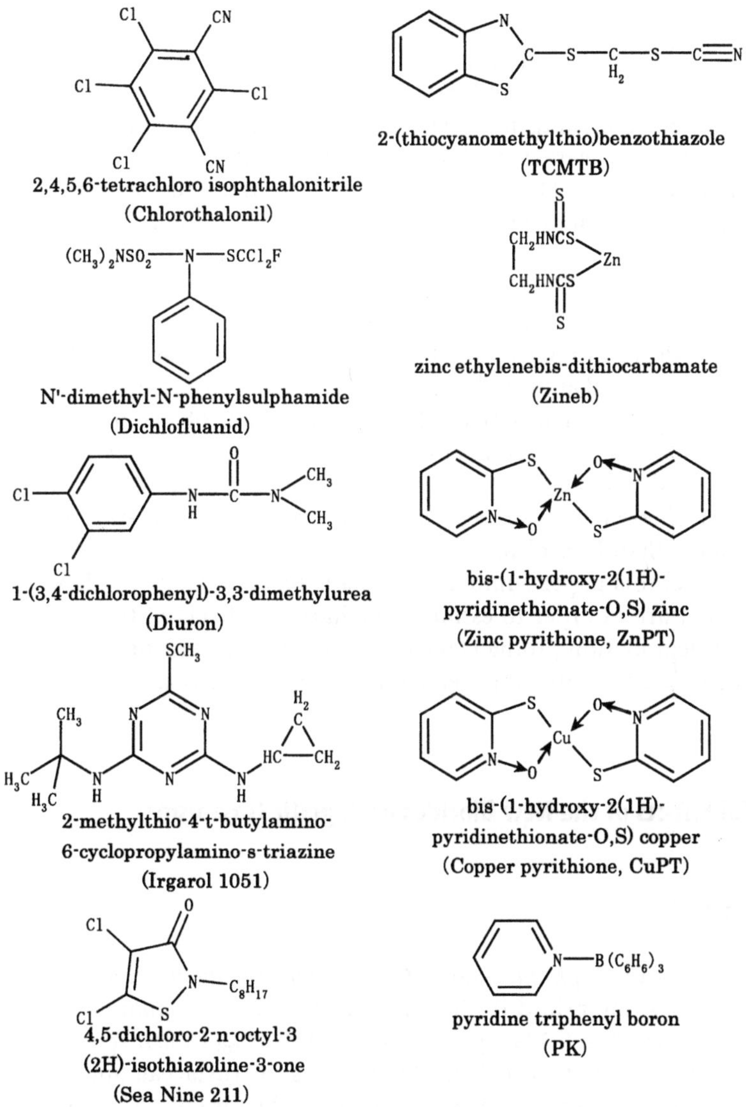

Fig. 1 The structure of representative antifouling biocides

Thomas [1] listed eight of these new biocides as shown in Fig. 1, namely 2,4,5,6-tetrachloro-isophthalonitrile (Chlorothalonil), *N*′-dimethyl-*N*-phenylsulphamide (Dichlofluanid), 1-(3,4-dichlorophenyl)-3,3-dimethylurea (Diuron), 2-methylthio-4-*t*-butylamino-6-cyclopropylamino-*s*-triazine (Irgarol 1051), 4,5-dichloro-2-(n-octyl)-4-isothiazoline (Sea Nine 211), 2-(thiocyanomethylthio)benzothiazole (TCMTB), bis-(1-hydroxy-2(1H)-pyridinethionate-O,S) zinc, (Zinc pyrithion, ZnPT), and zinc ethylene bis-(dithiocarbamate) (Zineb). Pyridinetriphenylboron (PK) and bis-(1-hydroxy-2(1H)-pyridinethionate-O,S) copper (Copper pyrithion, CuPT) are used in Japan as new biocides [2].

Concentrations of Chlorothalonil [3, 4], Dichlofluanid [5], Diuron [6, 7], Irgarol 1051, [8–16] and Sea Nine 211 [5, 7, 17] were determined in coastal waters including marinas and fishery harbors. The concentrations of Chlorothalonil were very low. However, Dichlofluanid, Diuron, Irgarol 1051, and Sea Nine 211 were detected in seawater, and the concentrations were higher in marinas and fishery harbors than in the coastal zones such as the beach. Liu et al. [18] demonstrated that Irgarol 1051 is degraded by the biological activities of white rot fungus, *Phanerachaete crysosporium*, to 2-methylthio-4-*t*-butylamino-6-amino-*s*-triazine (M1). Irgarol 1051 is also transferred to M1 by photolysis [19]. M1 is not degraded further by *P. crysosporium*, and appeared to be accumulated as an end degradation product in the aquatic environment. The concentration of M1 in seawater was determined by Okamura et al. [14] and Liu et al. [15] in the Seto Inland Sea. They reported that the M1 concentration tends to be higher than that of Irgarol 1051. These results suggest that M1 remains in the aquatic environment much longer than Irgarol 1051.

Therefore, this paper summarizes details on ecotoxicity available in published literature in order to estimate the harmful effects of the new biocides and their degradation products on aquatic organisms. Furthermore, the risks to the aquatic environment posed by these new biocides are also discussed.

2 Harmful Effects of the New Biocides on Aquatic Organisms

2.1 Chlorothalonil

Acute toxicity to *Daphnia magna*, *Onchorhynchus mykiss*, and *Gasterosteus aculeatus* are summarized in Table 1. The 48 h EC_{50} to *Daphnia magna* is in the range of 0.097 to 0.129 mg L^{-1}, and the 96 h LC_{50} to fish species is in the range of 0.069 to 0.076 mg L^{-1}. Acute toxicities to crustacean and fish species are significant, and are classified as "very toxic" by the OECD ecotoxicity classification guidelines [25].

2.2 Sea Nine 211

Sea Nine 211 exhibits acute toxicity to phytoplankton at concentrations of 0.0139–0.032 mg L^{-1}, to crustacean species at concentrations of 0.0047–1.312 mg L^{-1}, to bivalves at a concentration of 0.85 mg L^{-1}, and to fish species at concentrations of 0.0027–0.0205 mg L^{-1} (Table 1). From these data, it can be concluded that this compound exhibits weaker acute toxicity to bivalves than to other aquatic organisms such as phytoplankton and fish etc. On the basis of these acute toxicity values, it is classified as "very toxic–toxic" according to the OECD guidelines.

Data on the chronic toxicities of Sea Nine 211 to *Daphnia magna* and sheepshead minnow are reported to be 0.0012 mg L^{-1} and 0.006 mg L^{-1} [26], respectively (Table 2). The 28-day LC_{50} to rainbow trout larva is 0.014 mg L^{-1} [26], leading to fears that the chronic toxicity of Sea Nine 211 is equivalent to that of other antifouling compounds.

2.3 ZnPT and CuPT

As summarized in Table 1, the published literature contains very little data on the acute toxicity of ZnPT and CuPT to aquatic organisms. The respective toxicity values are reported to be 0.028 mg L^{-1} for phytoplankton, 0.029–0.034 mg L^{-1} for *Daphnia magna*, and 0.0026–0.4 mg L^{-1} for fish species. On the other hand, CuPT exhibits acute toxicity to phytoplankton growth, immobilization of *Daphnia magna*, and the survival of fish species at concentrations of 0.0028–0.035 mg L^{-1}, 0.022 mg L^{-1}, and 0.0043–0.011 mg L^{-1}, respectively. Although the obtained toxicity data are limited, CuPT tends to exhibit greater acute toxicity to fish species than ZnPT. Evaluated on the basis of the OECD classification guidelines, the toxicological data on ZnPT and CuPT correspond to the "very toxic" definition, suggesting that both compounds are highly toxic to aquatic organisms.

ZnPT has significant effects on the embryogenesis of fish species, and an early life-stage toxicological study in which fertilized eggs were exposed to ZnPT, has confirmed deformities in the spinal curvature of fish larva after hatching [27]. Meanwhile, Okamura et al. [26] obtained 28-day LC_{50} values of 0.0046 mg L^{-1} and 0.0013 mg L^{-1} for ZnPT and CuPT, respectively, in a study using rainbow trout larva (Table 2). These toxicity values concern that both compounds may exhibit high chronic toxicity to fish species.

ZnPT and CuPT are degraded to pyridine sulfonic acid in the aquatic environments [28]. As shown in Table 1, the 120 h LC_{50} to freshwater algae, 48 h EC_{50} to *Daphnia magna* and 96 h LC_{50} to fish species was 28.9 mg L^{-1}, > 122 mg L^{-1} and 57.1 –> 127 mg L^{-1}, respectively. Acute toxicities of pyridine sulfonic acid were weaker than the parent compounds, ZnPT and CuPT. These results suggest that toxicities of ZnPT and CuPT decrease in the process of degradation.

Table 1 Acute toxicity of alternative antifouling biocides to aquatic organisms

Aquatic organisms	Endpoint	Acute toxicity ZnPT[a] ($mg\,L^{-1}$)	CuPT[b] ($mg\,L^{-1}$)	Pyrizine-sulfonic acid ($mg\,L^{-1}$)	Pk[c] ($mg\,L^{-1}$)	Sea Nine 211[d] ($mg\,L^{-1}$)	Chlorothalonil[e] ($mg\,L^{-1}$)
Micro algae							
Selenastrum capricornutum	72–120 h EC_{50}	0.028 [21][f]	0.035 [21]		0.032 [23]		
Skeletonema costatum	72 h EC_{50}	0.00206 [21]	0.0284 [21]		0.00215 [20]	0.0139 [23]	
freshwater algae	120 h LC_{50}			28.9 [21]			
Crustacean							
Daphnia magna	48 h EC_{50}	0.029–0.034 [21]	0.022 [21]	>122 [21]			0.097–0.129 [24]
mysid shrimp	96 h LC_{50}	0.0063 [21]		71.6 [21]		0.0047 [23]	
brown shrimp	96 h LC_{50}					0.0124 [23]	
Uca pugilator	96 h LC_{50}					1.312 [23]	
Penaeus japonicus	96 h LC_{50}	1.78 [21]	0.0436 [21]		0.149 [20]	0.0126 [20]	
Tigriopus japonicus	24 h EC_{50}	0.16 [22]	0.031 [22]		0.016 [22]	0.03 [21]	
Tigriopus japonicus	24 h LC_{50}	>0.5 [22]	0.041 [22]		0.11 [22]	0.077 [21]	
Mussel							
bay mussel	96h LC_{50}					0.85 [23]	
Fish							
Lepomis macrochirus	96 h LC_{50}					0.014 [23]	
Pimephales promelas	96 h LC_{50}	0.0026 [21]	0.0043 [21]	68.5 [21]			
Oncorhynchus mykiss	96 h LC_{50}	0.0032 [21]		57.1 [21]		0.0027 [23]	0.069–0.076 [24]
Cyprinodon variegatus	96 h LC_{50}	0.4 [21]		>127 [21]		0.0205 [23]	
Pagrus major	96 h LC_{50}	0.273 [21]	0.00767 [21]		0.242 [20]		
Gasterosteus aculeatus							0.073 [24]

[a] bis-(1-Hydroxy-2(1H)-pyridinethionate-O,S)zinc. [b] bis-(1-Hydroxy-2(1H)-pyridinethionate-O,S)copper
[c] Pyridinetriphenylboron. [d] 4,5-Dichloro-2-n-octyl-3-isothiazolone(Sea Nine-211)
[e] 2,4,5,6-Tetrachloro-isophthalonitrile. [f] Figure in brackets exhibits the reference number

Table 2 Chronic toxicity of alternative antifouling biocides to aquatic organisms

Aquatic organisms	Endpoint	Chronic toxicity ZnPT[a] ($mg\,L^{-1}$)	CuPT[b] ($mg\,L^{-1}$)	PK[c] ($mg\,L^{-1}$)	Sea Nine 211[d] ($mg\,L^{-1}$)	Refs.
Crustacean						
Daphnia magna	21 day chronic EC_{50}				0.0012	[23]
Fish						
Cyprinodon variegatus	early life NOEC				0.006	[23]
Oncorhynchus tshawytcha	7 day LC_{50}	0.0084	0.0076	0.140	0.014	[26]
	14 day LC_{50}	0.0056	0.003	0.084	0.014	
	21 day LC_{50}	0.0049	0.0017	0.061	0.014	
	28 day LC_{50}	0.0046	0.0013	0.042	0.014	

[a] bis-(1-Hydroxy-2(1H)-pyridinethionate-O,S)zine
[b] bis-(1-Hydroxy-2(1H)-pyridinethionate-O,S)copper
[c] Pyridinetriphenylboron
[d] 4,5-Dichloro-2-n-octyl-3-isothiazolone (Sea Nine 211)

2.4 PK

PK is a new chemical substance that has been developed in recent years for use as a biocide in antifouling paints. Consequently, there are almost no reports on the harmful effects of this compound on aquatic organisms. As shown in Table 1, the only data that has been reported concerns its acute toxicity to phytoplankton, prawns, and red sea bream. The 72 h EC_{50} to *Skeletonema costatum*, the 96 h LC_{50} to prawns, and the 96 h LC_{50} to red sea bream, are reported to be 0.00215 $mg\,L^{-1}$, 0.149 $mg\,L^{-1}$, and 0.242 $mg\,L^{-1}$, respectively. Although the acute toxicity of PK is weaker than that of ZnPT and CuPT, it is classified as "very toxic" according to the OECD toxicity guidelines.

The 28-day LC_{50} of PK was found to be 0.042 $mg\,L^{-1}$ based on a 28-day rearing experiment using rainbow trout larvae carried out by Okamura et al. [26]. This value is equivalent to that of ZnPT and CuPT. These results cause great concern regarding the harmful effect of this compound on aquatic organisms.

2.5 Irgarol 1051

The acute toxicity of Irgarol 1051 and its degradation product 2-methylthio-4-*t*-butylamino-6-amino-*s*-triazine (M1) to various aquatic organisms are summarized in Table 3. Irgarol 1051 does not affect bacterial growth even at a concentration of 50 mg L^{-1} [14]. The 24 h LC_{50} of Irgarol 1051 to crustacean species is reported to be in the range of 5.7 to 12 mg L^{-1} [14]. It is found to inhibit mobilization (48 h EC_{50}) at a concentration of 8.1 mg L^{-1} [29]. The 48 h LC_{50} to oyster larvae is reported to be 3.2 mg L^{-1}, whilst the 96 h LC_{50} to fish species is reported to be in a range from 0.79 mg L^{-1} for rainbow trout to 4 mg L^{-1} for zebra fish [31].

On the basis of the OECD ecotoxicity classification guidelines [25], these acute toxicity values correspond to the definition of "toxic". Furthermore, the EC_{50} determined by the inhibition of phytoplankton growth is reported to be in the range of 0.0001 to 0.0023 mg L^{-1}. Irgarol 1051 is very toxic to seaweed, the germination and development of spores is inhibited in the range from 0.0006 to 0.0059 mg L^{-1} [30]. The chemical structure of Irgarol 1051 resembles that of triazine herbicide; therefore, it seems reasonable that Irgarol 1051 is remarkably more toxic to phytoplankton and seaweed, so-called aquatic vegetation, compared to marine animals such as crustaceans and fish species.

On the basis of the OECD ecotoxicity classification guidelines [25], the acute toxicity values to phytoplankton and aquatic animals (crustacean and fish) correspond to the definition of "very toxic" and "toxic", respectively.

As shown in Table 3, the 72 h EC_{50} of M1 to phytoplankton growth is reported to range between 0.0019 and 0.046 mg L^{-1}. The 96 h EC_{50} of M1 to the germination and development of the spores of seaweed is reported to be in the range of 0.0017 to 0.13 mg L^{-1}, whilst it exhibits acute toxicity to crustacean (*Daphnia pulex*) species at concentrations from 5.7 to 16 mg L^{-1}. The acute toxicity values of M1 are 10 times greater than those of the parent compound and it is obvious that the toxicity of Irgarol 1051 decreases as the compound is degraded. However, according to the OECD ecotoxicity classifications, the acute toxicity of M1 is classified as "toxic".

The chronic toxicity of Irgarol 1051 was evaluated by an early life-stage toxicity test using rainbow trout (embryo to larval stage) (Table 4), and the "no observed effect concentration" (NOEC) determined by the inhibition of larval growth during the 60 days after hatching was 0.00402 mg L^{-1} [31]. The chronic toxicity and acute toxicity values reveal that the acute/chronic toxicity ratio of Irgarol 1051 to rainbow trout is in the range of 3–10^2. This ratio is similar to that of numerous other chemical substances such as Lindane, PCP, DDT, and Dieldrin etc.

Despite the paucity of data on the toxicity of the new biocides to aquatic organisms, it is suggested from the available published data that Irgarol 1051 is toxic to phytoplankton and seaweed and that CuPT and ZnPT are highly toxic to marine animals. The acute toxicity of these new biocides was found to be in the order of μg L^{-1}. In other words, it is equivalent to that of the TBT compounds [32–34]. There is much evidence that they are highly toxic to aquatic organisms.

Table 3 Acute toxicity of Irgarol 1051 and its biodegradation products (M1) to aquatic organisms

Aquatic organisms Class	Species	Endpoint	Toxicity index	Acute toxicity ($mg\,L^{-1}$) Irgarol 1051[a]	M1[b]	Refs.
Bacteria	*Vibrio fisheri*	bioluminescence	30 min.EC_{50}	> 50	> 50	[14]
Micro algae	*Selenastrum capricornutum*	biomass	72 h EC_{50}	0.0016	0.019	[14]
	Selenastrum capricornutum	growth rate	72 h EC_{50}	0.0023	0.046	[14]
	Senedesmus sp.		72 h EC_{50}	0.0014		[29]
	Raphidocelis subcapitata		EC_{50}	0.001		[29]
	Anabaena flos-aquae		EC_{50}	0.002		[29]
	Navicula pelliculata		EC_{50}	0.0001		[29]
	Skeletonema costatum		EC_{50}	0.0004		[29]
Seaweed	*Porphyra yezoensis*	conchospore survival	96 h LC_{50}	5	6.5	[30]
		conchospore germination	96 h EC_{50}	0.0041	0.13	[30]
		conchospore growth	96 h EC_{50}	0.0006	0.017	[30]
	Eisenia bicylis	gamatephyte growth	96 h EC_{50}	0.0059	> 0.032	[30]
Crustacean	*Daphnia magna*	lethality	24 h LC_{50}	16	17	[14]
			8 h LC_{50}	8.3	11	[14]
	Daphnia pulex	lethality	24 h LC_{50}	5.7	27	[14]
	Thamnocepharus platyurus	lethality	24 h LC_{50}	12	19	[14]
	Daphnia magna		48 h EC_{50}	8.1		[29]
	Mysidopsis bahia	lethality	96 h LC_{50}	0.04		[31]
	Artemia salina	lethality	24 h LC_{50}	> 4	> 4	[14]
Mussel	oyster larvae	lethality	48 h LC_{50}	3.2		[29]
Fish	Cyprinodon variegatus	lethality	96 h LC_{50}	3.5		[31]
	Barachydanio rerio	lethality	96 h LC_{50}	4		[29]
	Oncorhynchus mykiss	lethality	96 h LC_{50}	0.79		[31]
	Menidia beryllina	lethality	96 h LC_{50}	1.58		[31]
	Lepomis macrochirus	lethality	96 h LC_{50}	2.6		[31]

[a] 2-Methylthio-4-*t*-butylamino-6-cyclopropylamino-*s*-triazine. [b] 2-Methylthio-4-*t*-butylamino-6-amino-*s*-triazine

Table 4 Chronic toxicity of Irgarol 1051 to shrimp and fish

Aquatic organisms Class	Species	Endpoint	Exposure period	Chronic toxicity ($mg L^{-1}$)	Refs.
Crustacean	*Mysidopsis bahia*	growth	28 day	0.11 ([a] LOEC: 0.26)	[31]
	shrimp	survival	28 day	0.26 (LOEC: 0.49)[a]	[31]
Fish	*Oncorhynchus mykiss*	hatchability		0.184	[31]
		survival	60 day after hatch	0.184	[31]
		growth	60 day after hatch	0.00402 (LOEC : 0.00914)[a]	[31]
		reproduction	28 day	0.26 (LOEC: 0.49)[a]	[31]
	Oncorhynchus tshawytcha	LC_{50}	28 day	0.088	[26]

[a] LOEC: lowest observed effect concentration

No literature has been found concerning toxicities of Dichlofluanid, Diuron, TCMTB, and Zineb. Furthermore, very little is known about the chronic toxicities of these new biocides. In order to estimate the exact PNEC value, ecotoxicological information such as acute chronic ratios should be elucidated by performing detailed experiments on the ecotoxicity of these new biocides.

3
Preliminary Risk Assessment

The predicted "no effect concentration" (PNEC) was tentatively estimated from the lowest acute toxicity value by the OECD method. According to OECD, the uncertainty factor of 100 was used to calculate PNEC. The values of PNEC for several alternative biocides are shown in Table 5. The highest seawater concentrations determined by individual survey are also shown in Table 5.

As shown in Table 5, the PNEC of Irgarol 1051, M1, Sea Nine 211, ZnPT, CuPT, and PK was 690 $ng\,L^{-1}$, 16 $ng\,L^{-1}$, 190 $ng\,L^{-1}$, 27 $ng\,L^{-1}$, 26 $ng\,L^{-1}$, 22 $ng\,L^{-1}$, and 22 $ng\,L^{-1}$, respectively. The PNEC of Chlorothalonil was two orders of magnitude higher than the highest concentration found in natural waters. Although the available data on the Chlorothalonil concentration in natural waters is very limited, these results suggest that its hazardous impact on the aquatic environment is considered small. The PNECs of Irgarol 1051, M1, and Sea Nine 211 were one order magnitude lower than the concentrations of these biocides in the seawater of marinas. We are concerned with these findings which suggest that Irgarol 1051, M1, and Sea Nine 211 are hazardous to the aquatic ecosystem especially in marinas and fishery harbors. The concentrations of Irgarol 1051, M1, and Sea Nine 211 in seawater collected in coastal areas such as beaches were significantly lower when compared to the PNEC values. The hazardous impact of these new biocides seems to be restricted to particular areas such as marinas and fishery harbors where many pleasure boats and/or fishing boats are moored.

Harino et al. [7] did not detect pyrithion compounds (ZnPT and CuPT) in the seawater of the port of Osaka, Japan. Because of the results from the port of Osaka it is assumed that any hazardous impacts from the pyrithion compounds can be ignored at present. The toxicity of pyridine sulfonic acid, which is a degradation product of ZnPT and CuPT was lower than that of the parent compounds. These results suggest that the toxicity of ZnPT and CuPT decreases during the degradation process.

Published data on the concentrations of PK have not been found. Therefore, any potential hazardous impacts on the marine ecosystem from PK cannot be evaluated at present and further studies are needed.

Table 5 Preliminary risk assessment of several alternative biocides to aquatic organisms by comparing PNEC with the highest concentration in natural waters

Biocide	Aquatic organism	Lowest value of acute toxicity data ($mg\,L^{-1}$)		PNEC [a] ($ng\,L^{-1}$)	Highest concentration in natural water ($ng\,L^{-1}$)
Chlorothalonil	*D. magna*	EC_{50}	0.097	690	1.38 (Black water estuary) [4][b]
	O. mykiss	LC_{50}	0.069		
Irgarol 1051	*S. capricornutum*	EC_{50}	0.0016		85.3 (marina, Japan) [14]
	D. magna	EC_{50}	8.1	16	296 (fishery harbor, Japan) [14]
	O. mykiss	LC_{50}	0.791		700 (marina, France) [8]
					264 (port, France) [9]
					682 (marina, south coast of UK) [11]
					440 (marina, the Baltic Sea) [13]
					400 (marina, Sweden) [9]
M1	*S. capricornutum*	EC_{50}	0.019	190	1270 (marina, Japan) [14]
	D. magna	EC_{50}	27		1210 (fishery harbor, Japan) [14]
Sea Nine 211	*S. capricornutum*	EC_{50}	0.032		260–3700 (Mediterranean Sea, Spain) [17]
	Mysid shrimp	LC_{50}	0.0047	27	< 6.3–49 (Greece) [5]
	O. mykiss	LC_{50}	0.0027		< 0.3–0.55 (Osaka port, Japan) [2]
Zn PT	*S. capricornutum*	EC_{50}	0.028		Is not detected in Osaka port, Japan [7]
	D. magna	EC_{50}	0.029	26	The contamination is not exactly
	P. promelas	LC_{50}	0.0026		demonstrated
Cu PT	*S. capricornutum*	EC_{50}	0.035		Is not detected in Osaka port, Japan [7]
	D. magna	EC_{50}	0.022	22	The contamination is not exactly
	P. promelas	EC_{50}	0.0043		demonstrated
PK	*S. costatum*	EC_{50}	0.0022		No available information on
	T. japonicus	EC_{50}	0.016	22	the concentration in seawater
	P. major	LC_{50}	0.242		

[a] PNEC was tentatively estimated from the lowest acute toxicity value by the OECD method using an uncertainty factor of 100
[b] Figure in brackets exhibits the reference number

4 Conclusions

(1) The toxicities of the new biocides are classified as "toxic" or "very toxic" by the OECD classification guidelines, and show almost the same toxicity as TBT. However, these new biocides are not stable in comparison with the previously used antifouling biocides [tributyltin compounds (TBT)]. It can be concluded from these results that the new biocides do not persist for a long time in aquatic environments, and that the environmental impact of these biocides is smaller than that of TBT.

(2) Chlorothalonil, Dichlofluanid, Diuron, Irgarol 1051, M1, and Sea Nine 211 have been detected in natural waters, the concentrations were higher in marinas and small fishery harbors than in the coastal waters. The only research carried out in Japan has been on the occurrence of ZnPT and CuPT in seawater and was by Harino et al.; however, pyrithion compounds were not detected. A detailed survey is needed to clarify the present situation regarding contamination by the new antifouling biocides.

(3) The reported concentrations of Irgarol 1051, M1, and Sea Nine 211 in natural waters exceed the PNEC values of these new biocides especially in marinas and fishery harbors. Irgarol 1051, M1, and Sea Nine 211 are causing a hazardous impact on the aquatic ecosystem in marinas and fishery harbors. In order to assess the ecotoxicological risk of the new biocides in detail, further research is needed to clarify the toxicity and to develop the methods of estimating concentrations in natural waters.

Acknowledgements I am grateful to Dr. Kazufumi Takayanagi (Seikai National Institute of Fisheries Science, Fisheries Research Agency, Japan) for his advice and critical readings of this paper. Thanks are also due to Mr. Toshimitsu Onzuka for his assistance in preparing a figure.

References

1. Thomas KV (2001) Biofouling 17:73
2. Harino H (2004) Coast Mar Sci 29:28
3. Thomas KV, McHugh M, Waldock M (2002) Sci Total Environ 293:117
4. Voulvoulis N, Scrimshaw MD, Lester JN (2000) Mar Pollut Bull 40:938
5. Sakkas VA, Konstantinou IK, Lambropoulou DA, Albanis TA (2002) Environ Sci Pollut Res 9:327
6. Thomas KV, Fileman TW, Readman JW, Waldock MJ (2001) Mar Pollut Bull 42:677
7. Harino H, Kitano M, Mori Y, Mochida K, Kakuno A, Arima S (2005) J Mar Biol Ass UK 85:33
8. Readman JW, Kwing LLW, Grondin D, Bartocci J, Villeneuve JP, Mee LD (1993) Environ Sci Technol 27:1940
9. Tolosa I, Readman KJW, Bllaevote A, Ghilini S, Bartocci J, Horvat M (1996) Mar Pollut Bull 32:335

10. Gough MA, Forthergill J, Hendrie JD (1994) Mar Pollut Bull 28:613
11. Zhou JL, Fileman TW, Evans S, Donkin PD, Mantoura RF, Rowland S (1996) Mar Pollut Bull 32:599
12. Scarlett A, Donkin ME, Fileman TW, Donkin P (1997) Mar Pollut Bull 34:64
13. Biselli S, Bester K, Huhnerfuss H, Fent K (2000) Mar Pollut Bull 40:233
14. Okamura H, Aoyama I, Liu D, Maguire RJ, Pacepavicius GJ, Lau YL (2000) Water Res 34:3523
15. Liu D, Pacepavicius GJ, Maguire RJ, Lau YL, Okamura H, Aoyama I (1999) Water Res 33:2833
16. Albanis TA, Lambropoulou DA, Sakkas VA, Konstantinou IK (2002) Chemosphere 48:475
17. Martinez K, Ferrer L, Hernando MD, Fernandez-Alba AR, Marce RM, Borull F, Balcelo D (2001) Environ Technol 22:543
18. Liu D, Maguire RJ, Lau YL, Pacepavicius GJ, Okamura H, Aoyama I (1997) Water Res 31:2363
19. Okamura H (2002) Chemosphere 48:43
20. Shipbuilding Research Association of Japan (1999) Technical report on the prevention of marine pollution by antifouling paints – 1997 edn, p 17 (in Japanese)
21. Shipbuilding Research Association of Japan (2001) Technical report on the prevention of marine pollution by antifouling paints – 1999 edn, p 48 (in Japanese)
22. Shipbuilding Research Association of Japan (2002) Technical report on the prevention of marine pollution by antifouling paints – 2001 edn, p 24 (in Japanese)
23. Shade WD, Hurt SS, Jacobson AH, Reinert KH (1993) Environ Toxicol Risk Assess ASTP STP 1216:381
24. Ernst W, Doe K, Jonah P, Young J, Julien G, Hennigar P (1991) Arch Environ Contam Tocicol 21:1
25. OECD (1996) Proposal for a harmonized classification system based on acute toxicity
26. Okamura H, Watanabe T, Aoyama I, Hasobe M (2002) Chemosphere 46:945
27. Goka K (1999) Environ Res Section A 81:81
28. Turley PA, Fem RJ, Ritter JC (2000) Biofouling 15:175
29. Cibba-Geigy Co. (1995) Irgarol 1051 in antifouling paints. Technical Information Bulletin
30. Okamura H, Aoyama I, Takami T, Maruyama T, Suzuki Y, Matsumoto M, Katsuyama I, Hamada J, Beppu T, Tanaka O, Maguire RJ, Liu D, Lau YL, Pacepavicius GJ (2000) Mar Pollut Bull 40:754
31. Hall Jr LW, Giddings JM, Solomon KR, Balcomb R (1999) Cri Rev Toxicol 29:367
32. Koyama K, Shimizu A (1992) In: Satomi Y, Shimizu M (eds) Organotin Pollution and its Effects on Aquatic Organisms. Koseisha Koseikaku, Tokyo, p 86 (in Japanese)
33. Horiguti T, Shimizu M (1992) In: Satomi Y, Shimizu M (eds) Organotin Pollution and its Effects on Aquatic Organisms. Koseisha Koseikaku, Tokyo, p 99 (in Japanese)
34. IMO (1997) MEPC41/INF, Call for the worldwide ban on every use of organotin-based antifouling paints for ship bottoms

Hdb Env Chem Vol. 5, Part O (2006): 227–262
DOI 10.1007/698_5_057

Published online: 22 December 2005

General Aspects of Natural Products Antifoulants in the Environment

Iwao Omae

Omae Research Laboratories, 335-23, Mizuno, Sayama, 350-1317 Saitama, Japan
um5i-oome@asahi-net.or.jp

Abstract Many marine lives prevent the surface of their bodies with antifouling substances without causing serious environmental problems. Some terrestrial plants also have the same kinds of substances. Therefore, these substances may be expected to be utilized as new environmentally friendly antifouling agents, especially those having high anesthetic, repellent, settlement deterrent, settlement inhibitory properties, etc., without having biocidal properties. These natural products are classified into five kinds of compounds, i.e., terpenes, nitrogen-containing compounds, phenols, steroids and others.

Many compounds such as sesquiterpenes, diterpenes, trigonelline, primidines, macrocarpals and higher fatty acids, showed the same level of antifouling properties as that of $CuSO_4$, and some compounds such as gramines, cycloviolacin O2, shogaols, alkylphenols, kaempferols and bufalin, showed the similar high antifouling properties as those of the organotin antifoulants. The mixtures of these compounds are expected to be much better antifoulants than the organotin compounds since some of them showed synergistic properties.

Keywords Antifouling · Marine environment · Natural products · Settlement inhibition · Terpenes

Abbreviations

A Ala	Alanine
C Cys	Cysteine
D Asp	Aspartic acid
E Glu	Glutamic acid
F Phe	Phenylalanine
G Gly	Glycine
H His	Histidine
I Ile	Isoleucine
K Lys	Lysine
L Leu	Leucine
M Met	Methionine
MIC	Minimum inhibitory concentration
N Asn	Aspergine
PSA	6-Pentadecylsalicylic acid
Q Gly	Glutamine
R Arg	Arginine
S Ser	Serine
T Thr	Threonine
TBG	5,6-tribromo-1-methylgramine
TBT	Tributyltin compounds
TBTCl	Tributyltin chloride
TBTF	Tributyltin fluoride
TBTO	Tributyltin oxide
TPT	Triphenyltin compounds
V Val	Valine
W Trp	Tryptophan
Y Tyr	Tyrosine

1 Introduction

The use of organotin antifouling paints for small vessels was banned in many countries in the late of 1980s because the deformity of oysters was found in 1970s. Since then, especially, in mid 1990s, many articles reported the decrease of concentration of organotin antifoulants, on the other hand, the high concentrations of Irgarol 1051 as an alternative antifoulant were reported in the harbors or marinas in the world. Therefore, the many users of ship bottom paints expect to have new antifouling paints which are environmentally friendly and also have the similar level of both high antifouling properties and long life of usage to those of the organotin antifoulants.

Recent new researches regarding environmentally friendly alternative methods of antifouling treated natural products, fouling-release coating and electrical antifouling systems [1]. One promising approach to find such antifouling method is to explore natural products occurring in the marine environment, especially those produced by sessile fouling-free marine organisms. This article reports natural product antifoulants.

2
Natural Products Antifoulants

Natural products include both inorganic compounds such as asbestos, mica, graphite, etc., and organic compounds. However, generally, the natural products mean the natural organic compounds formed in living organisms. For example, formic acid is formed in the body of ants. There are natural dyes such as arotenoids, quinones, flavonoids, chlorophylls, hemoglobins, and anthocyanins; and biopolymers such as starch, cellulose, proteins, DNA and RNA. There are terpenes having isoprene units in plants, alkaloids, steroids, vitamins, hormones, insect pheromones, arachidonic acid, prostaglandins, penicillins and streptomycins as antibiotics. Further, there are fats, oils, wax, lipids, saccharides, pyrimidine bases and purine bases. In these natural products, some of them may be expected to be as environmentally friendly antifoulants.

Among aquatic fouling organisms in seawater, 1746 kinds of protists, animals, and plants are shown in Table 1 [2]. These prominent marine organisms are barnacles, algae, ascidians, shellfishes, seamosses and hydrozoa. Marine lives such as corals, sponges, marine plants, and dolphins, etc., prevent the surface of their bodies with antifouling substances without causing serious environmental problems. Therefore, these substances may be expected to be used, as new environmentally friendly antifouling agents, especially those having highly anesthetic, repellent, and settlement inhibitory properties, etc., without showing biocidal properties, are desirable. Many of the antifouling substances are found from these marine animals, marine plants and microorganisms. Moreover, many of the antifoulants are also found in terrestrial plants such as a green tea, a wasabi and an oak tree leaf.

The fouling of manmade structure involves molecular bonds and biological adhesions with abiotic or biotic substances. These biotic foulings are formed by microfouling, e.g., those of bacteria and diatoms, and by macrofoulings, e.g., algae and vertebrates. These natural antifoulants include

Table 1 Marine organisms to settle on the structures immersed in seawater [2]

Organism	Number
Bacteria	37
Fungi	14
Diatom	111
Algae	452
Protozoa	99
Porifera	33
Coelenterata	286
Annelida	108
Tentaculata	139
Arthropoda	292
Echinodermata	19
Chordata	127
Other invertebrates	29
Total	1746

not only toxins but also anesthetics, growth inhibitors, metamorphosis inhibitors, settlement inhibitors, settlement deterrent agents, attachment inhibitors, repellants, anti-mussel agents, anti-barnacle agents, general antibacterial agents, various antifungal agents and stimulants [3].

It is well known that many marine invertebrates such as sponges and corals remain remarkably free from settlement by fouling organisms, suggesting that they have biologically active compounds preventing the larvae of other marine organisms from settlement and growth on their body. So far, several compounds with such activities have been found among marine invertebrates. These compounds are considered to play an important role in the antifouling mechanism of marine organisms [4].

Natural products antifoulants consist mainly of five kinds of compounds such as terpenes, nitrogen-containing compounds, phenols, steroids and others. These are produced from sponges, corals, starfishes, mussels, algae, terrestrial plants, etc.

3 Terpenes

3.1 Introduction

Terpenes were the most widespread compounds in nature, mainly in plants such as pine trees and citrus fruits, as constituents of essential oils. But, some larger and more complex terpenes (e.g. aqualene and lanosterol) occur in animals. Terpenes typically have very strong odors in low concentrations. Many terpenes are hydrocarbons, but oxygen-containing compounds such as alcohols, aldehydes, ketones, acids, esters and epoxides (terpeneoids) are also found. Their building block is a hydrocarbon, i.e., isoprene. Terpene hydrocarbons therefore are expressed by a molecular formula: $(C_5H_8)_n$, and they are classified according to the number of isoprene units: monoterpenes (number of isoprene units is two: $n = 2$), sesquiterpenes ($n = 3$), diterpenes ($n = 4$), sesterterpenes ($n = 5$), triterpenes ($n = 6$) and tetraterpenes ($n = 8$). Some steroids are also classified into triterpenes. Terpenes are considered to be biodegrable. Many of these terpenes are biologically active substances such

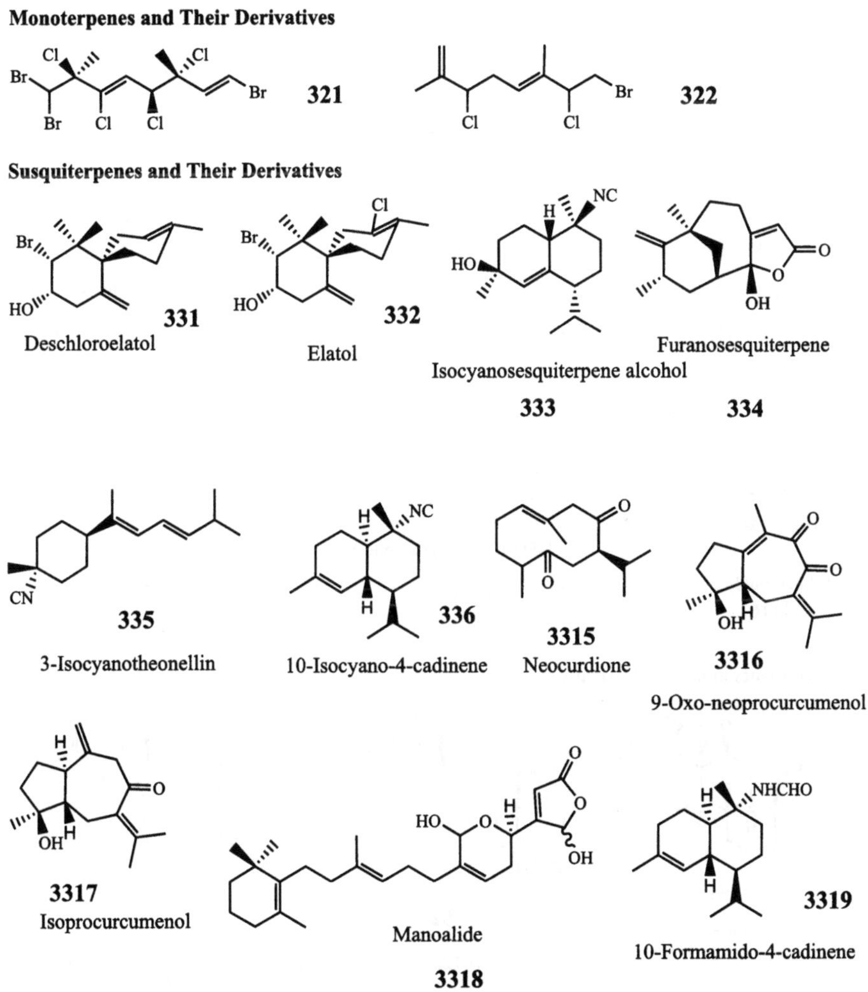

Scheme 1

as antibacterials, plant growth inhibitors, biocides, insect antifeedants and antitumor agents.

Diterpenes and Their Derivatives

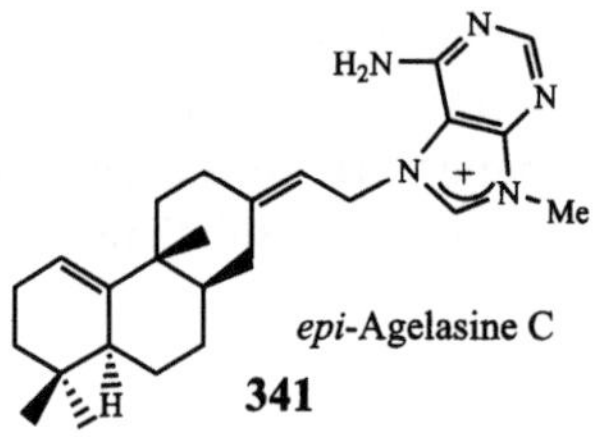

epi-Agelasine C

341

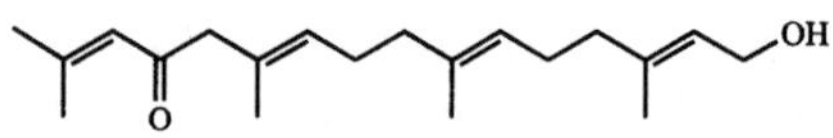

Elenganolone

342

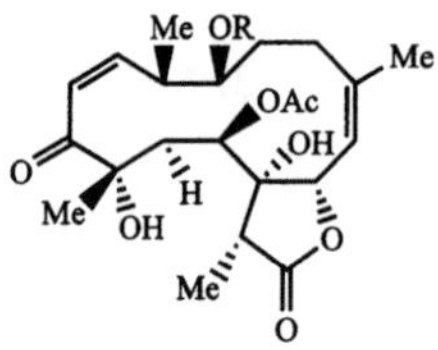

Renillafoulins

3411 **3412** **3413**

R = Ac, CH_3CH_2CO-, $CH_3CH_2CH_2CO$-

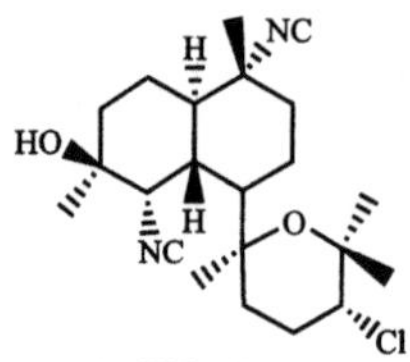

Kalihinol A

3414

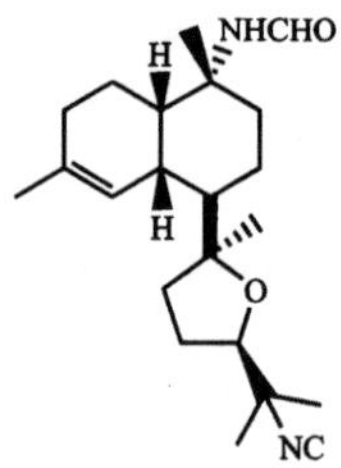

10-Formamidokalihinene

3415

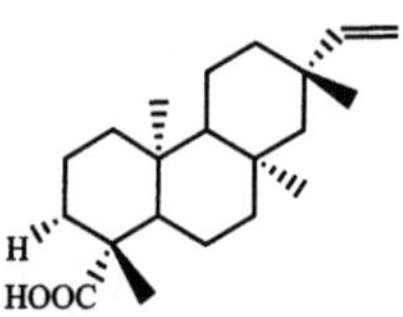

ent-Pimara-8(14),15-dien-19-oic acid

3416

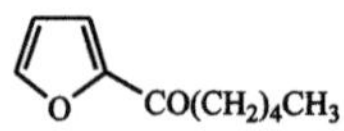

1-Furyl-*n*-pentyl ketone

3417

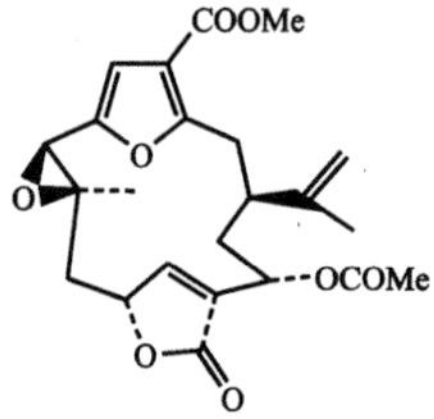

Pukalide **3418**

Sesterterpenes and Their Derivatives

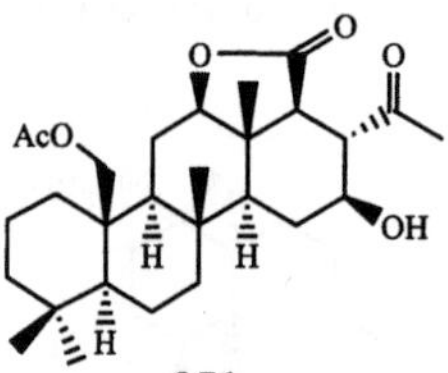

351

22-Acetoxy-16b-hydroxy-24-methyl-24-oxoscalarano-25,12b-lactone

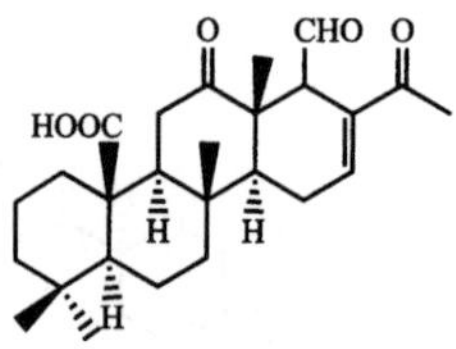

352

24-Methyl-12,24,25-trioxoscalar-16-en-22-oic acid 352

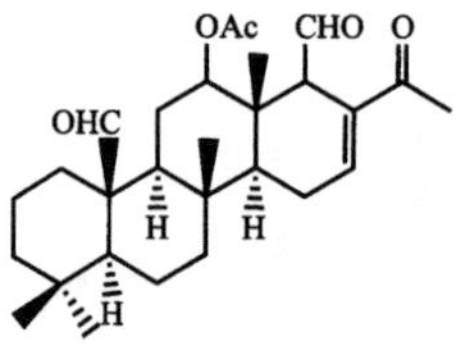

353

12a-Acetoxy-24-Methyl-24-oxoscalar-16-en-22,25-dial

Scheme 2

Table 2 Terpenes and Their Derivatives

Compound	Living thing Species	Refs.
3.2 Monoterpenes and Their Derivatives		
(3*R*, 4*S*, 7*S*)-trans,trans-3,7-Dimethyl-1,8,8-tribromo-3,4,7-trichloro-1,5-octadiene **321**	temperate red alga *Plocamium costatum* *Plocamium cartilagineum*	[5–7]
1-Dibromo-2,6-dichloro-3,7-dimethyl-7-octene **322**	temperate red alga *Plocamium costatum*	[5]
3.3 Susquiterpenes and Their Derivatives		
Deschloroelatol **331** Elatol **332**	marine red alga *Laurencia rigida*	[8]
Isocyanosesquiterpene alcohol (1*S**, 4*S**, 7*R** 10*S**)-10-isocyano-5-cadinen-4-ol **333**	nudibranch *Phyllidia pustulosa*	[9]
2-Hydroxy-9,11-dimethyl-10-methylene-3-oxatricyclo[7.3.1.02,6]tridec-5-en-4-one **334**	Palauan marine sponge *Dysidea herbacea*	[10]
3-Isocyanotheonellin **335** 10-Isocyano-4-cadinene **336**	nudibranchs (family *Phyllidiidae*)	[11]
Neocurdione **3315**	rhizomes of *Curcuma* *Curcuma aromatica*	[12]
9-Oxo-neoprocurcumenol **3316**	rhizomes of *Curcuma* *Curcuma aromatica*	[12]
Isoprocurcumenol **3317**	rhizomes of *Curcuma* *Curcuma zedoaria*	[12]
Manoalide **3318**	marine sponge *Smenospongia* sp.	[13]
10-Formamido-4-cardinene **3319**	marine sponge *Acanthella cavernosa*	[14]
3.4 Diterpenes and Their Derivatives		
epi-Agelasine C **341**	marine sponge *Agelas mauritiana*	[15]
Eleganolone **342**	common brown alga *Bifurcaria bifurcata*	[16]
Renillafoulins **3411–3413**	Atlantic sea pansy *Renilla reniformis*	[17]
Kalihinenes (Kalihinol A) **3414**	marine sponge *Acanthella cavernosa*	[18–20]
10-Formamidokalihinene **3415**	marine sponge *Acanthella cavernosa*	[18–20]
ent-Pimara-8(14),15-dien-19-oic acid **3416**	annual herb *Gnaphalium gaudichaudianum*	[21]
2-Furyl-*n*-pentyl ketone **3417**	octacorals	[22]

Table 2 (continued)

Compound	Living thing Species	Refs.
(Analogues diterpene marine product antifoulants to pukalide and renillafoulin)	*Renilla reniformis*	
	Leptogorgia virgulata	
3.5 Sesterterpenes and Their Derivatives		
22-Acetoxy-16β-hydroxy-24-methyl-24-oxoscalarano-25, 12β-lactone **351**	palauan marine sponge	[23]
24-Methyl-12,24,25-trioxoscalar-16-en-22-oic acid **352**	*Lendenfeldia chondrodes*	

Terpenes, terpeneoids and their derivatives such as glucosides are investigated for using them as antifouling agents. These agents are shown in Table 2 [5–23], and Schemes 1 and 2.

3.2 Monoterpenes and Their Derivatives

Halogenated monoterpene, 1,10,10-tribromo-3,4,7-trichloro-3,7-dimethyl-1,5-octadiene **321**, is obtained from sea hare, *Aplysia californica* [6, 7]. It showed an activity to deter the settlement of barnacle larvae, suggesting a potential ecological role of its terpene [5].

3.3 Susquiterpenes and Their Derivatives

Susquiterpenes, deschloroelatol (**331**) and elatol (**332**) extracted from a temperate marine red alga, *Laurencia rigida*, exhibited a wide spectrum of activity against fungi, bacteria and algae such as *Ustilago violace*, *Bacillus megaterium* and *Chlorella fusca* [8].

New isocyanosesquiterpene alcohol (**333**) was obtained from a nudibranch *Phyllidia pustulosa*, possessing an almost the same potent antifouling activity (0.17 $\mu g\,mL^{-1}$) as $CuSO_4$ (0.15 $\mu g\,mL^{-1}$) against the larvae of a barnacle *Balanus amphitrite* with very weak lethality [9].

Furanosesquiterpene (**334**) which is 2-hydroxy-9,11-dimethyl-10-methylene-3-oxatricyclo[7.3.1.$0^{2,6}$]tridec-5-en-4-one, isolated from a Palauan marine sponge, *Dysidea herbacea*, showed a weak repellent activity against a blue mussel, *Mytilus edulis galloprovincialis*. A 100% antifouling activity was observed at a concentration of 100 ppm. However, the tested mussels showed no reactivity at a concentration less than 10 ppm [10].

Fusetani et al. [11] isolated three new sesquiterpene isocyanates from a nudibranchs of Phyllidiidae family along with a new sesquiterpene peroxide and six known sesquiterpenes (**335–3314**). These compounds showed potent antifouling activities against the larvae of a barnacle *Balanus amphitrite* as shown in Table 3 [11] and Scheme 3 [11]. Especially, 3-isocyanotheonellin **335** and 10-isocyano-4-cadinene **336** showed the potent antifouling activities with IC_{50}'s of 0.13 and 0.14 $\mu g\,mL^{-1}$, respectively, while no toxicity was found at these concentrations [24]. Their activity was comparable with that of $CuSO_4$ (IC_{50} 0.15 $\mu g\,mL^{-1}$).

Neocurdione **3315**, isoprocurcumenol **3316** and a new sesquiterpene, 9-oxo-neoprocurcumenol **3317**, were isolated from the fresh rhizomes of *Curcuma aromatica* and *Curcuma zedoaria* (Zingiberaceae). These compounds showed an attachment-inhibiting activity that was two times stronger than that of standard $CuSO_4$ against the blue mussel [12].

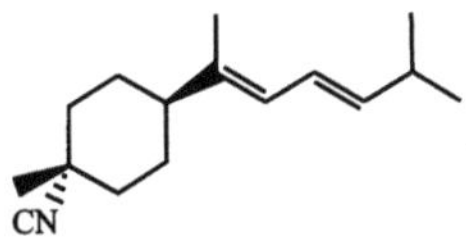

3-isocyanotheonellin

335

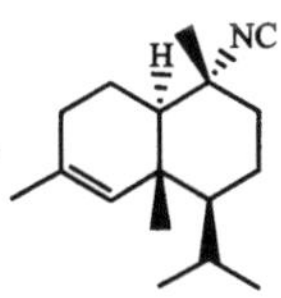

10-isocyano-4-cadinene

336

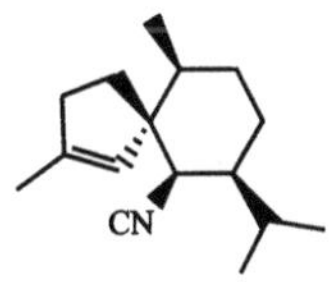

10-*epi*-axisonitrile-3

337

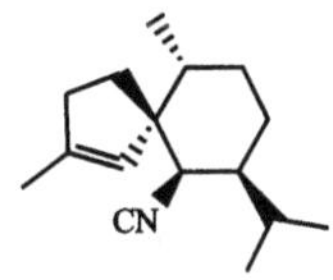

Axisonitrile-3

338

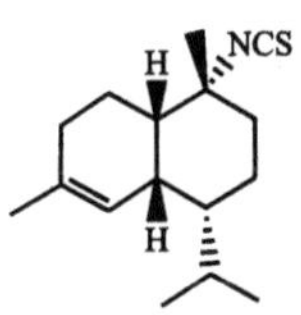

(-)-10-isothiocyano-4-amorphene

339

2-thiocyanatoneopupukeanane

3310

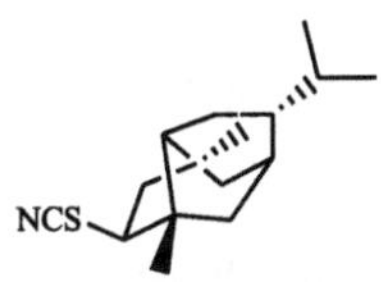

4-thiocyanatoneopupukeanane

3311

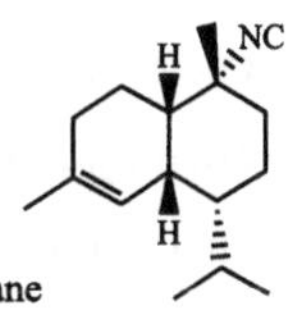

10-isocyano-4-amorphene

3312

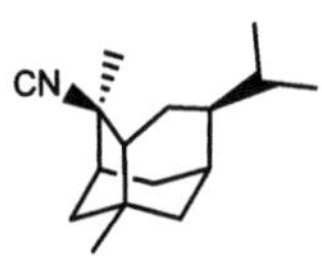

2-isoxyanotrachyopsane

3313

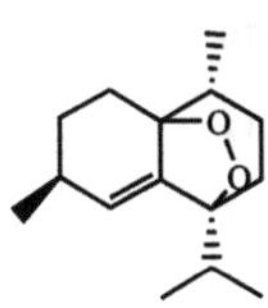

17-epidioxy-5-cadinene

3314

Scheme 3

Manoalide **3318** together with a steroid, isolated from a marine sponge *Smenospongia* sp., showed an antifouling activity but no toxicity against the cypris larvae of a barnacle *Balanus amphitrite*, with an EC_{50} value of 0.24 μg mL^{-1} ($CuSO_4$: 0.15 μg mL^{-1}) [13].

10-Formamido-4-cadinene **3319** obtained from a sponge, *Acanthella cavernosa*, also inhibited a larval settlement at a concentration of 0.5 μg mL^{-1} [14].

The sixteen secondary metabolites of a green alga *Caulerpa prolifera* were isolated. They are shown as compounds **C3320–C3335**. The antifouling activity of the algal extracts in laboratory assays against two of the major groups of fouling organisms. The extracts exhibited moderate to significant activities against two Gram-positive and one Gram-negative marine bacteria. The measured inhibition zone varied from 17 to 83% of that observed with the biocide TBTO. On the other hand, the growth inhibitory effect of the *Caulerpa prolifera* extract was evaluated against a marine microalga *Phaedactylum tri-*

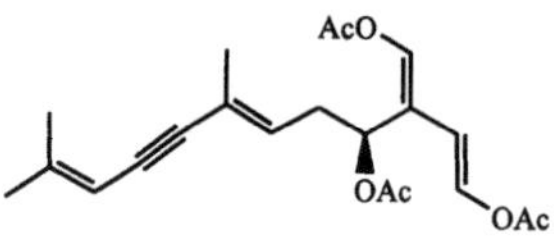

3320

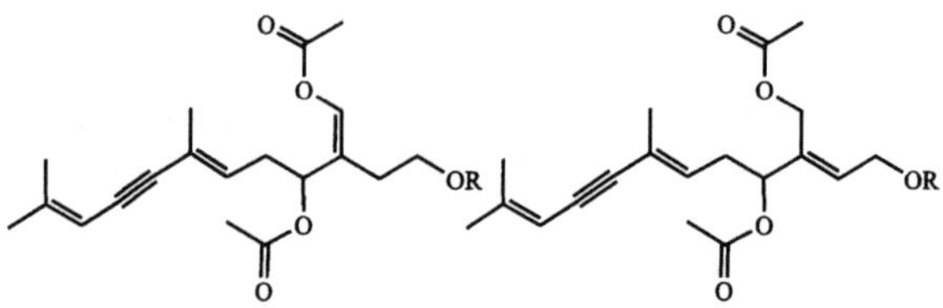

R = CH_3-$(CH_2)_4$-$(CH{=}CH$-$CH_2)_2$-$(CH_2)_6CO$-	**3321**	**3322**
R = CH_3-$(CH_2)_{14}CO$-	**3323**	**3324**
R = CH_3-CH_2-$(CH{=}CH$-$CH_2)_3$-$(CH_2)_4CO$-	**3325**	**3326**
R = CH_3-CH_2-$(CH{=}CH$-$CH_2)_3$-$(CH_2)_6CO$-	**3327**	
R = CH_3-CH_2-$(CH{=}CH$-$CH_2)_3$-$(CH_2)_2CO$-	**3328**	
R = CH_3-$(CH_2)_{16}CO$-	**3329**	
R = CH_3-$(CH_2)_7$-$(CH{=}CH)$-$(CH_2)_9CO$-	**3330**	
R = CH_3-$(CH_2)_4$-$(CH{=}CH$-$CH_2)_4$-$(CH_2)_2CO$-	**3331**	
R = CH_3-CH_2-$(CH{=}CH$-$CH_2)_4$-$(CH_2)_3CO$-	**3332**	
R = CH_3-$(CH_2)_{12}CO$-		**3333**
R = CH_3-$(CH_2)_5$-$(CH{=}CH)$-$(CH_2)_7CO$-		**3334**
R = CH_3-$(CH_2)_7$-$(CH{=}CH)$-$(CH_2)_7CO$-		**3335**

Scheme 4

Table 3 Sesquiterpenes from nudibranchs of the family phyllidiidae and their antifouling activity [11]

Compound	Antifouling activity IC_{50} ($\mu g\, mL^{-1}$)
3-Isocyanotheonellin **335**	0.13
10-Isocyano-4-cadinene **336**	0.14
10-*epi*-Axisonitrile-3 **337**	10
Axisonitrile-3 **338**	3.2
(–)-10-Isothiocyano-4-amorphene **339**	7.2
2-Thiocyanatoneopupukeanane **3310**	4.6
4-Thiocyanatoneopupukeanane **3311**	2.3
10-Isocyano-4-amorphene **3312**	0.7
2-Isoxyanotrachyopsane **3313**	0.33
17-Epidioxy-5-cadinene **3314**	> 50
Cu_2SO_4	0.15

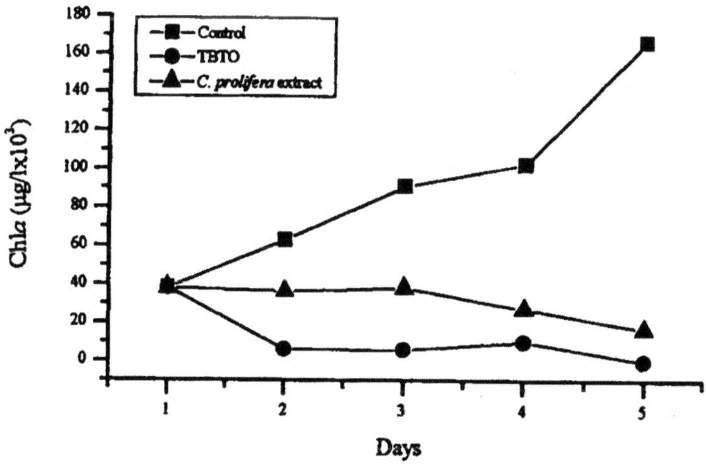

Fig. 1 Inhibition effect on the growth of *Phaedactylum tricornutum* (as presented by chlorophyll concentration) [26]

cornutum which is a representative of primary surface colonizers [25]. The extract proved to be as effective as TBTO against the development of the fouling microalga as shown in Fig. 1 [26].

3.4 Diterpenes and Their Derivatives

Linear diterpenes (**342–348**) isolated from *Bifurcaria bifurcata*, a common brown alga of the Atlantic shores of Europe, and their derivatives (**349, 3410**) were tested in laboratory assays against the representative species of major groups of fouling organisms, *viz.* bacteria, fungi, diatoms, the spores and zygotes of macroalgae and the blue mussels *Mytilus edulis*. Several components showed promising levels of the activity. Especially, two diketonyl diterpenes (**345, 3410**) showed almost the same bioassays of antibacterial and antifungal activities as those of TBTO and also showed almost the same minimum inhibitory concentration of 8 μg mL^{-1} for marine bacteria and marine fungi with those of TBTO, and elegandiol (**344**) showed high bioassay activities for the inhibition of microalgal growth, and the development of macroalgal spores and zygotes [16].

Three renillafoulins **3411–3413** from the Atlantic sea pansy *Renilla reniformis*, inhibited barnacle settlement with EC_{50} values ranging from 0.02 to 0.2 μg mL^{-1} [17, 27]. Renillafoulins and pukalide **3418** isolated from oc-

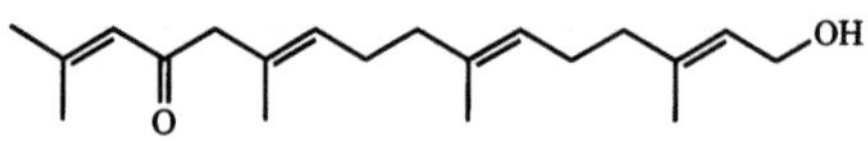

342

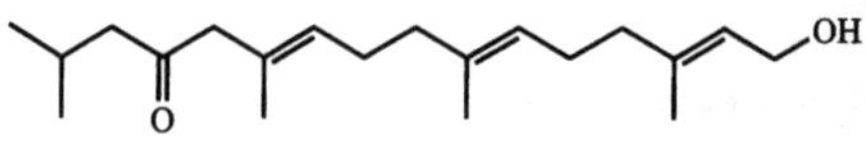

343

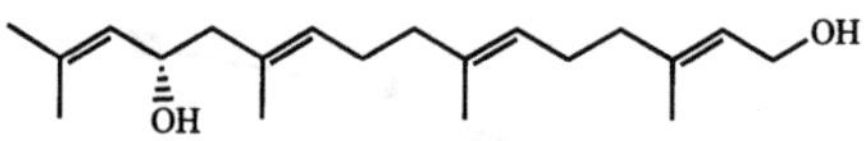

344

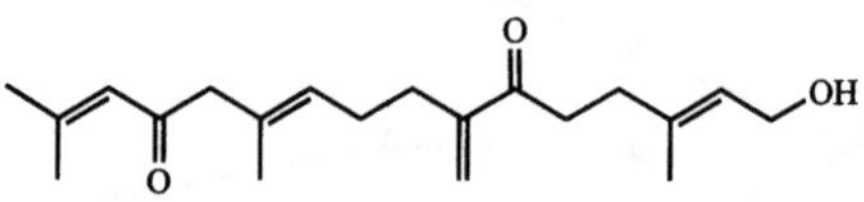

345

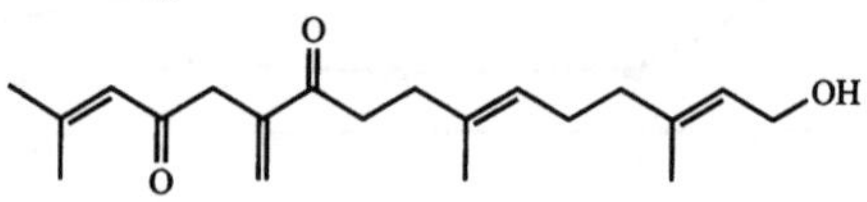

346

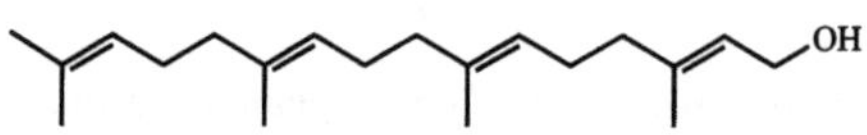

347

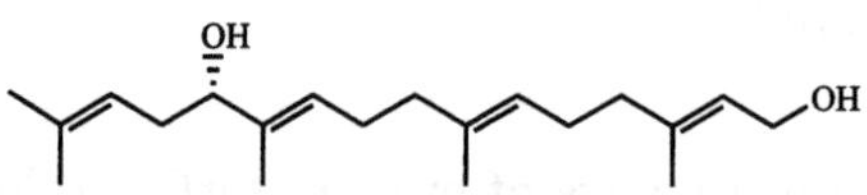

348

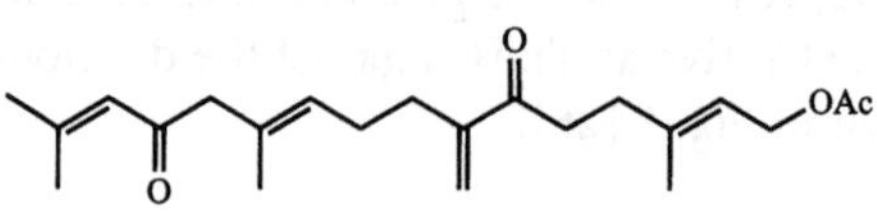

349

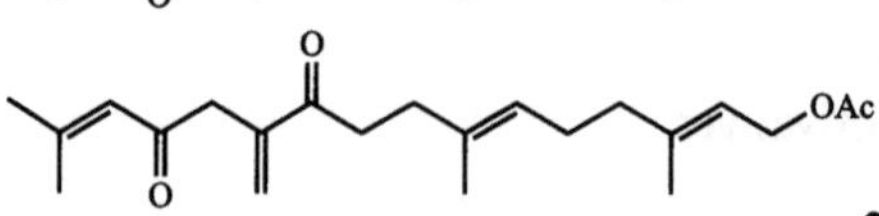

3410

Fig. 2 Chemical formulae of diterpenes isolated from *Bifurcaria bifurcata* [16]

Table 4 Diterpenes from marine sponge *Acanthella carvernosa* and their antifouling activity [18–20]

Compound	Antifouling activity IC_{50} ($\mu g\,mL^{-1}$)
Kalihinene X **3419**	0.49
Kalihinene Y **3420**	0.45
Kalihinene Z **3421**	1.1
Kalihipyrans A **3422**	1.3
Kalihipyrans B **3423**	0.85
Kalihinol A **3414**	0.087
10-Formamidokalihinene **3415**	0.095
15-Formamidokalihinene **3424**	0.14
Bifora-4,9,15-triene **3425**	4.6
10-Formamido-5-isocyanatokalihinol-A **3426**	ca. 0.05
10-Formamido-5-isothiocyanatokalihinol-A **3427**	ca. 0.05
Cu_2SO_4	0.15

tacorals *Renilla reniformis* and *Leptogorgia virgulata*, are potent inhibitors of barnacle settlement, however, these are comparatively complex and thus are not amenable to commercial exploitation. Clare et al. [22] examined about 20 analogues, bases on the functional groups of lactone and furan rings in the parent molecules, such as 2-furyl-*n*-pentyl ketone, khellin and γ-decanolactone for anti-settlement activity and toxicity. Especially, 2-furyl-*n*-pentyl ketone **3417** (EC_{set50} 0.002 μM) was more active than copper ($CuSO_4 \cdot 5H_2O$) in preventing the settlement of diatom, *Nitzschia* spp. [22].

Fusetani et al. [18] isolated two new diterpene formamides along with seven known ditepenes from a marine sponge *Acanthella cavernosa* at Yakushima Island in Japan. These nine diterpenes showed potent antifouling activities against larvae of the barnacle *Balanus amphitrite* as shown in Table 4 [19, 20]. The new kalihipyrans A (**3422**) and B (**3423**) inhibited the larval settlement and metamorphosis of the barnacle *Balanus amphitrite* with IC_{50} of 1.3 and 0.85 $\mu g\,mL^{-1}$, respectively. These activities are comparable to those of kaihinene X–Z (IC_{50}: **3419** = 0.49, **3420** = 0.45, **3421** = 1.1 $\mu g\,mL^{-1}$), whereas the corresponding isocyano compounds are more active (IC_{50}: **3414** = 0.087, **3415** = 0.095, **3424** = 0.14 $\mu g\,mL^{-1}$) [18]. Further, 10β-formamido-5-isocyanatokalihinol-A **3426** and 10β-formamido-5-isothiocyanatokalihinol-A **3427** were highly antifouling with the IC_{50} of only ca. 0.05 $\mu g\,mL^{-1}$ [20]. It should be noted that these 10β-formamido-5-kalihinens **3426** and **3427** were more active than $CuSO_4$ (IC_{50}: 0.15 $\mu g\,mL^{-1}$), and their toxicities were quite low [24].

3.5
Sesterterpenes

Three sesterterpenes **351–353** together with epidioxy sterols isolated from a Palauan marine sponge, *Lendenfeldia chondrodes*, showed a repellent activity against a blue mussel *Mytilus edulis galloprovincialis*. In three sesterterpenes, 22-acetoxy-16β-hydroxy-24-methyl-24-oxoscalarano-25,12β-lactone

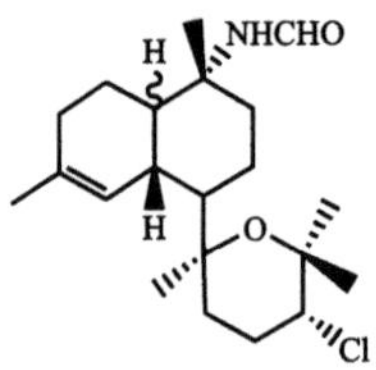

Kalihinene

1β-H, 14-α-Cl **3419**
1α-H, 14-α-Cl **3420**
1β-H, 14-β-Cl **3421**

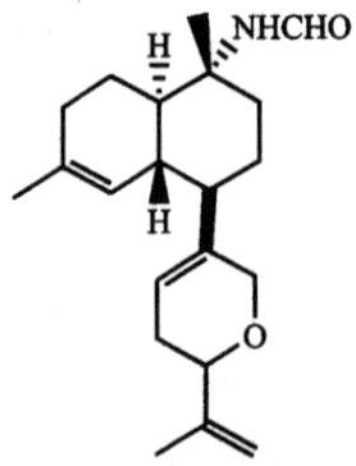

Kalihipyrans A

3422

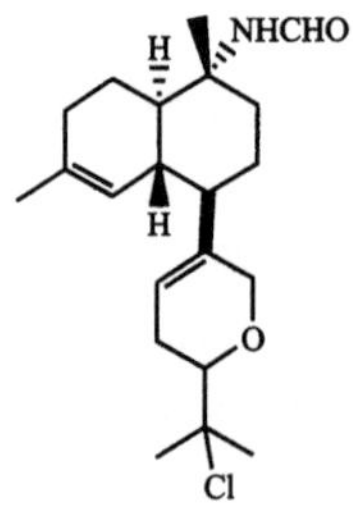

Kalihipyrans B

3423

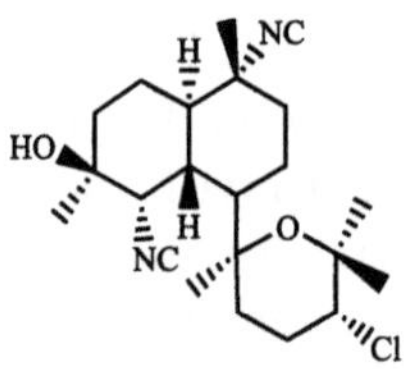

Kalihinol A

3414

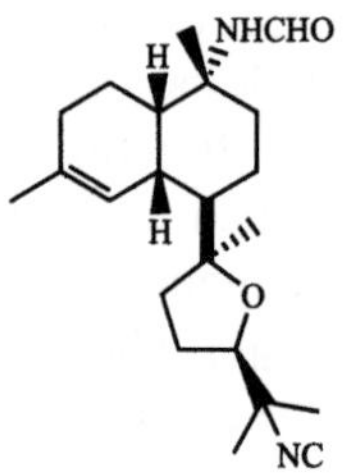

10-Formamidokalihinene

3415

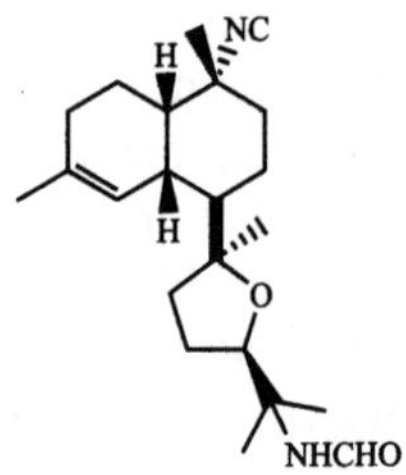

15-Formamidokalihinene

3424

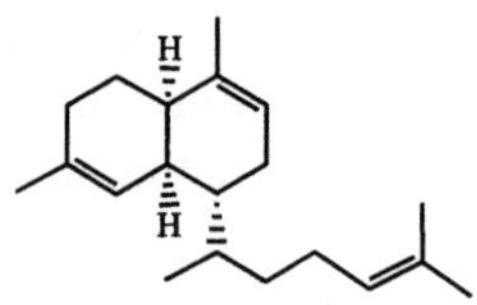

Bifora-4,9,15-triene

3425

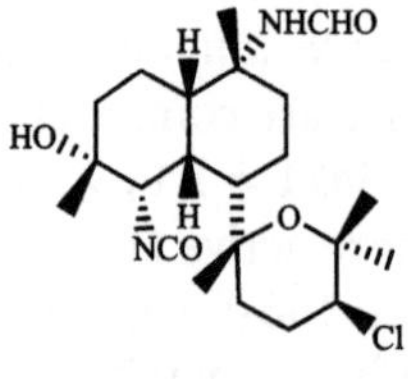

10 -Formamido-5-isocyanatokalihinol-A

3426

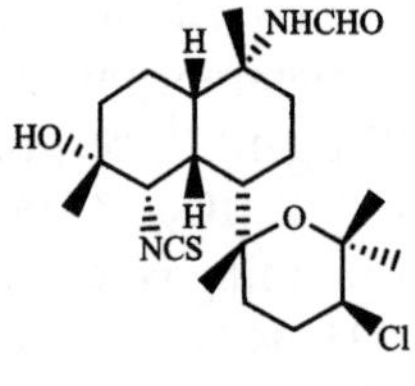

10 -Formamido-5-isothiocyanatokalihinol-A

3427

Scheme 5

351 and 24-methyl-12,24,25-trioxoscalar-16-en-22-oic acid **352** showed potent antifouling activities. The antifouling activities of lactone **351** and aldehyde **352** show 90% and 95% at the concentration in 1 ppm. The value of the other aldehyde **353** is 10% [23].

4 Nitrogen-containing Compounds

4.1 Introduction

One of representative natural products is an alkaloid which has a pharmacological effect. They are found in nature, usually in plants. They are nitrogenous organic molecules. The name derives from the word alkaline; originally, the term was used to describe any nitrogen-containing base (an amine in modern terms). Some kinds of these alkaloids and the other nitrogen-containing compounds also showed antifouling activities. They are heterocyclic compounds (e.g., gramines, pyrrols, and pyrazoles), amides, carbamates, primary amines, peptides, etc. as shown in Table 5 [4, 28–47].

4.2 Hetrocyclic Compounds

4.2.1 Gramines

Many kinds of heterocyclic compounds such as gramines, pyrrols, and pyrazoles, show antifouling activities. Especially, 2,5,6-Tribromo-1-methylgramine **421** isolated from a marine bryozoan, *Zoobotryon pellucidum*, showed antifouling activities against a barnacle *Balanus amphitrite* and a blue mussel *Mytilus edulis*. The antifouling activities of 2,5,6-tribromo-1-methylgramine and $CuSO_4$ against the blue mussels *Mytilus edulis* are shown in Table 6 [4], and minimum inhibitory concentration for larval settlement and IC_{30} values of 2,5,6-tribromo-1-methylgramine and TBTO for reared cyprids are shown in Table 7 [4]. The antifouling activity of 2,5,6-tribromo-1-methylgramine is higher than that of $CuSO_4$ and its inhibitory activity was 6 times as strong as that of TBTO, while its toxicity to the cyprid larvae was one-tenth of that of TBTO. This substance also inhibited the settlement of the blue mussel *Mytilus edulis* [4].

Recently, Kawamata et al. [28] synthesized approximately one hundred kinds of gramine compounds. Of these compounds, 5,6-dichlorogramine **422** showed much higher antifouling activity than that of TBTO as shown in Table 8 [28].

Table 5 Nitrogen-containing Compounds

Compound	Living thing species	Refs.
4.2 Heterocyclic Compounds		
4.2.1 Gramines		
2,5,6-Tribromo-1-methylgramine **421**	marine bryozoan *Zoobotryon pellucidum*	[4]
5,6-Dichlorogramine **422**		[28]
4.2.2 Pyrrols		
Pseudoceratidine **424**	marine sponge *Pseudoceratina purpurea*	[29–31]
1,7-Diaminoheptane derivative **425**		
Mauritiamine **426**	marine sponge *Agelas mauritiana*	[32]
4.2.3 Pyrazoles		
5-Bromoverongamine **427**	marine sponge	[33]
5-Oxo-5,6,7,8-tetrahydroimidazo[1,5-c] pyrimidine **428**	*Pseudoceratina sp.*	
4.2.4 Others		
3,4,5,6-Tetrahydro-6-hydroxymethyl-3,6-dimethyl-4-pyrimidine-carboxylic acid **429**	Palauan sponge *Protophlitaspongia aga*	[48]
Zooanemonin **4210**		
Polymeric 3-alkylpridium salt **4211**	Mediterranean sponge *Reniera sarai*	[34]
Phenazine-1-carboxylic acid **4212**	marine epibiotic bacteria,	[35]
2-*n*-Heptylquinol-4-one **4213**	*Pseudomonas* sp.	
4.3 Amides		
Ceratinamide A **431**	marine sponge *Pseudoceratina purpurea*	[36]
Haliclonamides	marine sponge	[37]
Waiakeamides	*Haliclona* sp.	[38]
Trigonelline **432**	octacoral *Dendronephthya* sp.	[39]
Nicotinamide **433**	mallotus *Mallotus japonicus* Mueller-Arg.	[40]
Cyanoformamide **434**	marine sponge *Pseudoceratina purpurea*	[41]
N-Docosanoyl-*D*-erythro-(2*S*, 3*R*)-16-methyl-heptadecasphing-4(E)-enine **435** (C22 Ceramide)	marine sponge *Haliclona koremella*	[42]
Cyclotides	terrestrial plant	[43, 44]
Cycloviolacin O2	*Viola odorata*	
4.4 Other Nitrogen-containing Compounds		
Moloka'iamine **441**	marine sponge *Pseudoceratina purpurea*	[45, 46]
Ethyl *N*-(2-phenylethyl)carbamate **442**	marine bacteria *Cytophaga* sp.	[47]
Ethyl *N*-[2-(4-nitrophenyl)ethyl]carbamate **443**		

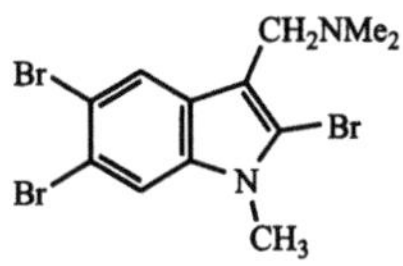

2,5,6-Tribromo-1-methylgramine

421

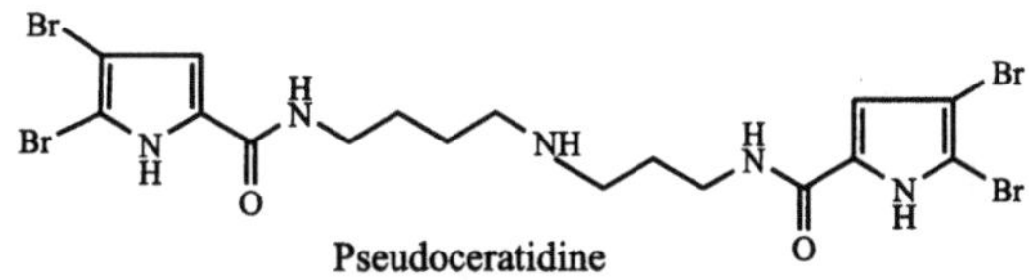

Pseudoceratidine

424

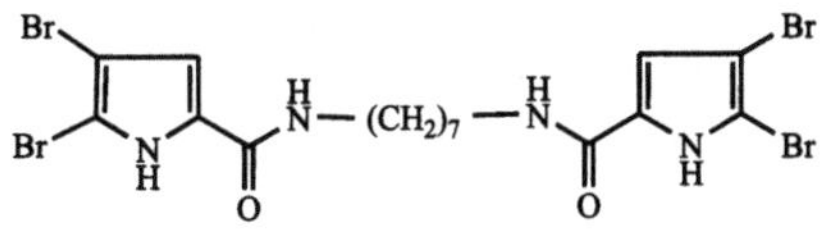

1,7-Diaminoheptane derivative

425

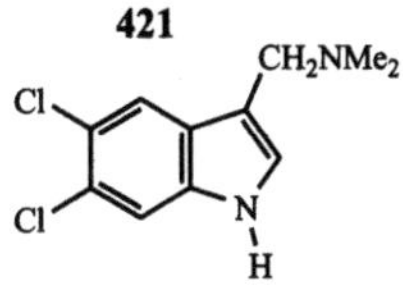

5,6-Dichlorogramine

422

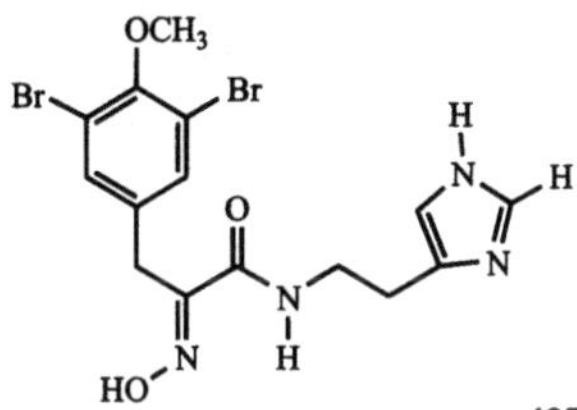

427

5-Bromoverongamine

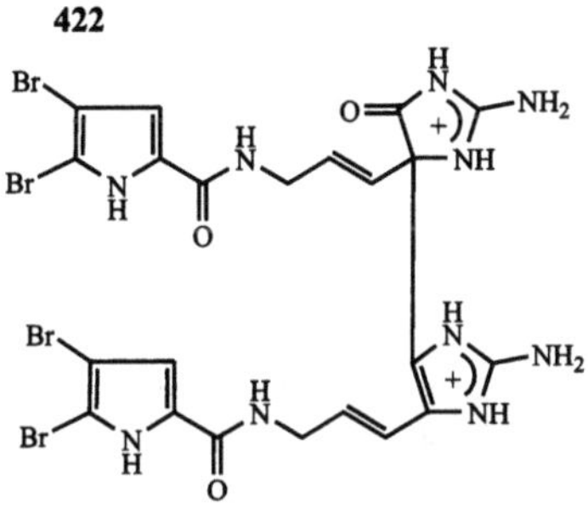

Mauritiamine

426

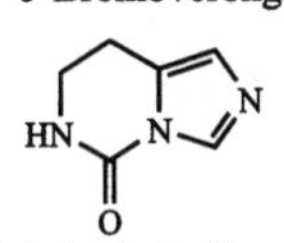

428

5-Oxo-5,6,7,8-tetrahydroimidazo[1,5-c]pyrimidine

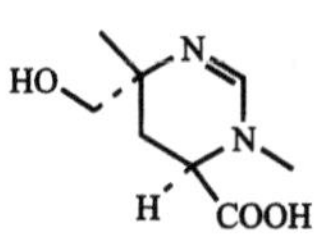

4-Pyrimidine carboxylic acid

429

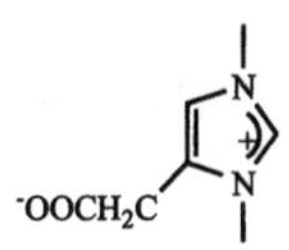

Zooanemonin

4210

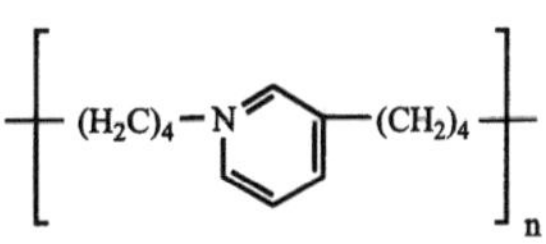

Polymeric 3-alkylpridium salt

4211

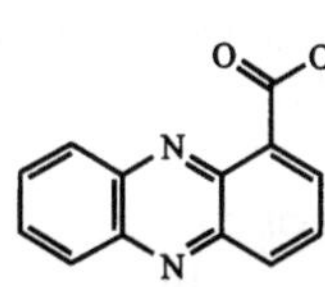

Phenazine-1-carboxylic acid

4212

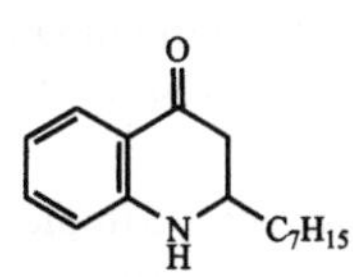

2-*n*-Heptylquinol-4-one

4213

Scheme 6

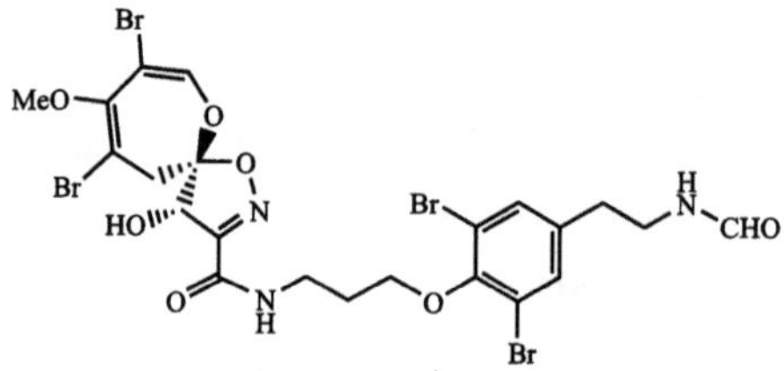

Ceratinamide A

431

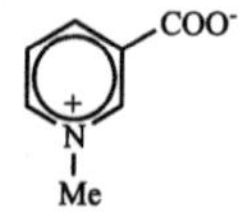

Trigonelline

432

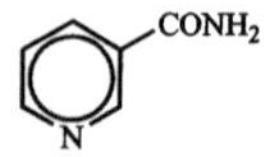

Nicotinamide

433

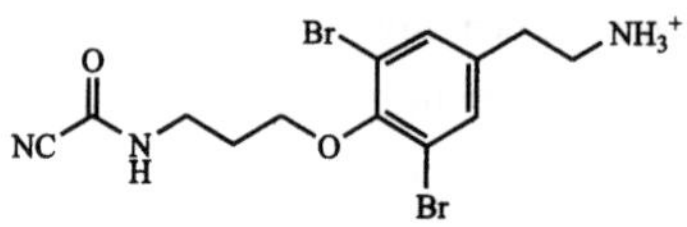

Cyanoformamide

434

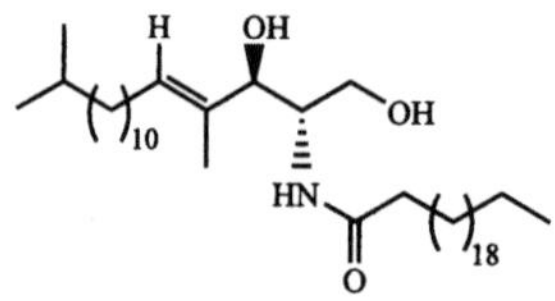

C22 Ceramide

435

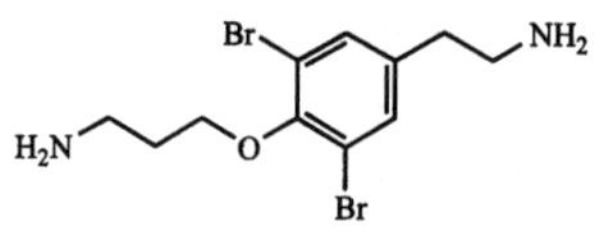

Moloka'iamine

441

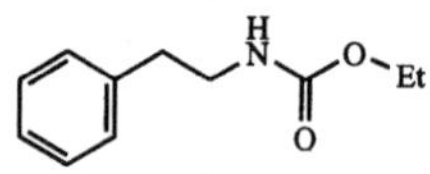

Ethyl *N*-(2-phenylethyl)carbamate

442

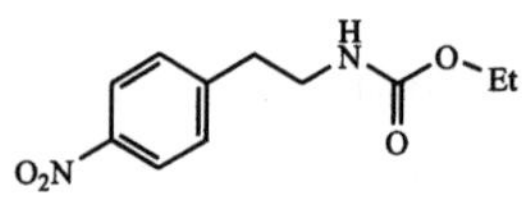

Ethyl *N*-[2-(4-nitrophenyl)ethyl]carbamate

443

Scheme 7

Table 6 Antifouling activity of 2,5,6-tribrom-1-methylgramine (TBG) and $CuSO_4$ against the blue mussels *Mytilus edulis* [4]

Substance	Sample amount (ng cm^{-2})					Unit [a]
	80	30	16	8	1.6	
TBG	++	++	++	++~+	+~±	680–1360
$CuSO_4$	++	+				100

[a] $\text{Unit} = 100 \times \frac{CuSO_4\text{(minimum dose for ++ activity, }\mu\text{mol cm}^{-2})}{\text{Sample (minimum dose for ++ activity, }\mu\text{mol cm}^{-2})}$

Table 7 Minimum inhibitory concentration (MIC) for larval settlement and 30% lethal concentration (LC_{30}) values of 2,5,6-tribrom-1-methylgramine (TBG) and TBTO (ppm) for reared cyprids [4]

Substance	MIC	LC_{30}
TBG	0.03	0.60
TBTO	0.20	0.06

Table 8 Antifouling activity of gramine compounds [28]

Compound	Antifouling activity (ppm)
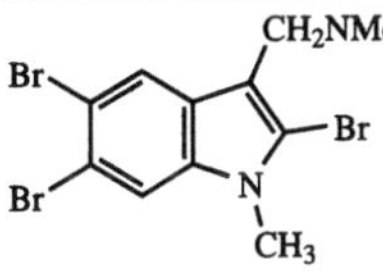 **421**	0.063
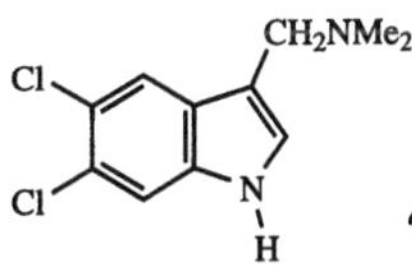 **422**	0.008
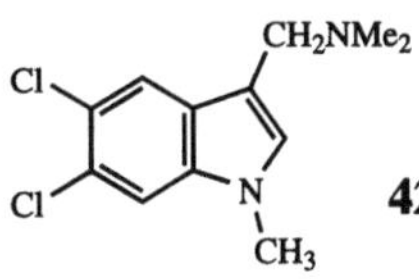 **423**	0.063
TBTO	0.127
Cu_2O	1.25

4.2.2 Pyrrols

Pseudoceratidine **424** having 4,5-dibromopyrrole-2-carbamyl units, isolated from a marine sponge *Pseudoceratina purpurea*, inhibited the larval settlement and metamorphosis of a barnacle *Balanus amphitrite* with an ED_{50} value of 8.0 $\mu g\,mL^{-1}$ [29]. Further, in order to explore structure-activity relationships responsible for antifouling and antimicrobial activity, Ponasik et al. [30, 31] synthesized its several analogs. Of these compounds, 1,7-diaminoheptane diamide (**425**) showed a much higher antifouling activity with ED_{50} value of 0.1 $\mu g\,mL^{-1}$.

Another pyrrole compound, mauritiamine **426** isolated from a marine sponge *Agelas mauritiana*, also showed moderate antifouling activity, and it inhibited larval metamorphorsis of a barnacle *Balanus amphitrite* with ED_{50} values of 15 $\mu g\,mL^{-1}$ [32].

4.2.3
Pyrazoles

5-Bromoverongamine **427**, a novel antifouling tyrosine alkaloid has been isolated together with 5-oxo-5,6,7,8-tetrahydroimidazo[1,5-c]pyrimidine **428** from the Caribbian Sea specimen of a marine sponge *Pseudoceratina* sp. [33]. 5-Bromoverongamine **427** inhibited the settlement of barnacle larvae at 10 $\mu g\,L^{-1}$ (EC_{50} = 1.03 $\mu g\,L^{-1}$) and was not toxic at this concentration [33].

4.2.4
Others

Pyrimidine derivative, 3,4,5,6-tetrahydro-6-hydroxymethyl-3,6-dimethyl-4-pyrimidinecarboxylic acid **429** and azooanemonin **4210** were isolated from a marine sponge *Protophitaspongia aga*. Their antifouling activity against the barnacle is shown in Table 9 [48]. The activities of these compounds were comparable to that of $CuSO_4$. These compounds proved to be lethal to the cyprids at 50 ppm [48].

Table 9 IC_{50} Values (ppm) of primidine derivative **429**, zooanemonin **4210**, and $CuSO_4$ against barnacle [48]

Compound	24 hours	48 hours
Primidine derivative **429**	5.0	6.7
Zooanemonin **4210**	7.5	7.8
Cu_2SO_4	4.0	5.2

Polymeric 3-alkylpridium salt **4211** from a Mediterranean sponge, *Reniera sarai*, was less effective for inhibiting the cyprid larvae settlement as compared with the organic booster biocides, copper pyrithione or zinc pyrithione, and their effects were non toxic and reversible, as measured by very low toxicity bioassays [34].

Other heterocyclic compounds such as phenazine **4212** and alkylquinol **4213**, isolated from marine epibiotic bacteria, *Pseudomonas* sp., showed the inhibition of settlement of barnacle larvae, *Balanus amphitrite* and lagal spores of *Ulva lactuca*, and the growth of *Ulva lactuca* [35].

4.3 Amides

Ceratinamide A **431** of bromotyrosine derivative, isolated from a marine sponge *Pseudocertina purpurea*, inhibited the larval settlement and metamorphosis of a barnacle *Balanus amphitrite* with ED_{50} value of 0.10 $\mu g\,mL^{-1}$ [45, 48].

Sera et al. [37, 38] isolated cyclic peptides, haliclonamides and waiakeamides, from a marine sponge, *Haliclona* sp. They showed a weak repellent activity against a blue mussel, *Mytilus edulis galloprovincialis*.

Trigonelline **432** occurs widely in mammalian urine, leguminous plants and marine invertebrates. It is isolated from an octacoral *Dendronephthya* sp. It showed the same level of settling-inhibitory activity against the barnacle larvae with that of $CuSO_4$ (minimum inhibitory concentrations for the larval settling were 2–10 ppm for trigonelline and 5 ppm for $CuSO_4$). The conventional antifoulants such as TBTO and $CuSO_4$ have strong toxicities (the values of LC_{50} are 0.3 and 10, respectively). On the other hand, trigonelline is considered to be safer because the substance is very commonly found in leguminous plants, marine invertebrates, macroalgae and microalgae and are often eaten as foods [39]. Very simple compound, nicotinamide **433** (313, $CuSO_4 = 100$), also shows a high repelling activity against the blue mussel, *Mytilus edulis* [40].

Cyanoformamid, ceratinamine **434**, from a marine sponge, *Pseudoceratina purpurea*, shows antifouling activity against the cyprids of a barnacle *Balanus amphitrite* with EC_{50} value of 5.0 $\mu g\,mL^{-1}$ and cytotoxic against P388 murine leukemia cells with IC_{50} value of 3.4 $\mu g\,mL^{-1}$ [41].

Ceraminde *N*-docosanoyl-*D*-*erythro*-(2*S*, 3*R*)-16-methyl-heptadecasphing-4(*E*)-enine (C22 ceramide) **435** as another amide, isolated from a marine sponge *Hliclona koremella*, also showed an antifouling activity against the macrolagae [42].

Cycloviolacin O2, a plant peptide of a cyclotide family, isolated from the terrestrial plant, *Viola odorate*. The sequence was determined to be cyclo-(VWIPCISSAIGCSCKSKVCYRNGIPCGESC) [43, 44]. It showed inhibition against fouling barnacles to be well within the same range as that for TBTO; its inhibition was complete at a concentration of 0.25 μM. This degree of inhibition might be compared to an EC_{50} value of 0.15 μM (0.09 $\mu g\,mL^{-1}$) for TBTO in the same assay. The EC_{50} value for TBTO represents the concentration at which 50% of the cyprids are killed, whereas the effect of it in this assay was nontoxic and reversible, giving it a distinct advantage over TBTO and other commercially used marine antifouling agents [43].

4.4 Other Nitrogen-containing Compounds

Primary amine moloka'iamine **441**, isolated from a marine sponge, *Pseudeoceratina purpurea*, is a potential antifoulant against a zebra mussel, *Dreissena polymorpha*, (EC_{50} = 10.4 μM). The moloka'iamine **441** showed the absence of phytotoxic activity toward *Lemna pausicostata* and significant selectivity against macrofouling organisms such as the zebra mussels and was suggested to have the potential utility as a lead compound of naturally derived antifoulants [45, 46].

Ethyl *N*-(2-phenylethyl)carbamate **442** isolated from marine bacteria, *Cytophaga* sp. inhibited a biofilm formation. Furthermore, dithiocarbamate derivatives of 2-(4-nitrophenyl)ethylamine **443**, significantly inhibited the growth of marine attaching bacteria (bioassay IC_{50} 0.5 μg cm^{-2}) [47].

5 Phenols

5.1 Monophenol Derivatives

Phenol compounds consist of monophenols having one hydroxy group and polyphenols having two or more hydroxy groups on one phenyl group. They generally show biocidal properties. Thererfore, many of these compounds are used for pharmaceutical and agrochemical ingredients. These natural products are shown in Table 10 [49–57].

Trans-6-, 8-, and 10-shogaols (**513**) were obtained from a hexane extract of the roots of ginger, *Zingiber officinale*. Roscocoe showed highly potent attachment-inhibitors (three times more active than that of standard $CuSO_4$ in the blue mussel assay) from them. *Trans*-8-shogaol (**513**, $n = 8$) showed the highest antifouling activity comparable with that of tributyltin fluoride (TBTF) which is recognized as one of the most effective antifouling agents in a conventional submerged assay as shown in Table 11 [52].

Takasawa et al. [58] reported on many alkyl phenols with attaching repellent activities in the blue mussels assay and with antifouling activities in the conventional submerged assay as shown in Table 12 [58, 59]. Especially *p*-branched alkylphenols **516**, **517** show high repellent activities against the blue mussels, *Mytilus edulis* and the same antifouling activity as the organotin compounds [58, 59].

Table 10 Phenols and Their Derivatives

Compound	Living thing Species	Refs.
5.1 Phenol Derivatives (Stilben glucosides)		
Polydatin coumarate **511**	eucalyptus *Eucalyptus rubida*	[49]
Capsaicin **512**	pepper	[50, 51]
Shogaols **513**	ginger *Zingiber officinale* Roscoe	[52]
Zingerols **514** Zingerone **515**		[52]
5.2 Polyphenols (Tannins, Phloroglucinols, Kaempherol glucopyranosides)		
(+) Catechin **521** (–)-Epicatechin **522**	moutain cherry *Prunus jamasakura*	[49]
Kaempferol coumaroylglucopyranoside **523** (Kaempferol)	leaves of oak tree *Quercus dentata*	[49]
Polydatin 6″-*O*-(*E*)-*p*-coumarate **524** Raponticin 6″-*O*-(*E*)-*p*-coumarate **525**	mallee *Eucalyptus rubida*	[49]
Macrocarpal K **526** Isopentylphloglucinol-β-eudesmol adduct	leaves of mallee *Eucalyptus globulus*	[53, 54]
Epigallocatechingallate **527** (Tannins)	green tea	[55]
3,5-Diformyl phloroglucinols **528** Sideroxylonal A	mallee *Eucalyptus* p.	[56]
2,4-Dihydrooxy-3,5,6-trimethylbenzoate **529** Mortivinacin A **5210**	fungus *Mortierella vinacea*	[57]

Table 11 Antifouling activities of Shogaols, Zingerone, Zingerols and TBTF in the Conventional Submerged Assay [52]

Compound	Days 29 [a]	62	101	125
6-Shogaol (**513**) $n = 4$	○ [b]	○	×	×
8-Shogaol (**513**) $n = 6$	○	○	○	○
10-Shogaol (**513**) $n = 8$ [d]	○	○ [c]	×	×
Zingerone (**515**)	○	×	×	×
6-Zingerol (**514**) $n = 4$	○	×	×	×
8-Zingerol (**514**) $n = 6$	○	○	×	×
TBTF	○	○	○	○

[a] from 28 June 2001
[b] ○ activity was apparent; × the degree of fouling was the same as that in the blank zone. Each zone (5 cm in diameter) was coated with 300 mg of a sample
[c] 45 days [d] 140 mg/5 cm in diameter

Table 12 Repellent activity against the blue mussel, *Mytilus edulis* and the antifouling activity [58, 59]

Compound	Repellent activity [a] (Unit [b])	Antifouling activity [c] (50 µmol cm^{-2}) 15	30	60	90 days
Alkyl phenols:					
$HO-C_6H_4-(R)(-p)$					
$R=(CH_2)_6CH_3$	4	+	+	+	±
$R=C(CH_3)_2(CH_2)C(CH_3)_3$ **516**	30	+	+	+	+
$R=C_9H_{19}$(*o*- and *p*-(1 : 9)) **517**	40	+	+	+	+
Isothiocyanate					
6-Methylthiohexyl	10	+	+	+	+
Nicotinic acid derivative					
Nicotinamide	313	–	–	–	–
Naphthoquinone derivative					
1,4-Naphthoquinone	100	+	+	+	–
Antifouling agents					
TBTO	200	+	+	+	+
TPT acetate	150	+	+	+	+
$CuSO_4$[d]	100				

[a] Repellent activity is shown by unit, but compounds having no activity with 4 unit are shown by +, ±, or –

[b] Unit: $100 \times \frac{\text{minimum dose of } CuSO_4 \text{ for ++ activity } (\mu mol\, cm^{-2})}{\text{minimum dose of the sample for ++ activity } (\mu mol\, cm^{-2})}$

[c] +, no fouling organisms in the sample zone; ±, slightly active; –, the degree of fouling was the same as that in the blank zone

[d] Standard sample for repellent activity in the blue mussel assay

Table 13 Repellent activity against the blue mussel, *Mytilus edulis* [48]

Compound	Repellent activity (Unit [a])
(+) Catechin **521**	6
(–)-Epicatechin **522**	14
Kaempferol coumaroylglucopyranoside **523** (Kaempferol)	227
Polydatin 6″-*O*-(*E*)-*p*-coumarate **524**	185
Raponticin 6″-*O*-(*E*)-*p*-coumarate **525**	179
TBTO	200
TPT acetate	150
$CuSO_4$[b]	100

[a] Unit: $100 \times \frac{\text{minimum dose of } CuSO_4 \text{ for ++ activity } (\mu mol\, cm^{-2})}{\text{minimum dose of the sample for ++ activity } (\mu mol\, cm^{-2})}$

[b] Standard sample for repellent activity in the blue mussel assay: 100

Phenol Derivatives

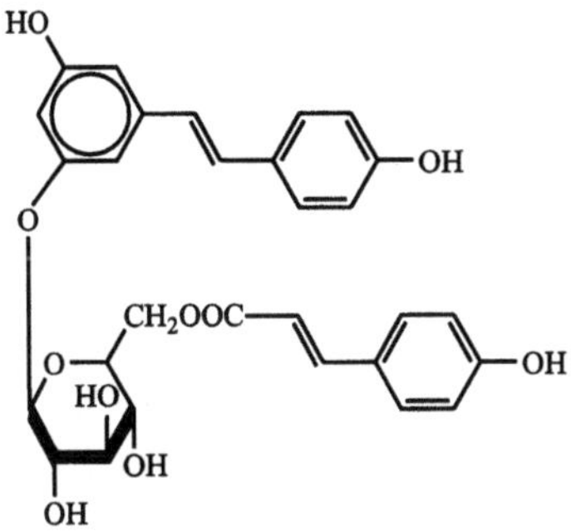

511

Polydatin 6"-*O*-(*E*)-*p*-coumarate

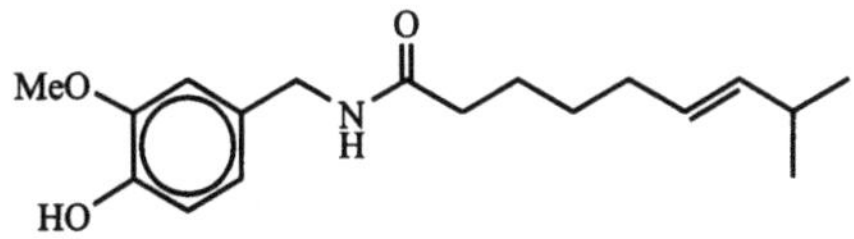

Capsaicin

512

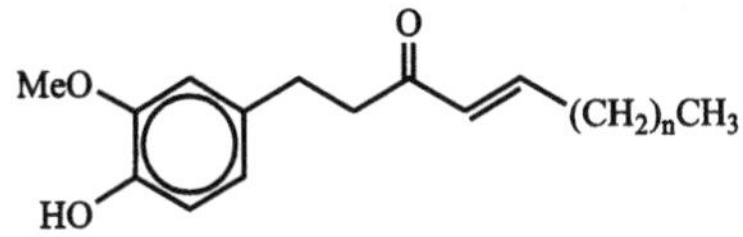

513 n = 4, 6, 8 Shogaols

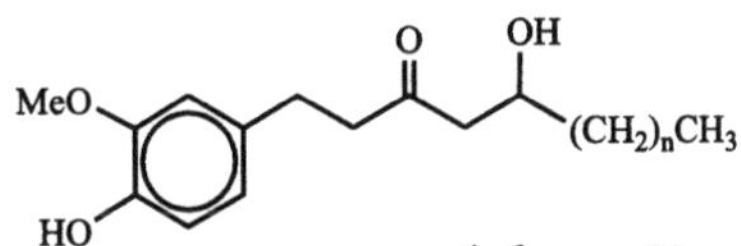

n = 4, 6 Zingerols

514

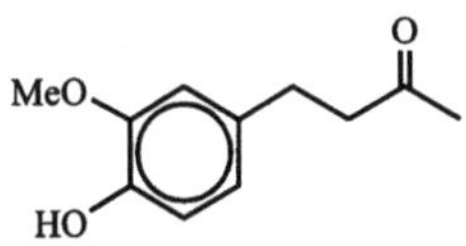

Zingerone

515

Scheme 8

5.2
Polyphenol Derivatives

Polyphenol derivatives of (+)-catechin **521** and (–)-epicatechin **522** extracted from a mountain cherry, *Prunus jamasakura*, kaempferol coumaroylglucopyranoside **523** extracted from the leaves of an oak, *Quercus dentata*, polydatin 6″-*O*-(*E*)-*p*-coumarate **524**, raponticin 6″-*O*-(*E*)-*p*-coumarate **525** extracted from mallee, *Eucalyptus rubida*, showed repellent activities against the blue mussel, *Mytilus edulis* as shown in Table 13 [48].

A macrocarpal named macrocarpal K **526**, an isopentylphloglucinol-β-eudesinol adduct, was isolated from the leaves of mallee, *Eucalyptus globulus*, and also six macrocarpals A, B, E, am-1, H and K showed attachment-inhibiting activities against the blue mussel *Mytilus edulis galloprovincialis*. These five macrocarpals A, B, E, H and K showed potent attachment-inhibiting activities against the blue mussel, which was 3 times higher than that of the standard antifoulant $CuSO_4$ [53]. Further, four kinds of polyphenol compounds, 3,5-diformyl phloroglucinols (**528**) also showed the same high repellent activity to that of the standard antifoulant $CuSO_4$ [56] against the blue mussel.

Polyphenols

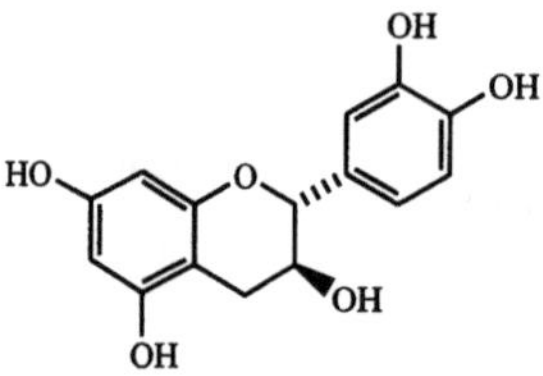

(+) Catechin

521

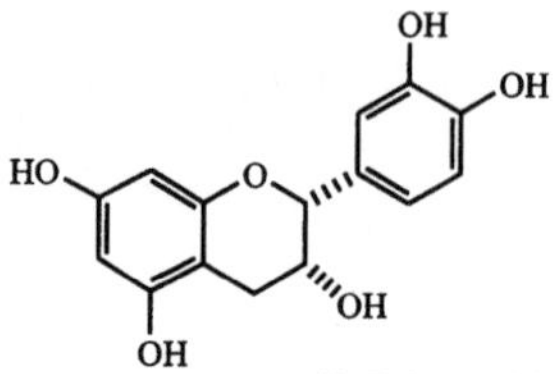

(-)-Epicatechin

522

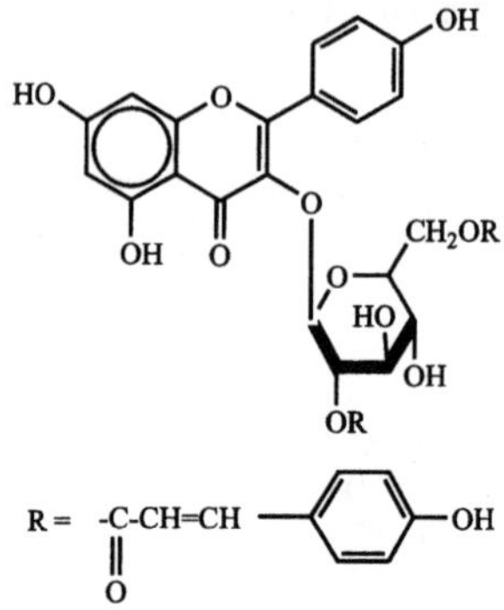

Kaempferol

523

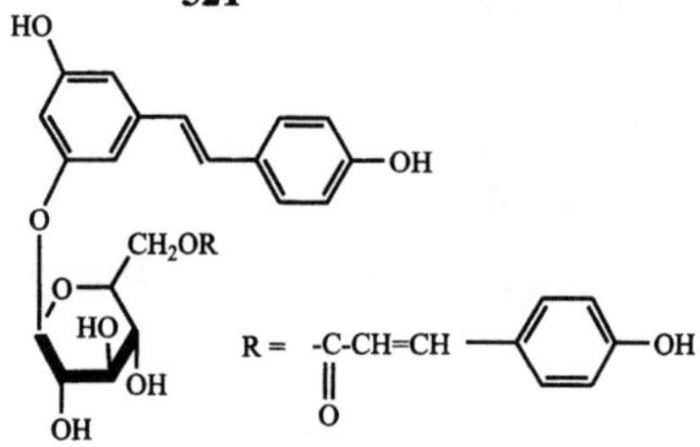

Polydatin 6"-O-(E)-p-coumarate

524

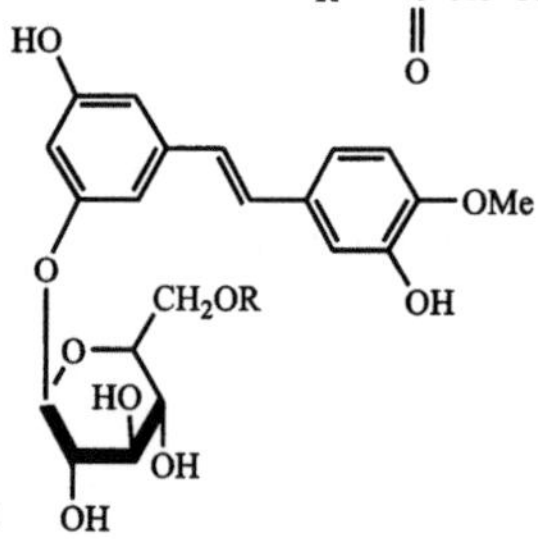

Raponticin 6"-O-(E)-p-coumarate

525

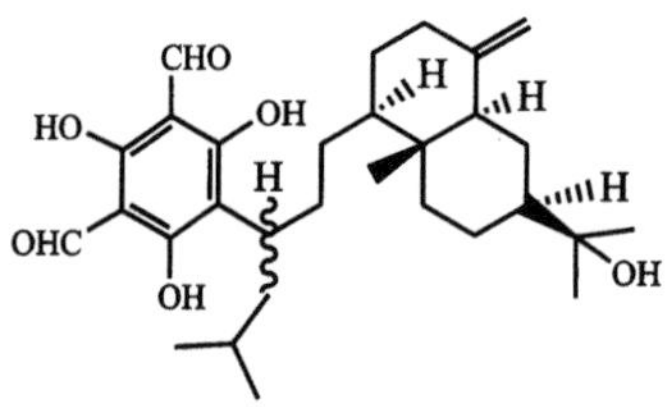

Macrocarpal K

526

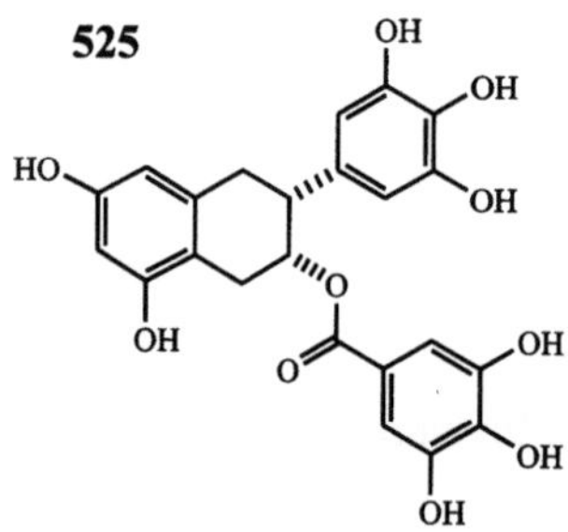

Tannins

527

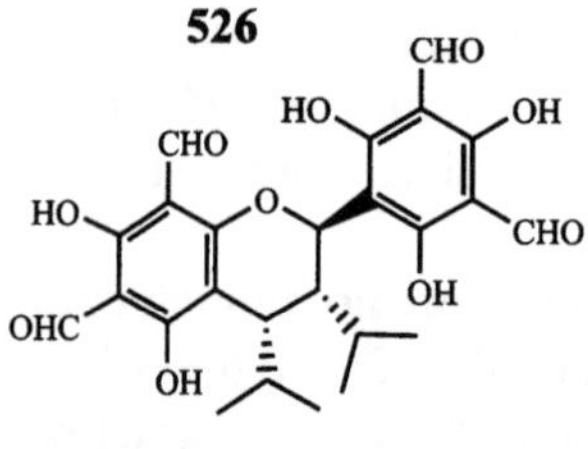

Sideroxylonal A

528

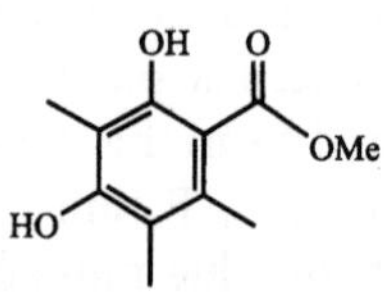

2,4-Dihydrooxy-3,5,6-trimethylbenzoate

529

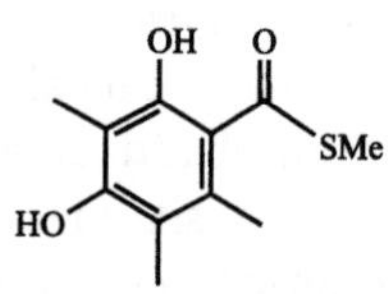

Mortivinacin A

5210

Scheme 9

Trimethylresorcinol dervatives 2,4-dihydroxy-3,5,6-trimethylbenzoate **529**, e.g. mortivinacin A **5210** and nicotinic acid ($C_5H_4N-COOH$) were extracted from fungus *Mortierella vinacea* of a solid-substrate fermentation culture. The resorcinol derivatives **529** and **5210** produces ca. 30-mm zones of inhibition in 48-h disk assays against *Fusarium verticillioides* by using a 250 μg $disk^{-1}$. Nicotinic acid caused a similar-sized zone of reduced growth at the above level. The resorcinol derivatives **529** and **5210** also showed antibacterial activities (10- to 15-mm zones of inhibition) in the standard disk assays against *Bacillus subtilis* and *Staphylococcus aureus* by using a 100 μg $disk^{-1}$ [57].

6 Steroids

Steroids are compounds that contain a cyclopenta[α]-phenathrene ring skeleton which when fully hydrogenated, is called gonan or steran **611** [60]. The

Table 14 Steroids

Compound	Living thing Species	Refs.
6.1 Epidioxy sterol		
5-α,8α-Epidioxycholest-6-en-3β-ol **612**	Palauan marine sponge *Lendenfeldia chondrodes*	[23]
6.2 Seco-steroids and Tetraene Steroids		
12α-Actoxy-13,17-*seco*-chlolesta-1,4-diene-3-ones **621**	Octocoral *Dendronephthya* sp.	[61]
3-Methoxy-19-norpregna-1.3,5(10), 20-tetraene **622**	octocoral *Alcyonium gracillimum*	[62]
6.3 Steroid Lycosides		
$\Delta^{9(11)}$ 3β,6α-Diydroxysteroidal aglycon **631**	Starfish *Henricia downeyae*	[63]
6.4 Muricins		
Muricins-1 Aglycone pregana-5, 20-dien-3β-ol **641**	Pacific gorgonians *Muricea californica* *Muricea fruticosa*	[64]
6.5 Steroidal Sulfated Glycosides		
Steroidal sulfated glycosides (Asterosaponins) Goniopectenosides A **651**	Starfish *Goniopecten demonstrans*	[65]

great variety of natural and synthetic steroidal compounds results from the presence of double bonds and various substituents on the ring skeleton and the side-chain. Trivial names are permitted for some steroids, mostly in natural steroids with high biological activities. Müller [54] reported that these

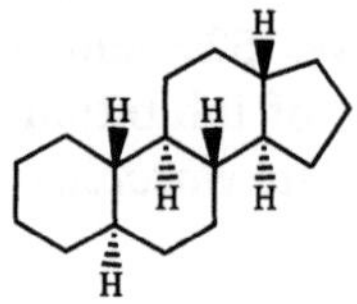

5α-Gonan or 5α-steran

611

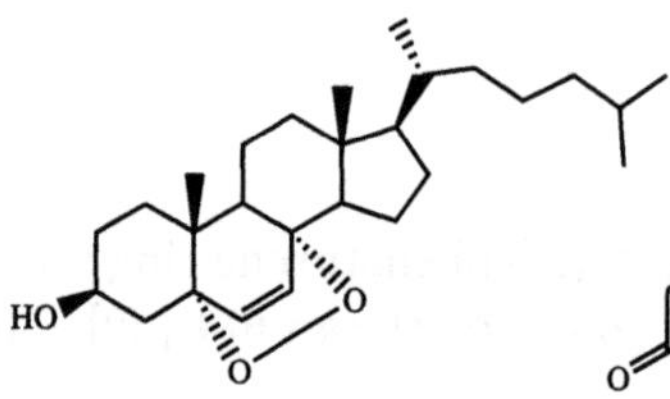

5-α,8α-Epidioxycholest-6-en-3β-ol

612

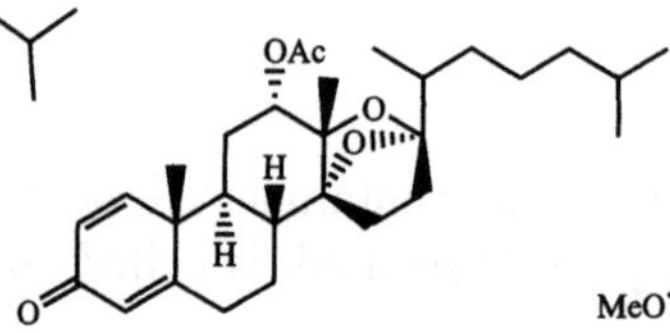

12α-Actoxy-13,17-seco-chlolesta-1,4-diene-3-ones

621

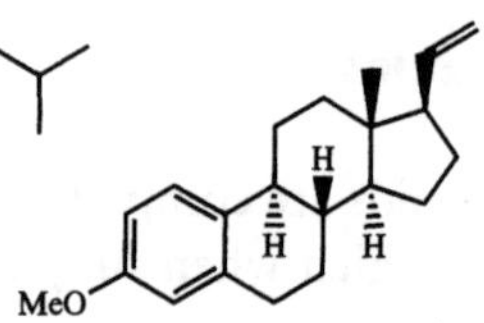

3-Methoxy-19-norpregna-1.3,5(10),20-tetraene

622

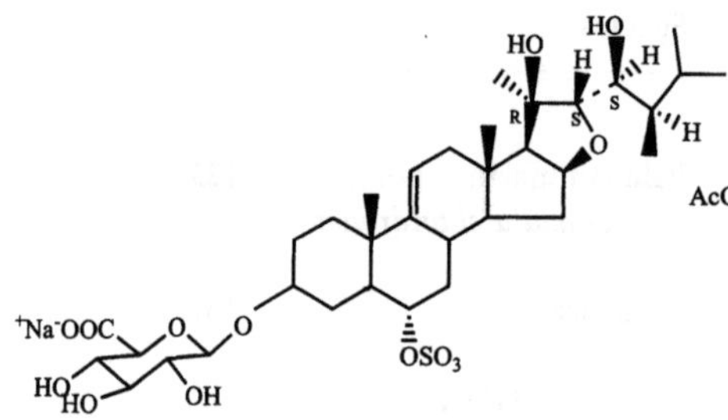

$\Delta^{9(11)}$3β,6α-Diydroxysteroidal aglycon

631

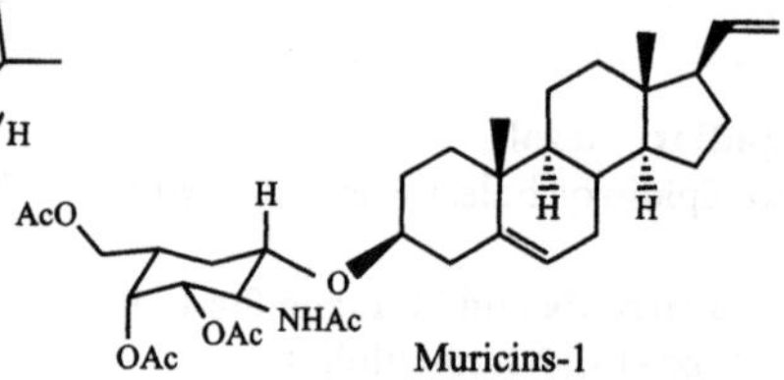

Muricins-1

641

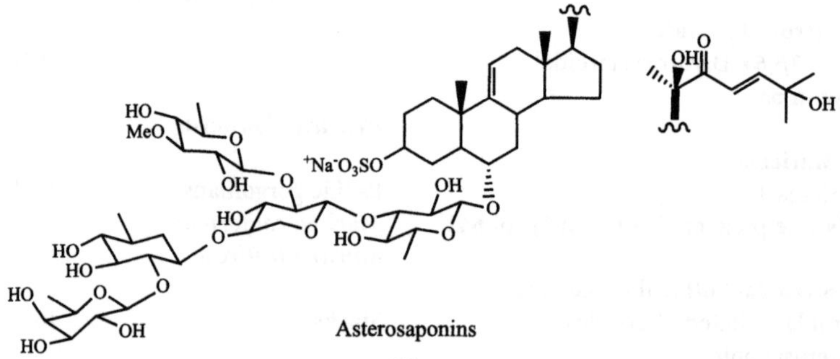

Asterosaponins

651

Scheme 10

steroids are classified into six types. They are (i) sterols isolated from the unsaponifiable parts of animal and plant oils and fats, e.g. cholesterol and related compounds, vitamin D, methyl sterols, (ii) bile acids of the oldest known steroid compounds, (iii) steroid hormones such as sex hormones, adrenal steroids, calcium-regulating sterols and insect skin-shedding hormones, (iv) sapogenins of plant glycosides that form a soapy foam in water, (v) steroid alkaloids occurs in both plants and animals and (vi) steroid lactones including the cardenolides, bufadienolides, holothurigenins, withanolides, and antheridiols.

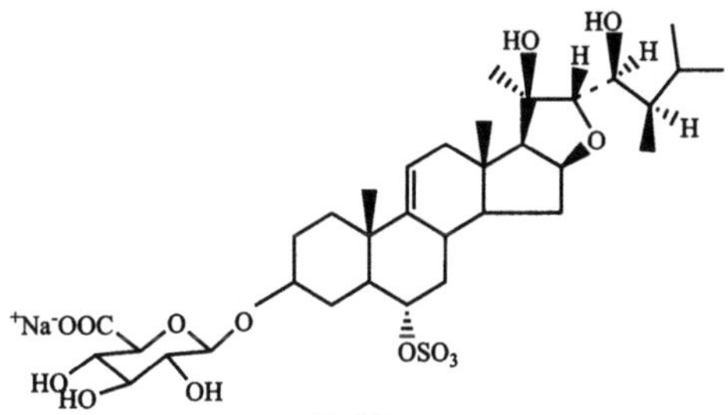

Bufalin

661

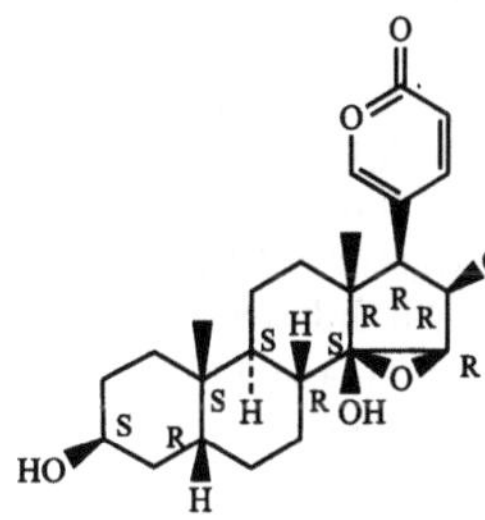

Cinobufagin

662

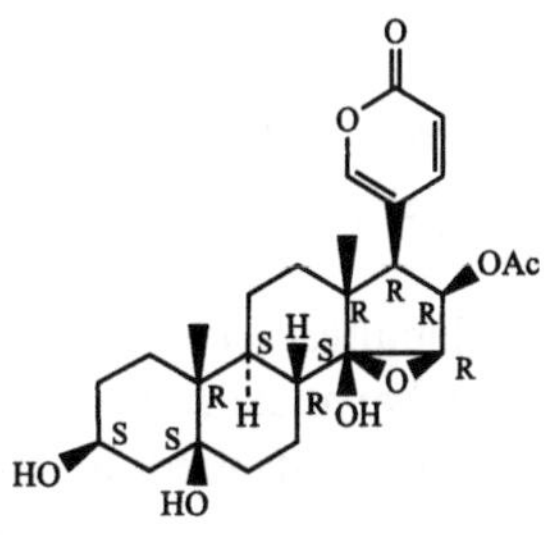

Cinobufatalin

663

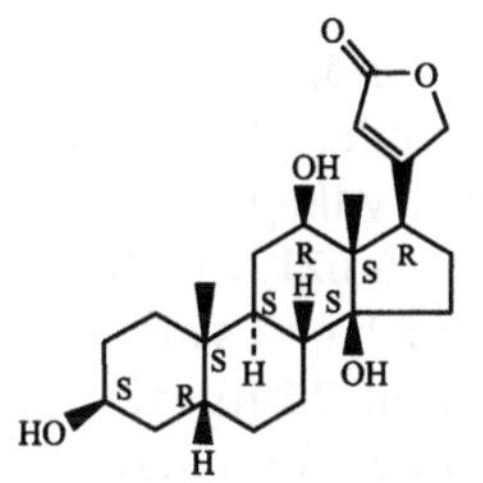

Digoxigenin

664

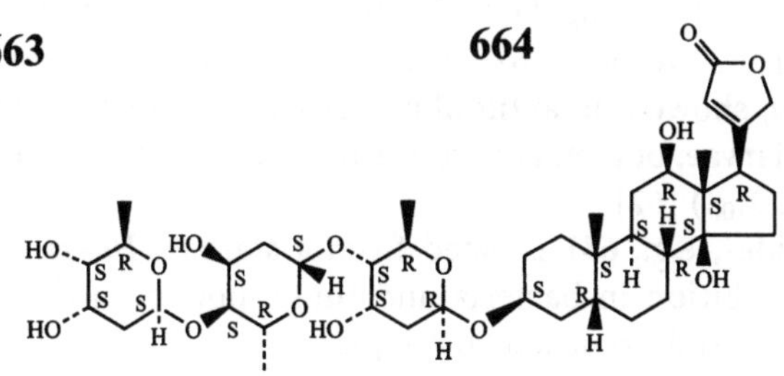

Digoxin

665

Scheme 11

Table 15 Antibarnacle activity of bufodienolids and tributyltin chloride [66, 67]

Antifoulants	Naupliar toxicity		Cyprid antisettlement	
	LD_{50} ng L^{-1}	Relative potency	EC_{50} ng L^{-1}	Relative potency
Bufalin [a] **661**	27	100.0	10	100.000
Cinobufagin [b] **662**	37	73.0	162	6.000
Cinobufatalin [c] **663**	837	32.0	2600	0.380
TBT Cl (n-Bu_3SnCl)	3400	0.8	66 000	0.001
Digoxigenin [d] **664**	4720	0.6	> 50 000	0.002
Digoxin [e] **665**	5450	0.5	6300	0.150

[a] Bufalin = 3,4-Dihydroxybufa-20,22 dienolide
[b] Cinobufagin = 16-(Acetyloxy)-14,15-epoxy-3 hydroxybufa-20,22-enolide
[c] Cinobufatalin = 16-(Acetyloxy)-14,15-epoxy-3,5-dihydroxybufa-20,22-enolide
[d] Digoxigerin = 3,12,14-Trihydroxycard-20(22)-enolide
[e] 3-[(O-2,6-dideoxy-β-*D*-*ribo*-Hexopyranosyl((1→4)-O-2,6-dideoxy-β-*D*-*ribo*-hexopyranosyl-(1→4)-2,6-dideoxy-β-*D*-*ribo*-hexopyranosyl)oxy)]-12,14-dihydroxycard-20(22)-enolide

In these steroids, sterols and their derivatives such as epidioxy sterols, *seco*steroids, tetraene steroids, steroid lycoside, muricins, and sulfated steroidal aglycones, etc. were applied for antifoulants as shown in Table 14 [23, 61–65], Schemes 10, 11.

5-α,8α-Epidioxycholest-6-en-3β-ol **612**, isolated from a Palauan marine sponge, *Lendenfeldia chondrodes*, showed a weak repellent activity against the blue mussel *Mytilus edulis galloprovincialis* [23].

*Seco*steroids, isogosterone A **621**, isolated from a Japanese *octocoral Dendronephthya* sp., inhibited the larval settlement of the barnacle *Balanus amphitrite* with an EC_{50} value of 2.2 μg mL^{-1}. A cyprids continued to swim without attaching to substrates for 7 days. These steroids were not lethal to the barnacle larvae even at 100 μg mL^{-1}, which was much less toxic than $CuSO_4$ (EC_{50} 0.15 μg mL^{-1}) [61]. On the other hand, 3-methoxy-19-norpregna-1.3,5(10),20-tetraene **622** isolated from another octocoral, *Alcyonium gracillimum*, showed no antifouling activity against the barnacle (*Balanus amphitrite*) larvae, but lethality to the barnacle larvae at a concentration of 100 μg mL^{-1} (LD_{100}) [62].

Steroid glycosides, e.g., **631** isolated from a starfish, *Henricia downeyae*, caused growth inhibition in bacteria and fungi, potent antifouling activity against the barnacle and bryozoan larvae [63].

Muricins-1 (3β-pregna-5,20-dienyl-2′-acetamido-2′-deoxy-3′,4′,6′-tri-O-acetyl-β-*D*-galactopyranosidel) **641**, isolated from Pacific gorgonians, *Muricea californi*a and *Muricea fruticasa*, showed the inhibition of growth of the marine diatoms, *Phaeodactylum tricornutum*. The muricins did not, however, possess the ichthyotoxic, cytotoxic, or antimicrobial effects [64].

Table 16 Others

Compound	Living thing *Species*	Refs.
7.1 Higher fatty acids		
Arachidonic acid **711**	marine sponge	[68]
Palmitoleic acid **712**	*Phyllospongia papyracea*	
7.2 Polyacetylenes		
Callytiol C **721**	marine sponge *Callyspongia truncata*	[69]
7.3 Polycyclic compounds		
Solenoide	marine sponge	[47]
Bis(deacetyl)solenoide D **731**	*Psammaplysilla purpurea*	
7.4 Isothiocyanates		
Alkyl isothiocyanates	wasabi	[49]
Phenylethylisothiocyate **741**	*Wasabi japonicas* Matsum	
7.5 Acylglucoglycerol derivatives		
Digalactosyldiacylglycerol **751**	wakame(sea weed) *Undaria pinnatifida*	[49]
7.6 Quinones		
p-Benzoquinones	fruits	[70]
Maesanin **761**		
Maesanol **762**		

Steroidal sulfate pentaglycosides (asterosaponins), goniopectenosides A, isolated from the polar extract of the starfish *Goniopecten demonstrans*, showed the significant inhibition of settlement of the biofouling marine brown macroalga, *Hincksia irregularis* [65].

Antifouling activities against the barnacles, the most troublesome fouling marine organisms, are shown in Table 15 [66, 67]. The most potent natural product tested, bufalin **661** was over 100 times more toxic than TBT and over 6000-fold more potent with respect to an antisettlement activity [66, 67]. The bufalin is a steroid of toad poison, and it is one of component in the secretory substance of *Bufo vulgaris*.

7
Others

Other compounds from the natural products are higher fatty acids, polyacetylenes, polycyclic compounds, isothiocyanates, acetylglucoglycerol derivatives, quinones, etc. These compounds are shown in Table 16 [47–49, 68–70] and Scheme 12.

Table 17 Antifouling activities of authentic fatty acids against the blue mussel *Mytilus edulis* and fatty acid composition of active fractions [72]

Fatty acids	Composition (%) [a]	Sample amount ($mg\,cm^{-2}$) 4.0	1.2	0.6	0.3	Unit [b]
C_{18}	13.8	+	+	–	–	< 5
$C_{18:3}$	3.0	++	++	++	+	50
$C_{20:5}$	0.6	++	++	++	+	53
$CuSO_4$						100

[a] Determined from each peak area
[b] Unit: $100 \times \frac{CuSO_4 \text{ (minimum dose for ++ activity } (\mu mol\,cm^{-2}))}{\text{Sample (minimum dose for ++ activity } (\mu mol\,cm^{-2}))}$

Table 18 Activity of maesanin **761** and maesanol **762** against marine crustaceans *Artemia salina* larvae [70]

	Test concentration ($mg\,L^{-1}$) 50	25	10	5	2.5	1	0.5
Maesanin **761**	3	3	3	2	2	1	1
Maesanol **762**	3	2	2	1	1	1	1
PSA[a]	3	2	2	1	1	1	1
TBTO	3	3	3	3	3	2	1

Evaluation criteria: 3: > 90% kill; 2: 10–90% kill; 1: < 10% kill; TBTO was used as reference compound
[a] PSA = 6-pentadecylsalicylic acid

Higher fatty acids showed the inhibition of settlement of the blue mussels, *Mytilus edulis*. Especially, arachidonic acid **711** and palmitoleic acid **712** isolated from a marine sponge, *Phyllospongia papyracea*, showed a high settlement inhibition to the mussels [68]. Goto et al. [71] pointed out the synergistic effect of higher fatty acids.

Unsaturated higher fatty acids, isolated from octocoral *Dendronephtyya* sp., showed high antifouling properties against the blue mussel *Mytilus edulis* as shown in Table 17 [72]. The above arachidonic acid **711** and palmitoleic acid **712** are also unsaturated higher fatty acids. A comparison of the antifouling activities of various fatty acids suggested that unsaturated fatty acids were thought to be responsible for the activity [72].

Polyacetylene derivatives, e.g. callytetraynetriol **721**, isolated from a marine sponge, *Callyspongia truncata*, showed a potent metamorphosis-inducing activity in the ascidian *Halocynthia roretzi* larvae, with ED_{100} values of 0.13 $\mu g\,mL^{-1}$ and show an antifouling activity against the barnacle *Balanus amphitrite* larvae with ED_{50} values of 0.24 $\mu g\,mL^{-1}$ [69].

Arachidonic acid

711

$CH_3(CH_2)_5$-CH=CH$(CH_2)_7$COOH

Palmitoleic acid

712

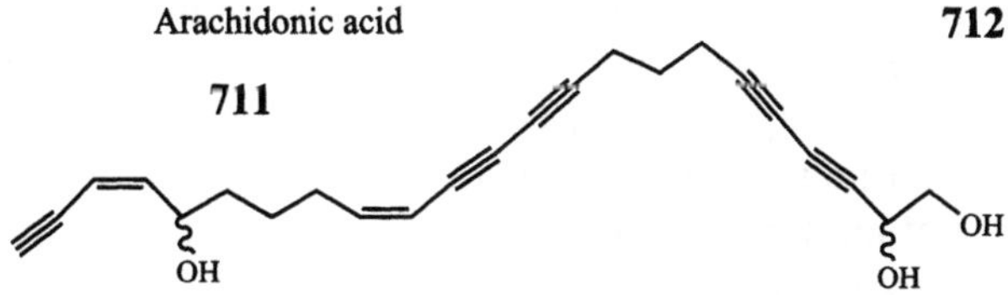

Callytiol C

721

Ph-CH_2-CH_2NCS

Phenylethylisothiocyanate

741

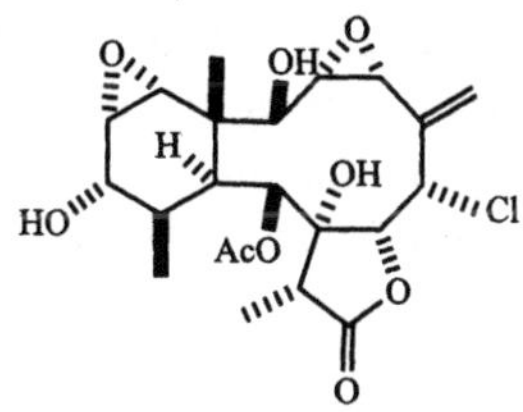

Bis(deacetyl)solenoide D

731

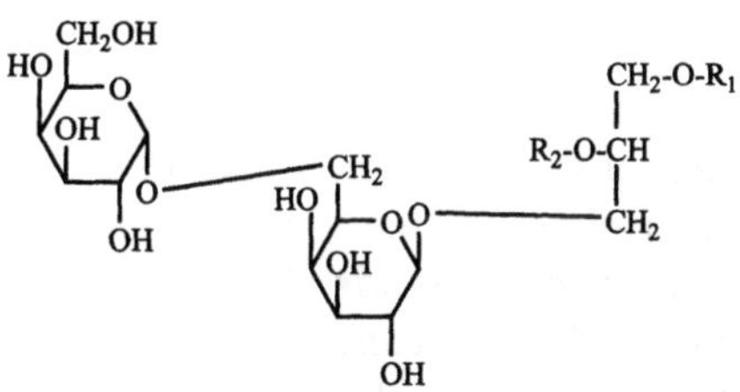

R_1, R_2 = C_{14}-C_{20} mixtures of saturated and unsaturated acyl groups

Digalactosyldiacylglycerol

751

Scheme 12

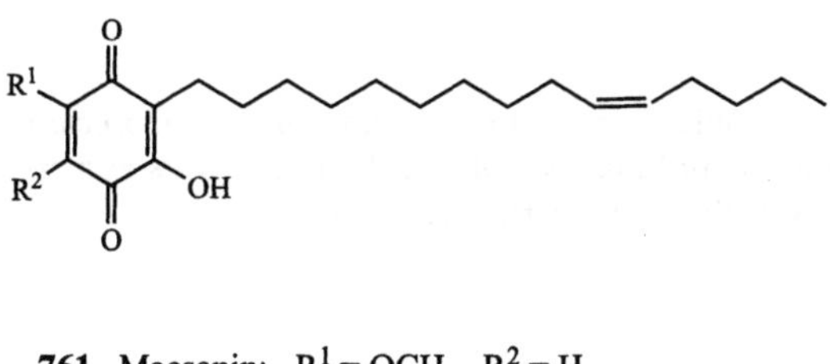

761 Maesanin: $R^1 = OCH_3$, $R^2 = H$
762 Maesanol: $R^1 = OH$, $R^2 = CH_3$

Scheme 13

A polycyclic compound, bis(deacetyl)solenolide D **731** was obtained from the marine sponge *Psammaplysilla purpurea*. It inhibited the biofilm formation [47].

Isothiocyanates isolated from a Japanese horse-radish, *Wasabi japonicas*, showed the moderate inhibition of settlement of the blue mussels *Mytilus edulis*, especially, $PhCH_2CH_2NCS$ **741** showed the highest value in the isothiocyanates. The mixture of $PhCH_2CH_2NCS$ **741** and $CH_3S(CH_2)_6NCS$ showed much higher synergistic antifouling activities [49].

Recently, a screening of antifouling activity from plant extracts led to the selection and further study of Myrsinaceae (*Maesa lanceolata* Forssk). Two *p*-benzoquinone compounds (maesanin **761**, maesanol **762**) were isolated from its fruits. They were found to be active against a representative marine crustaceans *Artemia salina* as shown in Table 18 [70]. Maesanin **761** and maesanol **762** showed the highest activity against *Artemia salina* among all the extracts, and were more active than 6-pentadecylsalicylic acid (PSA) isolated from *Ozoroa insignis*, but were less potent than TBTO. Hence, they are active against marine fouling organisms. The chemical modifications of these compounds were undertaken in order to obtain more active derivatives [70].

8 Conclusions

Five kinds of compounds, i.e., terpenes, nitrogen-containing compounds, phenols, steroids and others, were isolated as natural products antifouling substances. Of these substances, many compounds showed the same level of antifouling properties as $CuSO_4$, and some compounds also showed the same level of activities with the organotin antifoulants. However, these natural products showed generally no or very low toxicity, and are biodegradable in marine environment. Many simple natural products such as trigonelline, nicotinamide and ethyl *N*-(2-phenylethylcarbamate) and their synthetic derivatives had also high antifouling properties. Many of them showed synergistic properties. We expect to utilize some natural products, their synthetic derivatives or their mixtures as environmentally friendly antifouling agents in near future.

Acknowledgements I should like to express my sincere appreciation to Dr. Sumio Chubachi for reading the full manuscript, which enhanced its accuracy and clarity, and for providing much valuable constructive criticism.

References

1. Omae I (2003) Chem Rev 103:3431
2. Kishihara M (1998) Kagaku Kogyo 49:458

3. Clare AS (1996) Biofouling 9:211
4. Kon-ya K, Shimidzu N, Adachi K, Miki W (1994) Fish Sci 60:773
5. König GM, Wright AD, de Nys R (1999) J Nat Prod 62:383
6. Faulkner DJ, Stallard MO (1973) J Am Chem Soc 95:3413
7. Mynderse JS, Faulkner DJ (1975) Tetrahedron 31:1963
8. König GM, Wright AD (1997) J Nat Prod 60:967
9. Hirota H, Okino T, Yoshimura E, Fusetani N (1998) Tetrahedron 54:13971
10. Sera Y, Adachi K, Nishida F, Shizuri YJ (1999) Nat Prod 62:395
11. Okino T, Yoshimura E, Hirota H, Fusetani N (1996) Tetrahedron 52:9447
12. Etoh H, Kondoh T, Yoshioka N, Sugiyama K, Ishikawa H, Tanaka H (2003) Biosci Biotechnol Biochem 67:911
13. Tsukamoto S, Kato H, Hirota H, Fusetani N (1997) Biofouling 11:283
14. Nogata Y, Yoshimura E, Shinshima K, Kitano Y, Sakaguchi I (2003) Biolfouling 19:193
15. Hattori T, Adachi K, Shizuri Y (1997) J Nat Prod 60:411
16. Hello C, Thomas-Guyon H, Culioli G, Piovetti L, Bourgougnon N, Gal YL (2001) Biolfouling 17:189
17. Keifer PA, Rinehart KL Jr (1986) J Org Chem 51:4450
18. Okino T, Yoshimura E, Hirota H, Fusetani N (1996) J Nat Prod 59:1081
19. Okino T, Yoshimura E, Hirota H, Fusetani N (1995) Tetrahedron Lett 36:8637
20. Hirota H, Tomono Y, Fusetani N (1996) Tetrahedron 52:2359
21. Mergelman TL, Silva GL, Mongelli E, Gil RR (2003) Phytochemistry 62:569
22. Clare AS, Rittschof D, Gerhart DJ, Hooper IR, Bonaventura J (1999) Mar Biotechnol 1:427
23. Sera Y, Adachi K, Shizuri Y (1999) J Nat Prod 62:152
24. Fusetani N, Hirota H, Okino T, Tomono Y, Yoshimura E (1996) J Nat Toxins 5:249
25. Fletcher R (1980) Bot Mar 23:425
26. Smyrniotopoulos V, Abatis D, Tziveleka L-A, Tsitsimpikou C, Roussis V, Loukis A, Vagias C (2003) J Nat Prod 66:21
27. Rittschof D, Hooper IR, Costlow JD (1986) Bull Mar Sci 39:376
28. Kawamata M, Kaneko F, Fukunaga J, Fukushima K (2000) Taisei Kensetsu Gijyutsu Kenkyushoho 33:89
29. Tsukamoto S, Kato H, Hirota H, Fusetani N (1996) Tetrahedron Lett 37:1439
30. Ponansik JA, Conova S, Kinghorn D, Kinney WA, Rittschof D, Ganem B (1998) Tetrahedron 54:6977
31. Ponasik JA, Kassab DJ, Ganem B (1996) Tetrahedron Lett 37:6041
32. Tsukamoto S, Kato H, Hirota H, Fusetani N (1996) J Nat Prod 59:501
33. Thirionet I, Daloze D, Braekman JC (1998) Nat Prod Lett 12:209
34. Faimali M, Sepčić K, Turk T, Geraci S (2003) Biofouling 19:47
35. Burgess JG, Boyd KG, Armstrong E, Jiang Z, Yan L, Berggren M, May U, Pisacane T, Granmo Å, Adams DR (2003) Biolfouling 19:197
36. Tsukamoto S, Kato H, Hirota H, Fusetani N (1996) Tetrahedron 52:8181
37. Sera Y, Adachi K, Fujii K, Shizuri Y (2002) Mar Biotechnol 3:441
38. Sera Y, Adachi K, Fuji K, Shizuri Y (2003) J Nat Prod 66:719
39. Kawamata M, Kon-ya K, Miki W (1994) Fish Sci 60:485
40. Yamamori N (1990) Kagaku to Kogyo 43:1880
41. Tsukamoto S, Kato H, Hirota H, Fusetani N (1996) J Org Chem 61:2936
42. Hattori T, Adachi K, Shizuri Y (1998) J Nat Prod 61:823
43. Göransson U, Sjögren M, Svangård E, Claeson P, Bohlin L (2004) J Nat Prod 67:1287
44. Craik DJ, Daly NL, Bond T, Waine C (1999) J Mol Biol 294:1327
45. Schoenfeld RC, Ganem B (1998) Tetrahedron Lett 39:4147

46. Diers JA, Pennaka HK, Peng J, Bowling JJ, Duke SO, Hamann MT (2004) J Nat Prod 67:2117
47. Yamada A, Kitamura H, Yamaguchi K, Fukuzawa S, Kamijima C, Yazawa K, Kuramoto M, Wang G-Y-S, Fujitani Y, Uemura D (1997) Bull Chem Soc Jpn 70:3061
48. Hattori T, Matsuo S, Adachi K, Shizuri Y (2001) Fish Sci 67:690
49. Ina K (1991) Kagaku to Kogyo 44:666
50. Veech RL (1997) US Patent, 5,629,045 (1997) Chem Abstr 127:36004
51. Turgut C, Newby B, Cutright TJ (2004) ESPR-Environ Sci & Pollut Res 11:7
52. Etoh H, Kondoh T, Noda R, Singh IP, Sekiwa Y, Morimitsu K, Kubota K (2002) Biosci Biotechnol Biochem 66:1748
53. Singh IP, Umehara K, Etoh H (1999) Nat Prod Lett 14:11
54. Singh IP, Takahashi K, Etoh H (1996) Biosci Biotech Biochem 60:1522
55. Murakami R, Hachisako H (1998) Kobunsi Kako 47:66
56. Terada Y, Saito J, Kawai T, Singh IP, Etoh H (1999) Biosci Biotechnol Biochem 63:276
57. Soman AG, Gloer JB, Wicklow DT (1999) J Nat Prod 62:386
58. Takasawa R, Ogura C, Watanabe N, Etoh H, Sakata K, Ina K (1992) Biosci Biotech Biochem 56:1208
59. Takasawa R, Etoh H, Yagi A, Sakata K, Ina K (1990) Abri Biol Chem 54:1607
60. Müller R (2003) Ullmann's Encyclopedia of Industrial Chemistry, 6th ed. Wiley, vol 34, p 261
61. Tomono Y, Hirota H, Fusetani N (1999) J Org Chem 64:2272
62. Tomono Y, Hirota H, Imahara Y, Fusetani N (1999) J Nat Prod 62:1538
63. Palagiano E, Zollo F, Minale L, Iorizzi M, Bryan P, McChntock J, Hopkins T (1996) J Nat Prod 59:384
64. Bandurraga MM, Fenical W (1985) 41:1057
65. De Marino S, Iorizzi M, Zollo F, Amsler CD, Greer SP, McClintock JB (2000) Eur J Org Chem 4093
66. Rittschof D (2000) Biofouling 15:119
67. Gerhart DJ, Rittschof D, Bonaventura J (1993) US Patent 5,252,630
68. Kamiya H (1993) Japan Patent Kokai H5-4903 (1993) Chem Abstr 119:98086
69. Tsukamoto S, Kato H, Hirota H, Fusetani N (1997) J Nat Prod 60:126
70. He W, Van Puyvelde L, Bosselaers J, De Kimpe N, Van der Flaas M, Roymans A, Mathenge SG, Mudida FP (2001) Biofouling 17:221
71. Goto R, Kado R, Muramoto K, Hamiya H (1992) Biofouling 6:61
72. Mizobuchi S, Shimidzu N, Katsuoka M, Adachi K, Miki W (1993) Nippon Suisan Gakkaishi 59:1195

Subject Index

MIX
Papier aus verantwortungsvollen Quellen
Paper from responsible sources
FSC® C105338

If you have any concerns about our products,
you can contact us on
ProductSafety@springernature.com

In case Publisher is established outside the EU,
the EU authorized representative is:
Springer Nature Customer Service Center GmbH
Europaplatz 3, 69115 Heidelberg, Germany

Printed by Libri Plureos GmbH
in Hamburg, Germany